13th Conference on Theoretical Nuclear Physics in Italy 2011

AA002757

Journal of Physics: Conference Series Volume 336

Cortona, Italy
6-8 April 2011

ISBN: 978-1-62276-683-3
ISSN: 1742-6588

Printed from e-media with permission by:

Curran Associates, Inc.
57 Morehouse Lane
Red Hook, NY 12571

CURRAN ASSOCIATES INC. proceedings.com

Some format issues inherent in the e-media version may also appear in this print version.

Copyright© (2011) by the Institute of Physics
All rights reserved.

Printed by Curran Associates, Inc. (2013)

For permission requests, please contact the Institute of Physics
at the address below.

Institute of Physics
Dirac House, Temple Back
Bristol BS1 6BE UK

Phone: 44 1 17 929 7481
Fax: 44 1 17 920 0979

techtracking@iop.org

Additional copies of this publication are available from:

Curran Associates, Inc.
57 Morehouse Lane
Red Hook, NY 12571 USA
Phone: 845-758-0400
Fax: 845-758-2634
Email: curran@proceedings.com
Web: www.proceedings.com

13th Conference on Theoretical Nuclear Physics in Italy 2011

Journal of Physics: Conference Series Volume 336

Cortona, Italy
6-8 April 2011

ISBN: 978-1-62276-683-3
ISSN: 1742-6588

IOP Institute of Physics

In Memoriam of Alberto Dellafiore

Few-nucleon systems

Nuclear structure

Nuclear matter

Nuclear dynamics and relativistic heavy ion collision

Nuclear astrophysics

Nuclear physics with electroweak probes

Structure of hadrons and hadronic matter

Special topics

Journal of Physics: Conference Series **336** (2011) 011001 doi:10.1088/1742-6596/336/1/011001

Preface

These proceedings contain the invited and contributed papers presented at the 13th Conference on Theoretical Nuclear Physics in Italy held in Cortona, Italy, from April 6 to April 8, 2011. As usual, the meeting was held at *il Palazzone*, a 16th century castle owned by the *Scuola Normale Superiore di Pisa*. The aim of this biennial conference is to bring together Italian theorists working in various fields of nuclear physics to discuss their latest results and confront their points of view in a lively and informal way. This offers the opportunity to promote collaborations between different groups.

There were about 50 participants at the conference, coming from 15 Italian Universities and *Sezioni* of the *Istituto Nazionale di Fisica Nucleare* – INFN (Catania, Ferrara, Firenze, Genova, Lecce, Milano, Napoli, Padova, Pavia, Perugia, Pisa, Roma, Torino, Trento, Trieste). The program of the conference, prepared by the Organizing Committee (Ignazio Bombaci, Aldo Covello, Laura Elisa Marcucci and Sergio Rosati) focused on seven main topics: Few-Nucleon Systems, Nuclear Astrophysics, Nuclear Matter and Nuclear Dynamics, Nuclear Physics with Electroweak Probes, Nuclear Structure, Quark-gluon Plasma and Relativistic Heavy Ion Collisions, Structure of Hadrons and Hadronic Matter.

At variance with the previous meetings, the invited papers were not meant to give a comprehensive review of the research activities of the various Italian groups, as this appeared to be too heavy a burden for the speakers. Besides a dozen of invited papers, 15 contributed papers were presented, most of them by young participants.

The year 2011 has marked the hundredth anniversary of the discovery of the existence of the atomic nucleus by Rutherford, based on the results of the famous Geiger and Marsden's experiments. This memorable event was the subject of a fascinating talk given by Carlo Bernardini at the beginning of the meeting. In the last session of the conference, Marcello Lattuada gave an overview of the activities and perspectives of the *Laboratori Nazionali del Sud* of Catania. We warmly thank both Bernardini and Lattuada for having significantly contributed to widen the scope of our otherwise purely theoretical meeting. This Cortona meeting ended with a Round Table Discussion on "Future and perspectives of theoretical nuclear physics in Italy". A Bonaccorso, A Covello, M Di Toro, S Fantoni and M Giannini were on the panel, with G Orlandini acting as a moderator.

A prize established to honor the memory of Adelchi Fabrocini was assigned for the first time at this meeting. The prizewinner was Dr Arianna Carbone, to whom go our best wishes for a successful career.

This Cortona meeting was permeated with deep feelings of sorrow over the passing away on March 19 of our friend and colleague Alberto Dellafiore. Massimo Di Toro and Francesco Matera highlighted his human qualities and scientific achievements and we are pleased that the written text of their presentation is reported here. It is our wish that these proceedings be dedicated to Alberto's memory.

We gratefully acknowledge the financial support of INFN who helped make the conference possible.

I Bombaci, A Covello, L E Marcucci, S Rosati

Published under licence by IOP Publishing Ltd

List of participants

Barbaro Maria B (Università di Torino)
Benhar Omar (INFN Roma)
Bernardini Carlo (Università "La Sapienza" di Roma)
Bianco Davide (Università di Napoli)
Bombaci Ignazio (Università di Pisa)
Bonaccorso Angela (INFN Pisa)
Brenna Marco (Università di Milano)
Canton Luciano (INFN Padova)
Carbone Arianna (Universitat de Barcelona, Spain)
Catara Francesco (Università di Catania)
Colò Gianluca (Università di Milano)
Covello Aldo (Università di Napoli)
Cristoforetti Marco (ECT* Trento)
Di Toro Massimo (Università di Catania)
Faccioli Pietro (Università di Trento)
Fantoni Stefano (ANVUR Roma)
Gambacurta Danilo (Università di Catania)
Gargano Angela (INFN Napoli)
Gattobigio Mario (INLN, Université de Nice-Sophia Antipolis, CNRS, France)
Gensini Paolo (INFN Lecce)
Giannini Mauro (Università di Genova)
Girlanda Luca (Università del Salento)
Giusti Carlotta (Università di Pavia)
Greco Vincenzo (Università di Catania)
Idini Andrea (Università di Milano)
Illarionov Alexey (Università di Trento)
Itaco Nunzio (Università di Napoli)
Kievsky Alejandro (INFN Pisa)
Lattuada Marcello (INFN-LNS Catania)
Leidemann Winfried (Università di Trento)
Lo Iudice Nicola (Università di Napoli)
Lonardoni Diego (Università di Trento)

Vigezzi Enrico (INFN Milano)
Vitturi Andrea (Università di Padova)
Viviani Michele (INFN Pisa)
Losa Cristina (SISSA Trieste)
Lovato Alessandro (SISSA Trieste)
Mantovani Sarti Valentina (Università di Ferrara)
Marcucci Laura Elisa (Università di Pisa)
Matera Francesco (Università di Firenze)
Orlandini Giuseppina (Università di Trento)
Pacati Franco (Università di Pavia)
Pederiva Francesco (Università di Trento)
Radici Marco (INFN Pavia)
Ratti Claudia (Università di Torino)
Schulze Hans-Josef (INFN Catania)
Scopetta Sergio (Università di Perugia)
Shao Guo-Yun (INFN Catania)
Tinti Leonardo (Università di Firenze)
Traini Marco (Università di Trento)

Journal of Physics: Conference Series **336** (2011) 012001 doi:10.1088/1742-6596/336/1/012001

Remembering Alberto Dellafiore

Massimo Di Toro[1,2] and Francesco Matera[3]

[1] Laboratori Nazionali del Sud INFN, I-95123 Catania, Italy
[2] Physics and Astronomy Dept., University of Catania, Italy
[3] Physics and Astronomy Dept., University of Florence and INFN, Sezione di Firenze, Italy

Few weeks ago Alberto Dellafiore, senior researcher of INFN-Florence, passed away after a sudden acceleration of a cancer disease. We are still rather shocked and as very old friends of him we like to say a few words. Other comments from members of the Italian Nuclear Theory community can be find (in Italian) in the Appendix.

Personal memories

We both knew him since many years, in the research activity as well as in normal life.

- *MDT.* I met him at the end of the seventies in Oxford where he was working on a Ph.D. project with David Brink and I was visiting David for a collaboration. We started a common nuclear theory project (see later) and I was impressed for his very precise and exhaustive approach to the physics and mathematics problems. At the same time he was revealing a very pleasant character, with a deep sense of humor, but always joined to a very firm intellectual integrity. The children are naturally the best estimators of a person and my children (from 6 to 10 of age at that time) from the very first moment considered him one of their best friends. Back to Italy although living not very close (Florence vs Catania) we maintained our acquaintance for the work as well as for the families. Last time I have seen him was in November last year when he visited us with his wife Laura and their grand-daughter Sofia. I cooked some Chinese style noodles and we passed a very nice time together, although he was already suffering for the cancer effects, but he was very optimistic to be able to overcome it.
- *FM.* Alberto moved from Pisa to Florence in 1980, with an university research contract. He had a very important revitalizing effect on our small nuclear theory group, introducing the new research line on the dynamical response of finite Fermi systems treated in a semiclassical approach. Relevant results have been obtained, even with the close collaboration of David Brink from Oxford and Valery Abrimosov from Kiev. In all these years I had the chance to deeply appreciate the research motivation, the intellectual integrity and scientific rigor of Alberto. Rigor that sometimes was appearing as a kind of fussiness, but in fact it was always focused on physical contents of the problem and not on formal points. In this sense it has been always very rewarding to work with him. A great quality of Alberto was his deep natural annoyance in front of people misbehavior in science, management as well as in the everyday life. His indignation has always been focused on general features of the civil life or of the scientific research, never related to particular problems of personal interest. Although his academic career was not so smooth, as he would have deserved.

Published under licence by IOP Publishing Ltd

Alberto was close to retirement, but certainly he would have continued to give a valid contribution to the Florence nuclear theory group. With his untimely disappearance we have lost a precious collaborator, but above all a dear friend.

Few scientific notes

The main Alberto contribution to the many-body theory has been the development of a Semiclassical Linear Response for Finite Fermi systems. The idea comes back to the Oxford period when an analytical solution of the linearized Vlasov equation was worked out in a seminal paper [1]. This represents a kind of Phase Space Approach to the Random Phase Approximation, where collective motions are expressed in terms of natural quasiparticle orbits of the system. The connection to a fully quantum WKB approximation was clearly shown in a following work [2]. Moreover, within such frame a unified treatment of the giant resonances and low–energy levels in nuclei has been obtained [3].

The advantage of this approach has been the clear physical interpretation of the results (Classical Transition Densities, Landau Damping) and the simple extension to effects which go beyond the mean filed picture, like quasiparticle collisions and pairing correlations [4, 5].

Particularly interesting is the application to other Fermi systems, like metal clusters [6] and to more microscopic relativistic bosonic nuclear interactions [7]. The latter paper has been particularly important since the treatment of exchange terms has naturally led to the appearance of new degrees of freedom in the nuclear response. This can be of relevance also for other effective relativistic theories, like non-perturbative QCD models. Unfortunately Alberto is not anymore with us...

References

[1] Brink D M, Dellafiore A, Di Toro M 1986 *Nucl. Phys. A* **456** 205
[2] Dellafiore A, Matera F 1986 *Nucl. Phys. A* **460** 245
[3] Abrosimov V, Dellafiore A, Matera F 2003 *Nucl. Phys. A* **717** 44
[4] Dellafiore A, Brink D M, Matera F 1995 *Phys. Rev. A* **51** 914
[5] Abrosimov V, Brink D M, Dellafiore A, Matera F 2008 *Nucl. Phys. A* **800** 1
[6] Dellafiore A, Matera F 1990 *Phys. Rev. B* **41** 3488
[7] Dellafiore A, Matera F 1991 *Phys. Rev. C* **44** 2456

APPENDIX: Some Short Recollections from the Italian Nuclear Theory Community

- Come molti di noi, conoscevo Alberto da tanti anni. Anche senza una collaborazione scientifica diretta, avevamo una gradita occasione di incontro nelle conferenze organizzate periodicamente dal nostro gruppo di Napoli. In particolare, Alberto aveva partecipato con continuità alle ultime quattro, dal 2001 al 2010, e quello che non avrei immaginato fosse il nostro ultimo incontro è stato appunto a Vietri, nel maggio 2010. Circa un mese fa è stato pubblicato nel volume dei proceedings (Journal of Physics: Conference Series 267) il contributo di Alberto "Effects of self-consistency in semiclassical pairing theory", in collaborazione con Abrosimov, Brink e Matera. Questo rende ora più vivo il ricordo ed il rimpianto dell'amico scomparso. *Aldo Covello, Napoli*
- Mi unisco al ricordo di Alberto Dellafiore, che ho conosciuto come studente della Scuola Normale, e poi ho incontrato nel tempo in occasioni diverse. Ricordo la sua simpatia umana ed il suo equilibrio. Alla famiglia le mie condoglianze. È una perdita per tutti. *Adriano Di Giacomo, Pisa*
- Apprendo con molta tristezza la notizia della scomparsa di Alberto. Alberto faceva parte della nostra storia, della nostra comunità o almeno io avevo questa sensazione. Una persona a cui rivolgersi come collega e per avere consigli, anche di natura diversa da quella scientifica.

Journal of Physics: Conference Series **336** (2011) 012001 doi:10.1088/1742-6596/336/1/012001

E per questo, per questo suo essere parte di noi e quindi di me, non immaginavo possibile che non fosse più tra noi. Invece no, è possibile, e notizie come questa ti riportano alla realtà. Come purtroppo è stato per altri amici che ci hanno lasciato. Ora più che mai il suo volto, il suo parlare pacato, le cose che ha fatto mi ritornano in mente, perchè si fa maggiore attenzione a qualcosa che è tua ma che ti fa male. Abbiamo imparato da lui. E questo certo rimarrà. Ciao Alberto. *Stefano Fantoni, SISSA Trieste*

- Ho avuto modo di conoscere Alberto parecchi anni fa e ne ho apprezzato la competenza e gentilezza. Purtroppo non ci siamo frequentati molto, ma ne conservo un caro ricordo e la sua scomparsa mi ha colpito profondamente. Sono vicino alla famiglia in questo momento di dolore. *Mauro Giannini, Genova*
- Alberto era una carissimo amico, conservavamo una stima reciproca, sin dal tempo in cui avrebbe dovuto impartire un corso a Cagliari, alla fine degli anni 70, anche se non ci frequentavamo che raramente. Era nostro interesse comune lo studio delle correlazioni temporali fra nucleoni. Era sensibile, gentile e comprensivo, sempre disposto ad ascoltare i problemi degli altri. Ciao Alberto. *Piero Quarati, Cagliari*
- Conosco ed apprezzo Alberto dal lontano 1980, quando abbiamo pubblicato il frutto della nostra prima collaborazione. Una persona squisita e straordinaria nella sua sincerità e gentilezza, una di quelle persone che fanno amare di più il mestiere che ci è capitato e che facciamo con passione, come lo faceva lui. Non ha mai nascosto le difficoltà ed ha sempre aiutato ad affrontarle. Un caro saluto a lui ed un abbraccio a tutta la famiglia. *Marco Traini, Trento*
- Da vecchio *fiorentino*, la notizia della scomparsa di Alberto è stata per me come un pugno nello stomaco. Voglio estendere a tutti l'espressione del mio dolore, e sopratutto ai nostri colleghi di Firenze. *Paolo Gensini, Perugia*

Journal of Physics: Conference Series **336** (2011) 012002 doi:10.1088/1742-6596/336/1/012002

Scattering and resonances on p-shell nuclei

L Canton[1], P R Fraser[2], J P Svenne[3], K Amos[4], S Karataglidis[5] and D van der Knijff[4]

[1] Istituto Nazionale di Fisica Nucleare, Sezione di Padova, I-35131, Italy
[2] Instituto de Ciencias Nucleares, Universidad Nacional Autónoma de México, 04510 México D.F., Mexico
[3] Department of Physics and Astronomy, University of Manitoba and Winnipeg Institute for Theoretical Physics, Winnipeg, Manitoba, Canada R3T 2N2
[4] School of Physics, University of Melbourne, Victoria 3010, Australia
[5] Department of Physics, University of Johannesburg, P.O. Box 524, Auckland Park, 2006, South Africa

E-mail: luciano.canton@pd.infn.it

Abstract. We review recent results on low energy nucleon scattering from some light-mass nuclei and particularly consider resonances seen or expected in cross sections. We present results found using the multichannel algebraic scattering (MCAS) method which was developed to analyze such cross sections. The method gives both bound and resonant spectra of the compound nuclei. In 2006, using this method, we predicted nontrivial narrow resonances due to non-valence p-shell effects in the nucleus ^{15}F, a nucleus that lies beyond the proton drip-line. Subsequently, in 2009, those predictions were verified experimentally. Research on other such mirror systems is underway. In particular, the structure of ^{17}C has been used to define an appropriate MCAS nuclear interaction for the n+^{16}C system. Adding a Coulomb field to that interaction, an MCAS evaluation has been made for the mirror system p+^{16}Ne, to specify the low excitation spectrum of the yet unobserved, particle-unstable, ^{17}Na.
We also investigated how the scattering cross-section changes when the spectra of the colliding nuclei have low-excitation particle-emitting resonances. As a test case, we considered ^{8}Be, being particle-unstable, and analyzed neutron scattering cross sections and the spectra of the compound ^{9}Be system. If the ^{8}Be excited states have fixed-energy decay widths, we find that the bound states of the compound system are affected in an unphysical way. The shapes of the target resonances must vary from the usual Lorentzian in an energy-dependent way. Resonance width functions must go smoothly to zero at the elastic threshold. Ways of achieving this condition are being explored.

1. Introduction

The advent of radioactive ion beam (RIB) physics stimulates the consideration of new theoretical challenges involving weakly-bound systems. Radioactive nuclei, especially those close to the drip lines, can have quite low particle emission thresholds and consequently have low-lying resonance states in their spectra. Low-energy scattering of a RIB is a coupled-channel problem that involves such low-lying resonant states of the scattered ion. To address this, a Multi-Channel Algebraic Scattering (MCAS) formalism [1] is used, in which momentum space solutions of coupled Lippmann-Schwinger equations are found. The method also defines all negative-energy bound states. A finite-rank separable representation using an "optimal" set of sturmian functions [2] serves to construct an input matrix of nucleon-nucleus interactions. The MCAS

Published under licence by IOP Publishing Ltd

Journal of Physics: Conference Series **336** (2011) 012002 doi:10.1088/1742-6596/336/1/012002

method has the ability to locate all compound system resonance centroids and widths, regardless of how narrow. For full details see Ref. [1]. Also, by use of orthogonalizing pseudo-potentials (OPP) in generating Sturmians, it ensures the Pauli principle is not violated [3], even if the collective model formulation of nucleon-nucleus interactions is used. Otherwise, some compound nucleus wave functions contain spurious components.

In section 2, we briefly review the MCAS method [1, 3], and in section 3 we illustrate the fundamentals of the resonance identification. Section 4 is devoted to the discussion of results for ^{15}F, the mirror of ^{15}C, and section 5 for the preliminary results on ^{17}C, in view of future investigations on the mirror system ^{17}Na. Section 6 reviews the extension of the MCAS approach to unstable target states [4]. In section 7, we display and discuss results for neutron scattering from ^{8}Be, with and without the excited target states having widths different from zero. In section 8 we look briefly at the implications of letting the target-state widths have energy dependence, to ensure that these resonances do not extend to negative energies, where they would produce unphysical behaviour for bound, sub-threshold, states. Finally, section 9 gives our brief concluding remarks.

2. The MCAS formulation with stable collective excitations of the target

The MCAS method was developed to find solutions of coupled-channel, partial-wave expanded, Lippmann-Schwinger equations for the scattering of two nuclei. For each total system spin-parity (J^π), those equations are

$$T_{cc'}^{J^\pi}(p,q;E) = V_{cc'}^{J^\pi}(p,q) + \mu \left[\sum_{c''=1}^{\text{open}} \int_0^\infty V_{cc''}^{J^\pi}(p,x) \frac{x^2}{k_{c''}^2 - x^2 + i\varepsilon} T_{c''c'}^{J^\pi}(x,q;E)\, dx \right.$$
$$\left. - \sum_{c''=1}^{\text{closed}} \int_0^\infty V_{cc''}^{J^\pi}(p,x) \frac{x^2}{h_{c''}^2 + x^2} T_{c''c'}^{J^\pi}(x,q;E)\, dx \right] , \quad (1)$$

where the channels are denoted c and where $\mu = \frac{2\overline{m}}{\hbar}$, $\overline{m}$ being the reduced mass. There are two summations as the open and closed-channel components are separated, with wave numbers

$$k_c = \sqrt{\mu(E - \epsilon_c)} \text{ and } h_c = \sqrt{\mu(\epsilon_c - E)} , \quad (2)$$

for $E > \epsilon_c$ and $E < \epsilon_c$ respectively. ϵ_c is the energy threshold at which channel c opens. In the nucleon scattering case, they coincide with the, presumed discrete, excitation energies of the nucleus. Henceforth the J^π superscript is to be understood. Expansion of $V_{cc'}$ in terms of a finite number (N) of basis function leads to an algebraic representation of the scattering matrix [1]

$$S_{cc'} = \delta_{cc'} - \mathrm{i}^{l_{c'} - l_c + 1} \pi\mu \sum_{n,n'=1}^{N} \sqrt{k_c}\, \hat{\chi}_{cn}(k_c) \left([\boldsymbol{\eta} - \mathbf{G}_0]^{-1} \right)_{nn'} \hat{\chi}_{c'n'} \sqrt{k_{c'}} \, . \quad (3)$$

The indices c and c' refer now only to open channels, l_c is the partial wave in channel c and the Green's function matrix is

$$[\mathbf{G}_0]_{nn'} = \mu \left[\sum_{c=1}^{\text{open}} \int_0^\infty \hat{\chi}_{cn}(x) \frac{x^2}{k_c^2 - x^2 + \mathrm{i}\epsilon} \hat{\chi}_{cn'}(x)\, dx - \sum_{c=1}^{\text{closed}} \int_0^\infty \hat{\chi}_{cn}(x) \frac{x^2}{h_c^2 + x^2} \hat{\chi}_{cn'}(x)\, dx \right] . \quad (4)$$

Here, $\boldsymbol{\eta}$ is a column of weight values and $\hat{\chi}$ are a convenient set of expansion functions. Details are given in Ref. [1].

3. Resonance identification

Sturmians (also known as Weinberg states) represent an *alternative* way to formulate the Quantum Mechanical problem.

Consider a two-body like Hamiltonian $H = H_o + V$: then the Schrödinger equation is written in the standard (time-independent) way

$$(E - H_o)\Psi_E = V\Psi_E\,, \tag{5}$$

where E is the spectral variable, and Ψ_E is the eigenstate.

Sturmians, instead, are the eigensolutions of:

$$(E - H_o)\Phi_i(E) = \frac{V}{\gamma_i(E)}\Phi_i(E)\,, \tag{6}$$

where E is a parameter. The eigenvalue γ_i is the potential scale. Thus the spectrum consists of all the potential rescalings that give solution to that equation, for given energy E, and with well-defined boundary conditions.

The standard boundary conditions for Sturmians $\Phi_i(E)$ are:

- $E < 0$ Bound-state like; normalizable.
- $E > 0$ Purely outgoing/radiating waves; non-normalizable.

The spectrum of eigenvalues is purely discrete, and bounded absolutely. For short-range (nuclear-type) potentials, the eigenvalues can accumulate around 0 only.

Most interestingly, with the use of this sturmian representation, the coupled-channel S-matrix can be expressed in the form [2]:

$$S_{cc'}(E) = \delta_{cc'} - i^{(l_{c'}-l_c+1)}\pi\sqrt{k_c k_c'}\sum_i \Gamma_{ci}(E^{(+)};k_c)\frac{1}{1-\gamma_i(E^{(+)})}\Gamma_{c'i}(E^{(+)};k_{c'}) \tag{7}$$

where $\Gamma_{ci}(E^{(+)};k_c)$ denote the coupled-channel functions $V\Phi_i(E)$ in momentum-space representation.

It is remarkable that the following interpretation can be given to the last expression: the scattering process initiated in the asymptotic channel c is "captured" into Sturmians. Subsequently the Sturmian propagates freely in the interaction region and finally decays into the outgoing channels. This structure naturally reflects the description of the scattering process in terms of compound nucleus formation, and leads to an expression which is rather similar to that obtained in R-matrix formalism. However, in the Sturmian approach, such structure emerges directly from the Hamiltonian, while in the R-matrix formalism, the resonant (compound) structure is modeled phenomenologically in terms of specific boundary conditions given at the surface of an hypothetical R-space sphere.

Resonant structures can be obtained in terms of the properties of Sturmian eigenvalues. A resonance occurs when the eigenvalue, initially progressing along the the real axis, becomes complex before reaching the point (1,0). Such occurs for positive energies as the scattering threshold ($E = 0$) must be passed. The energy centroid of the resonance is the energy corresponding to the real part of sturmian eigenvalue matching the value 1. The width of the resonance can also be determined geometrically by the patterns of the sturmian trajectories, and relates to the imaginary part of the sturmian eigenvalue at the resonant energy. In figure 1 one observes this situation in a realistic case, namely $n+^{12}$C elastic scattering. The eigenvalue trajectories (eigenvalues labeled "(1)" and "(2)") produce two overlapping $\frac{3}{2}^+$ resonances that can be clearly seen in the experimental cross-section as well as in the theoretical calculation.

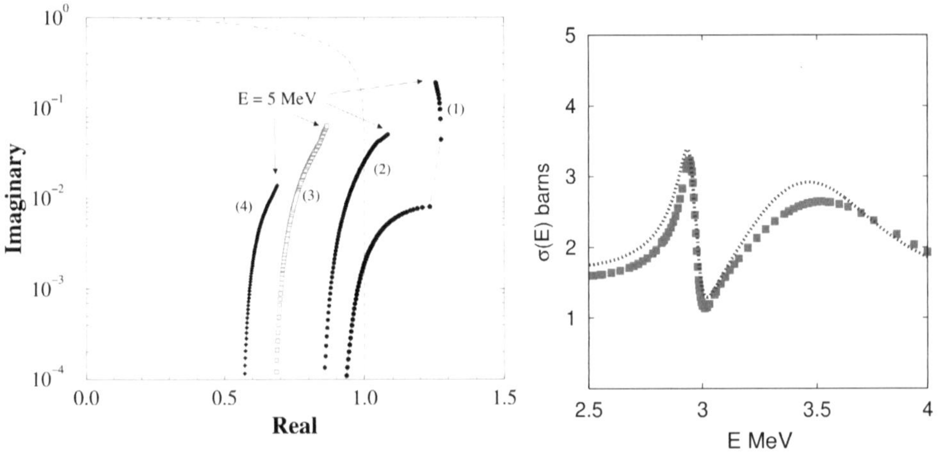

Figure 1. (Color online) n-^{12}C scattering and low-energy resonances in $\frac{3}{2}^+$ channel of the n-^{12}C system. Sturmian-eigenvalue patterns (left) and $\frac{3}{2}^+$ resonant cross-section (right). The dashed line in the left panel denotes the unit circle in the complex plane.

Table 1. The parameter values defining the nucleon + mass-14 potential.

		Odd parity	Even parity
V_{central} (MeV)		-44.2	-44.2
V_{ll} (MeV)		0.42	0.42
V_{ls} (MeV)		7.0	7.0
V_{ss} (MeV)		0.0	0.0
Geometry	$R_0 = 3.1$ fm	$a = 0.65$ fm	$\beta_2 = -0.5$
Coulomb	$R_c = 3.1$ fm	$a_c =$ —	$\beta_2 = 0.0$
	$0s_{1/2}$	$0p_{3/2}$	$0p_{1/2}$
0^+ $\lambda^{(OPP)}$	1000	1000	1000
2^+ $\lambda^{(OPP)}$	1000	1000	3.11
4^+ $\lambda^{(OPP)}$	1000	1000	3.87

4. The case of ^{15}F

The low-excitation spectrum of ^{14}O has a $0^+;1$ ground state followed by a band starting 0.545 MeV above the proton-^{13}N threshold and consisting of a $1^-;1$ state at 5.173 MeV, and then known proton-unstable resonances with spin-parities of $0^+;1$ at 5.920 MeV, of $3^-;1$ at 6.272 MeV, and of $2^+;1$ at 6.590 MeV [5].

Of the low energy spectrum, the 0_2^+ and 2^+ were selected for coupling, as in the mirror n+^{14}C system they generated the negative low-energy parity states in ^{15}C [6]. In this calculation, both are taken as zero-width (ignoring their known proton-decay widths) while in a more recent calculation [7], we took into account the upper limits of the known width of the two excited states 0_2^+ and 2^+. (The method for including couplings to resonant levels is explained later on in this work)

The coupled-channel nuclear interaction was generated assuming the collective model with rotor character for the structure [8]. Parameter values were chosen to be those used in Ref. [6]. For completeness, they are given in table 1. In this case, the proton distribution is taken as that of a uniformly charged sphere of radius 3.1 fm.

Table 2. Low-lying levels of ^{15}F. Only the lowest two states were known experimentally [10] at the time of the calculation [6]. The remaining three have been observed in Ref. [9] .

J^π	Theory $E, (\frac{1}{2}\Gamma)$	Experiment $E, (\frac{1}{2}\Gamma)$
$\frac{1}{2}^+$	1.31 (0.8)	1.47 (1.00)
$\frac{5}{2}^+$	2.78 (0.3)	2.77 (0.24)
$\frac{1}{2}^-$	5.49 (0.005)	4.9 (<0.2)
$\frac{5}{2}^-$	6.88 (0.01)	6.4 (<0.2)
$\frac{3}{2}^-$	7.25 (0.04)	
$\frac{1}{2}^+$	7.21 (1.2)	
$\frac{5}{2}^+$	7.75 (0.4)	7.8 (0.4) ?
$\frac{3}{2}^+$	7.99 (3.6)	?

The compound system, ^{15}F, is particle unstable and, until recently, only its ground and first excited resonance states were known. We used known properties of ^{15}C to predict [6] new states in ^{15}F, in particular three narrow resonances of negative parities $\frac{1}{2}^-$, $\frac{3}{2}^-$, and $\frac{5}{2}^-$ in the range 5-8 MeV of excitation.

Three years later, definite spin-parity assignments have been made for the lowest three experimentally known states while the fourth and fifth given in Ref. [9], are tentatively $\frac{3}{2}^-$ or $\frac{5}{2}^-$ for the 6.4 MeV state and $\frac{3}{2}^+$ or $\frac{5}{2}^+$ for the 7.8 MeV state. These resonance state centroids are compared with the spectra predicted from our MCAS calculations in figure 2. The measured states pair up quite reasonably with the calculated ones. Further comparison between the predictions and the experiments is made in table 2.

It should be noted that the parameters that produced this spectrum were fitted only to the observed $\frac{1}{2}^+$ and $\frac{5}{2}^+$ states, before the results of Ref. [9] were published.

5. The case of ^{17}C

We have made MCAS calculations of the n+^{16}C system using three states in ^{16}C; the 0^+ (ground), 2^+ (MeV), and 4^+ (MeV) states in a bid to improve an earlier calculation of the same system [11]. We assume that couplings to other states (presumed $0_2^+, 2_2^+$, 3^+) in the spectrum between 3 and 5 MeV excitation are not essential and that the rotational model, as defined previously [1], suffices in seeking the spectrum of ^{17}C. The parameter values required to get the results displayed in figure 3 are listed in table 3. The potential parameter values and Pauli blocking/hindrance weights are quite similar to the set used previously [11], now with inclusion of a small hexadecapole deformation to link the 4^+ state to the ground in first order, and with some Pauli hindrance of the $2s_{\frac{1}{2}}$ orbit in the connections to the 4^+ target state.

The spectrum that results is shown in figure 3 and labelled 'mcas(0+2+4)'. It is compared with the spectrum found previously from the 2-state MCAS evaluation ('mcas(0+2)'), and with the experimental one that is a combination of states listed in table 4 of Ref. [12] and in figure 5 of Ref. [13]. The latter are displayed by the dash-dot lines. Only positive parity states are known in the low excitation spectrum and (twice) their spins are given as the integers associated with the individual levels shown.

Clearly the three known subthreshold (n+^{16}C) states are now matched well in energy and spin-parity by the new MCAS results. The other known and uncertain spin-parity states also have matching MCAS partners in proximity of their excitation energies. Additionally, the uncertain states from Raimann *et al.* [13] seem to have possible matches with the first low

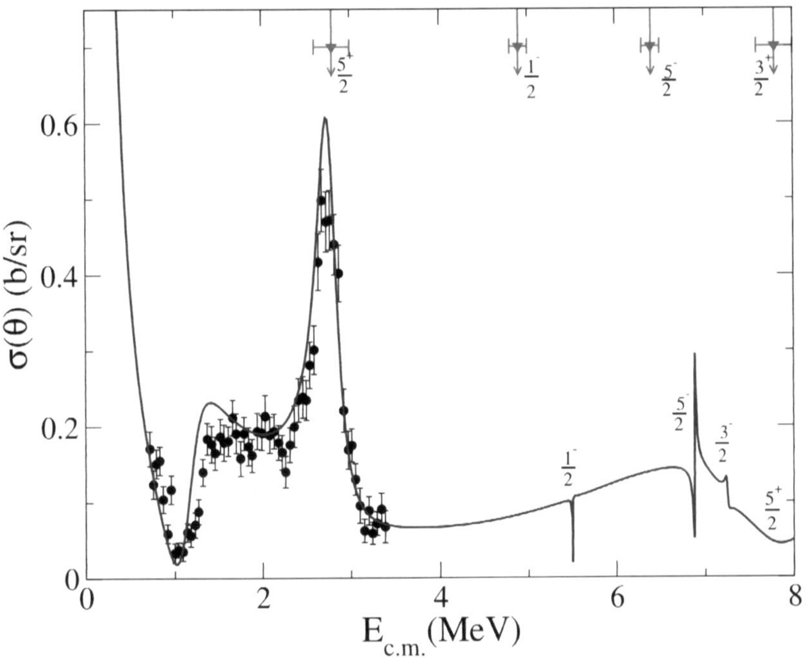

Figure 2. (Color online) MCAS calculation [6] of the differential cross section at $\theta = 147°$, c.m. The horizontal bars on the down arrows at the top indicate the widths (FWHM) of the experimental ^{15}F spectrum [9].

Table 3. The parameter values used to define the channel coupling properties of the n+^{16}C system. Energy units are MeV, length units are fm.

V_0	V_{ll}	V_{ls}	V_{Is}
-37.0	-2.0	9.0	1.6
R	a	β_2	β_4
2.9	0.8	0.33	0.1
state in ^{16}C		OPP λ_{lj}	
	$(1s_{\frac{1}{2}}, 1p_{\frac{3}{2}}, 1p_{\frac{1}{2}})$	$1d_{\frac{5}{2}}$	$2s_{\frac{1}{2}}$
0^+ (0.000)	10^6	2.7	0.0
2^+ (1.766)	10^6	2.7	0.0
4^+ (4.142)	10^6	0.0	2.0

lying state above threshold seeming to be a $\frac{7}{2}^+$ resonance. The MCAS spectrum has a number of aspects in common with that shown in figure 1 of Ref. [14]. Besides the three closely spaced and weakly bound sub-threshold states, there is a cluster of $\frac{3}{2}^+, \frac{5}{2}^+, \frac{7}{2}^+$, and $\frac{9}{2}^+$ states in the region around 3 MeV excitation and a second such cluster in the region of 5 MeV excitation. There is also a higher excited $\frac{1}{2}^+$ state found with both calculations above 6 MeV excitation, notable by being very broad ($\Gamma = 2.46$ MeV with MCAS). Our MCAS result shows more states, including one of spin-parity $\frac{11}{2}^+$ at 4.24 MeV excitation.

More specifics of the states in the ^{17}C spectrum are listed in table 4. Columns 1 to 3 display values of spin-parity J^π, excitation energy (or centroid) E, and widths Γ, of resonances

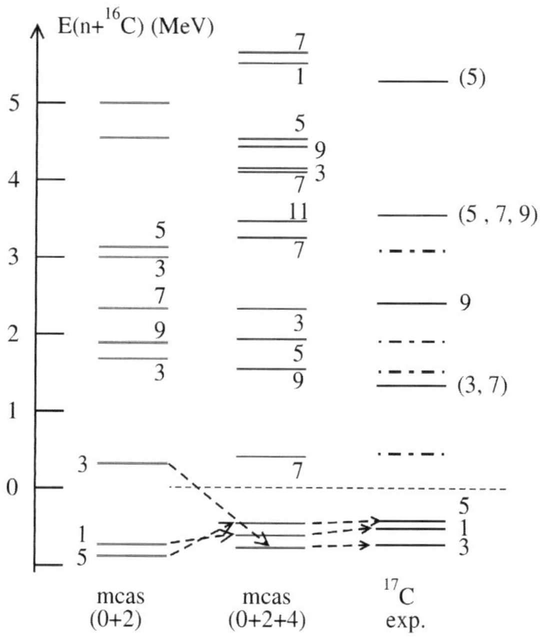

Figure 3. (Color online) The spectrum of ^{17}C compared with the MCAS results.

Table 4. Spectra for ^{17}C. Units are MeV.

	Ref. [12]		Ref. [13]		MCAS	
J^π	E	Γ	E	J^π	E	Γ
$\frac{3}{2}^+$	0.00	—	0.000	$\frac{3}{2}^+$	0.00	—
$\frac{1}{2}^+$	0.21	—	0.292	$\frac{1}{2}^+$	0.16	—
$\frac{5}{2}^+$	0.31	—	0.295	$\frac{5}{2}^+$	0.22	—
			(1.18)	$\frac{7}{2}^+$	1.17	10^{-8}
$(\frac{3}{2}^+ - \frac{7}{2}^+)$	2.06	0.25	(2.25)	$\frac{9}{2}^+$	2.31	3x10^{-5}
			(2.64)	$\frac{5}{2}^+$	2.70	0.018
$\frac{9}{2}^+$	3.10	0.10		$\frac{3}{2}^+$	3.01	0.096
			(3.82)	$\frac{7}{2}^+$	4.03	3x10^{-5}
$(\frac{5}{2}^+, \frac{7}{2}^+, \frac{9}{2}^+)$	4.25	0.14		$\frac{11}{2}^+$	4.24	2x10^{-5}
				$\frac{7}{2}^+$	4.88	0.010
				$\frac{3}{2}^+$	4.94	1.78

ascertained in a study [12] of three neutron transfer cross sections for ^{12}C scattering from ^{14}C. The excitation energies of states shown in figure 5 of Ref. [13] are listed in column 4, while the MCAS results are displayed in columns 5, 6, and 7. The three nucleon transfer reaction widths in general do not match those from the MCAS evaluation but the two sets are quite different; the latter being single nucleon removal values. The cluster of resonant states found using MCAS in the region of 3 MeV excitation have widths that agree well with the matching ones from the MCM evaluation of Timofeyuk and Descouvement [14]. Specifically their widths quoted in Ref. [14] are $\frac{7}{2}^+_1$ (10^{-12} MeV), $\frac{9}{2}^+_1$ (10^{-6} MeV), and $\frac{5}{2}^+_1$ (0.015 MeV), while the $\frac{3}{2}^+_1$

resonance width of 0.265 MeV is slightly larger than the MCAS value of 0.096.

Now, it is said that ^{17}C has a 'peculiar' structure connected perhaps with the neutron separation energy from the ground state being only 0.728 keV. That is typical of a halo nucleus. Indeed the coupled channel interaction we require to give the spectrum of ^{17}C reflects that with diffuseness being large. As well, features of this interaction are just as identified by Timofeyuk and Descouvement [14] as being required in a two-body potential model, *viz.* "*The bound* ^{17}C *spectrum cannot be understood in the two-body potential model with deformation and the* 2^+ *excitation of the* ^{16}C *core either, if standard sets of potentials are used. An* ℓ*-dependent and nonstandard spin-orbit* $n+^{16}$C *potential must be used for these purposes.*"

6. The MCAS formulation for unstable target states

The S-matrix equations in the MCAS methodology take the form given in Eq.(3). Traditionally, all target states are taken to have eigenvalues of zero width. Then the integrals in the (complex) Green's functions are evaluated using the method of principal parts, where, in the limit $\epsilon \to 0$, the Green's functions take the form given in Eq.(4). This method assumes time evolution of target states is given by

$$|x,t\rangle = e^{-\mathrm{i}H_0 t/\hbar}\,|x,t_0\rangle = e^{-\mathrm{i}E_0 t/\hbar}\,|x,t_0\rangle\,. \tag{8}$$

However, if states decay, they evolve as

$$|x,t\rangle = e^{-(\Gamma t/2\hbar)}\,e^{-\mathrm{i}E_0 t/\hbar}\,|x,t_0\rangle\,. \tag{9}$$

Thus, in the Green's function, channel energies become complex, as do the squared channel wave numbers,

$$\hat{k}_c^{\,2} = \mu\left(E - \epsilon_c + \mathrm{i}\tfrac{\Gamma_c}{2}\right)\ ;\ \hat{h}_c^{\,2} = \mu\left(\epsilon_c - E - \mathrm{i}\tfrac{\Gamma_c}{2}\right), \tag{10}$$

where $\frac{\Gamma_c}{2}$ is half the width of the target state associated with channel c. The Green's function matrix elements are then

$$\begin{aligned}[\mathbf{G}_0]_{nn'} &= \mu\left[\sum_{c=1}^{\text{open}}\int_0^\infty \hat{\chi}_{cn}(x)\frac{x^2}{k_c^2 - x^2 + \mathrm{i}\mu\frac{\Gamma_c}{2}}\,\hat{\chi}_{cn'}(x)\,dx\right.\\ &\left.\quad - \sum_{c=1}^{\text{closed}}\int_0^\infty \hat{\chi}_{cn}(x)\frac{x^2}{h_c^2 + x^2 - \mathrm{i}\mu\frac{\Gamma_c}{2}}\,\hat{\chi}_{cn'}(x)\,dx\right].\end{aligned} \tag{11}$$

When poles are moved significantly off the real momentum axis, integration along this axis is feasible.

7. The case of ^{9}Be

The low excitation spectrum of ^{8}Be has a 0^+ ground state that decays into two α-particles (width: 6×10^{-6} MeV), a 2^+ resonance state (centroid: 3.03 MeV; width: 1.5 MeV) and a 4^+ resonance state (centroid: 11.35 MeV; width: $\sim$3.5 MeV). In table 5 are given the parameters in the channel-coupling potentials, as well as the λ^{OPP} values giving the strengths of Pauli hindrance. Two evaluations of the $n+^8$Be cross section are obtained using this spectrum and set of parameter values; one with target-state widths set to zero, and the other taking into account the widths of the excited levels. The two results will be identified by the terms 'no-width' and 'width', respectively. Both cases are calculated with the same nuclear interaction, taken from a rotor model [7]. In table 6 , the resonance widths of the ^{9}Be states that result from these calculations are given and they are compared with experimental values.

Table 5. The parameter values defining the n+^{8}Be interaction.

	Odd parity		Even parity
V_{central} (MeV)	-31.5		-42.2
V_{ll} (MeV)	2.0		0.0
V_{ls} (MeV)	12.0		11.0
V_{ss} (MeV)	-2.0		0.0
Geometry	$R_0 = 2.7$ fm	$a = 0.65$ fm	$\beta_2 = 0.7$
	$0s_{1/2}$		$0p_{3/2}$
0^+ $\lambda^{(OPP)}$	1000		0.0
2^+ $\lambda^{(OPP)}$	2.0		0.2
4^+ $\lambda^{(OPP)}$	0.0		0.0

Table 6. Widths of resonances in n+^{8}Be scattering for the no-width and width cases, compared to experimental values.

J^π	$\Gamma_{exp.}$	$\Gamma_{no-width}$	Γ_{width}
$\frac{1}{2}^+$	0.217±0.001	—	1.595
$\frac{7}{2}^-$	1.210±0.230	2.08×10^{-5}	1.641
$\frac{1}{2}^-$	1.080±0.110	0.495	1.686
$\frac{5}{2}^+$	0.282±0.011	0.187	0.740
$\frac{3}{2}^-$	1.330±0.360	0.466	3.109
$\frac{5}{2}^-$	7.8×10^{-4}	0.060	2.772
$\frac{9}{2}^+$	1.330±0.090	0.386	2.498
$\frac{3}{2}^+$	0.743±0.055	3.286	5.162

Taking the excited states of ^{8}Be to be resonances gives the same spectral list as when they are treated as zero-width, but the evaluated widths of the compound nuclear states significantly increase, as reflected in the cross sections. These increases bring the theoretical ^{9}Be state widths closer, often significantly, to experimental values, as evidenced by comparison of the final columns of the two tables. Only the $\frac{5}{2}^-$, also poorly recreated in centroid, has a worse match with widths applied. The resultant elastic and reaction cross sections are shown in figure 4. The line labeled *no widths* depicts the results obtained without considering the decay widths in the target states, and the line labeled *widths* does take these decay widths into consideration, as constant values multiplied by the Heaviside (step) function: $H(E) = 0$ for $E \leq 0.0$; $H(E) = 1$ for $E > 0$. (For the line labeled $A = 0.5$, see section 8.)

Introducing target state widths, the resonances in ^{9}Be are suppressed but still present; their widths increasing and magnitudes decreasing so as all but the $\frac{5}{2}^+_1$, and arguably $\frac{5}{2}^+_2$, cannot be discerned from the background. The reaction cross section found in the no-width case is zero until 3.4 MeV (3.03 MeV in the centre of mass frame), the first inelastic threshold. When the target state has width different from zero the elastic cross section has flux loss from zero projectile energy upwards. There is an asymptotic behaviour as energy approaches zero. In the reaction cross section from the width calculation, there is a broad peak at 1.5 MeV, corresponding to

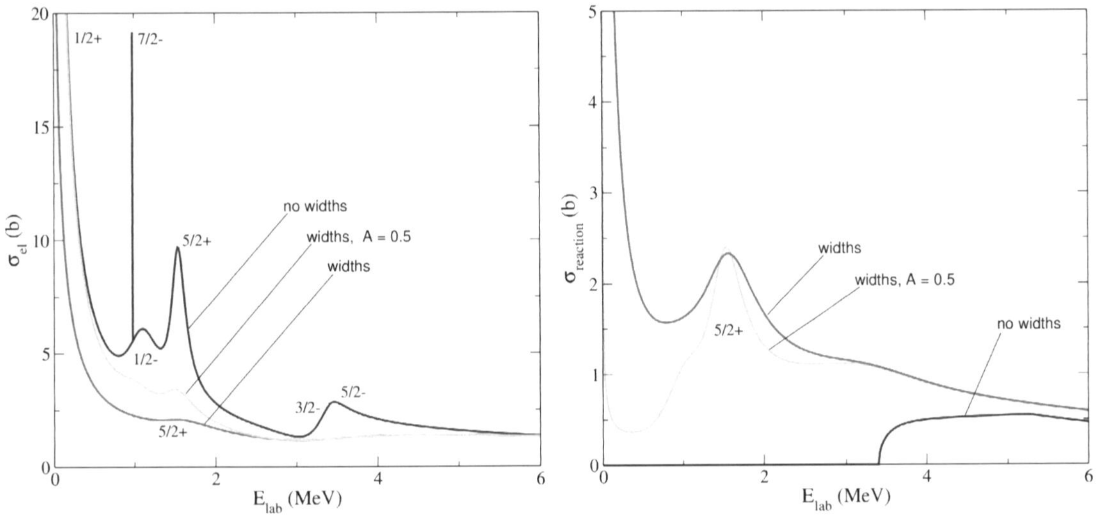

Figure 4. (Color online) Elastic (left panel) and reaction (right panel) cross section as a function of neutron laboratory energy with three treatments of the widths of target states (see text).

the $\frac{5}{2}^+$ resonance. While use of the Heaviside function ensures stability in the sub-threshold region of the compound system, it can generate 'ghost' levels where application of widths could move a (no-width) calculated sub-threshold state into the continuum. In such a case the action of the Heaviside function then creates two copies of the same state. An improvement of this technique is discussed in the next section.

8. Energy dependence of widths

A smooth description of the energy dependence of the decay width is needed to avoid any pathological behaviour caused by the Heaviside function. This function allows maintaining the stability of the compound system in the sub-threshold region, by setting to zero all widths for $E < 0$, but can generate "ghost" levels where the application of widths could move a sub-threshold state above $E = 0$, thus creating two copies of the same state. Also, a dramatic overestimation of the reaction cross-section occurs in the region close to threshold.

We therefore need to improve the model with a suitable parametrization of the energy-dependence of the decaying width for the excited levels, with the constraint that the width must be zero at the scattering threshold, and equal to the experimental width at the resonance centroid energy. We are presently exploring various forms of energy dependencies for the widths of these excited states ($\Gamma_c(E) = \Gamma_c \times U(E)$). Examples of such expressions are

$$U(E) = \frac{(1+A)}{(1+A)^2 - 1} \left\{ \frac{1+A}{A(1-x)^2+1} - \frac{1}{A(x^2)+1} \right\} H(E) \tag{12}$$

where $x = E/E_r$, $H(E)$ is the Heaviside function defined above and A is a parameter, and the Gaussian form

$$U(E) = e \left(\frac{E}{E_r} \right)^2 e^{-\left(\frac{E}{E_r}\right)} H(E). \tag{13}$$

In this paper we study only the first form. The second, and others, will be explored in future work. Also yet to be taken into account is causality, as needed when dealing with energy-dependent widths [15].

In figure 5 we give the induced energy dependence for the function $U(x = E/E_r)$ with respect to the renormalized energy x. The curves differ for different values of the parameter A. The line labeled $A = 0.5$ in figure 4 shows the effect of this energy-dependent width on elastic and reaction cross sections, respectively, when $A = 0.5$. Clearly, for this value of A, the asymptotic behaviour in the reaction cross section at $E = 0.0$ is suppressed, but not resolved. In principle, an energy

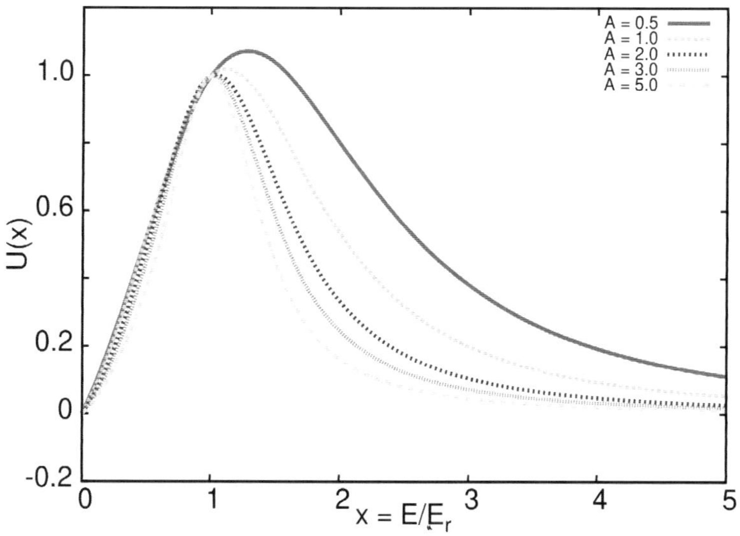

Figure 5. (Color online) The scaling function, $U(E/E_r)$, for various values of the parameter A.

dependence in the widths induces a corresponding energy-dependence in the positioning of the resonance centroids. This additional effect is a direct consequence of the principle of causality which demands that the real and imaginary part of any Green's function matrix elements be related by a dispersion integral. Such an aspect has been extensively discussed in the study of the properties of the nuclear optical potential [15]. Eq. (12) is particularly convenient in this regard because the corresponding dispersion integral leads to an analytical energy-dependent shift of the resonance centroids [16].

9. Conclusions

The MCAS approach has been used with isospin mirror mass-15 systems to define the spectroscopy of the particle-unstable nucleus, ^{15}F. The procedure involved first making an analysis of the neutron-rich ^{15}C system for which experimental information is known. Crucial to the description of the experimental spectrum was the concept of Pauli-hindrance of single-particle orbits coupled to the collective 0_2^+ and 2^+ excitations in the mass-14 nuclei. It leads to an appropriate description of the observed three low-lying negative-parity resonances. Then, by incorporating Coulomb interactions, the same nuclear force was used to analyze the proton-^{14}O case and thus to predict the spectroscopy of ^{15}F up to 8 MeV excitation. Experimental information concurs with those predictions.

As a prelude to a study of the low-excitation spectrum for the exotic nucleus ^{17}Na treated as a p+^{16}Ne system, we have used MCAS to define a nuclear interaction to give a reasonable low excitation spectrum of the mirror system, ^{17}C, when treated as a n+^{16}C coupled-channel problem. Three states, a 0^+ (ground) a first excited 2^+ ($\sim$ 1.7 MeV), and a 4^+ ($\sim$ 4 MeV), were taken as the target states. Both sub-threshold and low excitation resonances for ^{17}C with centroid energies in quite reasonable agreement with the limited set of known values were found. The ^{17}C spectrum contains both sharp and narrow resonances. By adding Coulomb interactions for the interaction between a proton and the individual states of ^{16}Ne, the spectrum of the nucleon emissive ^{17}Na should also contain narrow and broad resonances. However, this system involves target states that are resonances and so is a work in progress at this stage.

An extension to the MCAS formalism that considers particle-decay widths of target nucleus eigenstates has been applied to a range of light mass nuclear targets [7], in addition to the results for n+^{8}Be shown in this paper. To ensure that any resonance aspect of target states does not have influence below the nucleon-nucleus threshold in our formalism, we have used a Heaviside function to cut off tails of the positive-energy resonances below zero energy of the compound system. This procedure, however, is too simplistic since it introduces non-physical singularities in cross sections as the energy tends to zero, and may also incorrectly overestimate reaction cross just above the threshold.

A better energy-dependent scaling factor is needed: one with value one at the centroid energy and approaching zero for lower and higher energies. We propose here possible means of achieving such modified Lorentzians, and show, for one example, their effect on the elastic and reaction cross sections in the scattering of neutrons from ^{8}Be. Work is in progress to apply these and other forms to the light-nuclear systems considered in our previous work. Work is in progress, also, for preserving causality in the presence of energy-dependent widths by the use of appropriate dispersion relations.

Acknowledgments
This work is supported by the National Research Foundation, South Africa, the Natural Sciences and Engineering Research, Council, Canada, INFN, Italy and the Australian Academy of Science.

References

[1] Amos K *et al.* 2003 *Nucl. Phys. A* **728** 65
[2] Rawitscher G H and Canton L 1991 *Phys. Rev. C* **44** 60
[3] Canton L *et al.* 2005 *Phys. Rev. Lett.* **94** 122503
[4] Fraser P *et al.* 2008 *Phys. Rev. Lett.* **101** 242501
[5] Tilley D R *et al.* 2004 *Nucl. Phys. A* **745** 155
[6] Canton L *et al.* 2006 *Phys. Rev. Lett.* **96** 072502
[7] Fraser P R *et al.* 2011 *Rev. Mex. Fis.* **57** 20
[8] Tamura T 1965 *Rev. Mod. Phys.* **37** 679
[9] Mukha I *et al.* 2009 *Phys. Rev. C* **79** 061301 (R)
[10] Goldberg V Z *et al.* 2004 *Phys. Rev. C* **69** 031302
[11] Karataglidis S *et al.* 2008 *Nucl. Phys. A* **813** 235
[12] Bohlen H G *et al.* 2007 *Eur. Phys. J. A* **31** 279
[13] Raimann G *et al.* 1996 *Phys. Rev. C* **53** 453
[14] Timofeyuk N K and Descouvemont P 2011 *Phys. Rev. C* **81** 051301
[15] Mahaux C, Ngô H, and Satchler G R 1986 *Nucl. Phys. A* **449** 354
[16] Canton L *et al.* 2011 *Phys. Rev. C* **83** 047603

Study of parity violating observables in few-nucleon systems

M Viviani [a], A Baroni [b], R Schiavilla [b,c], L Girlanda [d], A Kievsky [a] and L E Marcucci [e,a]

[a]INFN-Pisa, 56127 Pisa, Italy
[b]Department of Physics, Old Dominion University, Norfolk, VA 23529, USA
[c]Jefferson Lab, Newport News, VA 23606
[d]Department of Physics, University of Salento & INFN-Lecce, 73100 Lecce, Italy
[e]Department of Physics, University of Pisa, 56127 Pisa, Italy

E-mail: michele.viviani@pi.infn.it

Abstract. Parity violation in few-nucleon systems is studied using a nucleon-nucleon parity-violating (PV) potential derived within an effective field theory framework at next-to-next-to-leading order. The potential includes one- and two-pion exchanges, contact interactions and relativistic corrections and depends on six low-energy constants: the pion-nucleon coupling constant h^1_π and five parameters multiplying the independent contact interaction terms (with one four-gradient). This potential is used to study the $\vec{p}$-p longitudinal asymmetry, the neutron spin rotation in $\vec{n}$-d scattering, and the longitudinal asymmetry in the ^{3}He$(\vec{n},p)^3$H reaction.

1. Introduction

A number of experiments aimed at studying parity violation in low-energy processes involving few nucleon systems are being completed or are in an advanced stage of planning at cold neutron facilities, such as the Los Alamos Neutron Science Center, the NIST Center for Neutron Research, and the Spallation Neutron Source at Oak Ridge. The primary objective of this program is to determine the fundamental parameters of the parity-violating (PV) nucleon-nucleon (NN) potential [1].

Until recently, the standard framework by which nuclear PV processes were analyzed theoretically was based on the use of potentials derived from meson-exchange mechanisms. In particular, the model proposed by Desplanques, Donoghue, and Holstein (DDH) [2] over thirty years ago consisted of π, ρ, and ω exchanges, and contained seven unknown meson-nucleon coupling constants.

On the other hand, in recent years, a new, more systematic, approach based on a model-independent field-theoretic treatment of nuclear forces has been developed. This approach uses the chiral symmetry exhibited by quantum chromodynamics (QCD) to severely restrict the form of the interactions of pions among themselves and with other particles [3]. In particular, the pion couples to nucleons by powers of its momentum Q, and the Lagrangian describing these interactions can be expanded in powers of Q/Λ_χ, where $\Lambda_\chi \sim 1$ GeV specifies the chiral-symmetry breaking scale. As a consequence, classes of Lagrangians emerge, each characterized by a given power of Q/Λ_χ and each involving a certain number of unknown coefficients, the so

Published under licence by IOP Publishing Ltd

called low-energy constants (LEC's), which are then determined by fits to experimental data (see, for example, the review papers [4, 5, 6, 7], and references therein).

This effective field theory (EFT) approach has been used also to describe the PV components in the NN interaction. One starts by assuming that these components derive from the weak interactions between the quarks inside the hadrons. As is well known, in the "Standard Model" the weak interactions contains both parity-conserving (PC) and PV components. The PC part contributes to the PC NN interaction and is of no interest, since its contribution is totally dominated by the strong (and electromagnetic) interaction. In contrast, the part of the weak interaction which violates parity (P) gives origin to a PV component in the NN potential. Note that the combination of charge conjugation and parity ($\mathcal{CP}$) is known to be violated to a much lesser extent than $\mathcal{P}$, and, therefore, it is reasonable to consider only $\mathcal{P}$-violating but $\mathcal{CP}$-conserving terms. Clearly, the weak Lagrangian at the quark level is not symmetric under chiral symmetry, however, one can construct the most general PV Lagrangian of nucleons and pions requiring that chiral symmetry be violated in the same way as is violated at quark level.

Following this approach, Kaplan and Savage in a pioneering work [8] wrote down a PV effective Lagrangian describing the interaction of pions and nucleons up to one derivative. This Lagrangian includes a "Yukawa" pion-nucleon interaction with no derivatives, multiplied by a parameter denoted as h_π^1 and known as the "weak pion-nucleon" coupling constant. It leads to a long-range, one-pion exhange (OPE) contribution to the PV NN interaction. Several PV experiments are devoted to reveal this component and to obtain a measurement of h_π^1. So far, however, the experimental evidence for its presence is inconclusive, for a review see Ref. [9]— see also a very recent study, using lattice simulations, to determine h_π^1 [10]. The Lagrangian derived by Kaplan and Savage includes five additional pion-nucleon interaction terms with one derivative, each of them multiplied by a different low-energy constant. However, they do not enter the PV NN interaction at either tree level or one loop [8].

After the work by Kaplan and Savage, there have been several studies of the PV NN interaction in EFT. The PV NN potential at next-to-next-to-leading ($\mathrm{N^2LO}$) was derived for the first time by Zhu and collaborators [11, 12]. This potential includes the long-range OPE component, medium-range components originating from two-pion exchange (TPE) processes, and short-range components deriving from *ten* four-nucleon contact terms (the Authors of Ref. [12] noted that at low energy their ten contact interactions collapse into *five* independent operators, corresponding to the five S–P low-energy PV amplitudes [13]). In a series of other papers, Desplanques and collaborators., have also derived the contribution of the TPE diagrams at $\mathrm{N^2LO}$ [14, 15, 16] to study, in particular, PV effects in the capture reaction $^1\mathrm{H}(\vec{n},\gamma)^2\mathrm{H}$. The expression of the TPE contribution obtained by Desplanques *et al.* is slightly different from that reported by Zhu *et al.* [12].

In the present contribution, we report on a new derivation of the PV NN potential at $\mathrm{N^2LO}$, aimed at clarifying the correct form of its TPE component. Moreover, the recent analysis of Ref. [17] has shown that there exist only *five* independent contact terms with one derivative. In view of these developments, it is worthwhile to perform new calculations of various PV observables which can be measured in $A \leq 5$ nuclear systems, using this updated PV potential.

Measurements are already available (or will become available in the near future) for the following PV observables: the longitudinal analyzing power in $\vec{p}$-p [18, 19, 20] and $\vec{p}$-α [21] scattering, the photon asymmetry and photon circular polarization in, respectively, the $^1\mathrm{H}(\vec{n},\gamma)^2\mathrm{H}$ [22]–[23] and $^1\mathrm{H}(n,\vec{\gamma})^2\mathrm{H}$ [24] radiative captures, and the neutron spin rotation in $\vec{n}$-α scattering [25, 26, 27]. There is also a set of experiments which are currently being planned, including measurements of the neutron spin rotation in $\vec{n}$-p [25] and $\vec{n}$-d [28] scattering, and of the longitudinal asymmetry in the charge-exchange reaction $^3\mathrm{He}(\vec{n},p)^3\mathrm{H}$, at cold neutron energies [29].

In this contribution we present a preliminary study of the $\vec{p}$-p longitudinal asymmetry, the

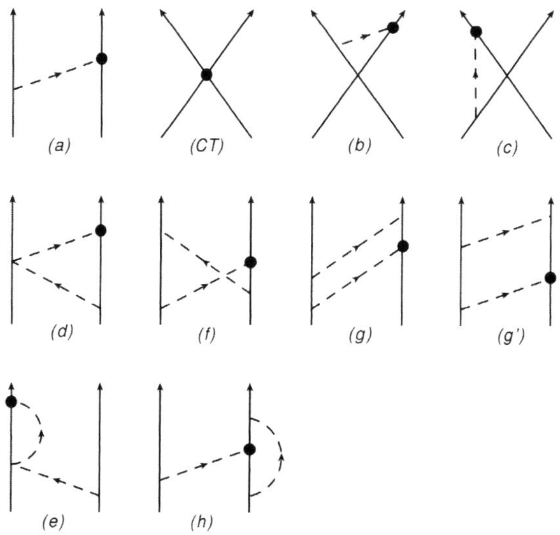

Figure 1. Time-ordered diagrams contributing to the PV potential (only one time ordering is given). Nucleons and pions are denoted by solid and dashed lines. The solid dot represents a PV vertex.

neutron spin rotation in $\vec{n}$-d scattering, and the longitudinal asymmetry in the ^{3}He$(\vec{n},p)^3$H reaction.

2. The parity-violating potential

As discussed above, the weak interaction between quarks induces a PV NN potential. This potential can be constructed starting from an effective pion-nucleon Lagrangian including all terms which violate chiral symmetry in the same way as at the quark level. The general expression for the Lagrangian describing PV pion-nucleon interactions up to one 4-gradient has already been given by Kaplan and Savage [8]. Beyond pion-nucleon interaction terms, the PV Lagrangian at $\mathcal{O}(Q)$ also involves four-nucleon contact terms, which represent the part of the potential due to the exchange of heavy mesons, the excitation of Δ-resonances, etc. The number of independent four-nucleon contact terms (up to one 4-gradient) is 5, as shown in Ref. [17].

More recently, we have started a program to identify the minimal set of both pion-nucleon and contact interaction terms containing two or more 4-gradients, by imposing the constraints arising from discrete symmetries and Fierz identities. A preliminary analysis indicates that *no* new terms enter the PV potential at N^{2}LO (further details of this study will be reported elsewhere [31]). Therefore, the PV potential (at N^{2}LO) depends on 6 unknown coupling constants: h^1_π plus the low-energy constants (LEC's) multiplying the 5 independent contact interactions with one 4-gradient.

For clarity, we report below the expressions of the "leading" πN interactions:

$$\mathcal{L}^{PC}_{\pi NN}(x) = -\frac{g_A}{2f_\pi}\overline{N}(x)\gamma^\mu\gamma^5\vec{\tau}\cdot\partial_\mu\vec{\pi}(x)N(x)+\dots\,, \qquad (1)$$

$$\mathcal{L}^{PV}_{\pi NN}(x) = \frac{h^1_\pi}{\sqrt{2}}\overline{N}(x)\Big(\vec{\tau}\times\vec{\pi}(x)\Big)_z N(x)+\dots\,, \qquad (2)$$

where $g_A \approx 1.26$, $f_\pi \simeq 93$ MeV is the pion decay constant, and h^1_π is the PV πN coupling constant. The different diagrams contributing to the PV NN potential at $\mathcal{O}(Q)$ are shown in figure 1. Note that in our notation, the chiral counting is given as

(i) leading πNN vertex: PC $\sim Q^{1/2}$, PV $\sim Q^{-1/2}$
(ii) leading $\pi\pi NN$ vertex: $\sim Q^0$ for both PC and PV vertices
(iii) leading $NN \to NN$ vertex: PC $\sim Q^0$, PV $\sim Q^1$
(iv) loops: Q^3
(v) energy denominators with pions $\sim Q^{-1}$
(vi) energy denominators without pions $\sim Q^{-2}$

The OPE diagram (a) gives the lowest order (LO) contribution (of order Q^{-1}). Explicitly

$$V^{(a),\mathrm{LO}}_{\alpha_1'\alpha_2',\alpha_1\alpha_2} = \left(\frac{g_A h^1_\pi}{\sqrt{2}F_\pi}\right)(\boldsymbol{\tau}_1 \times \boldsymbol{\tau}_2)_z \, \frac{i\boldsymbol{k}\cdot(\boldsymbol{\sigma}_1+\boldsymbol{\sigma}_2)}{k^2+m_\pi^2} \,, \tag{3}$$

where $F_\pi = 2f_\pi$ and $\alpha_i \equiv \{\boldsymbol{p}_i, s_i, t_i\}$ ($\alpha_i' \equiv \{\boldsymbol{p}_i', s_i', t_i'\}$) specifies the initial (final) momentum, and spin and isospin projections, respectively, of particle i. Above $\boldsymbol{k} = \boldsymbol{p}_1' - \boldsymbol{p}_1 = -(\boldsymbol{p}_2' - \boldsymbol{p}_2)$ and $(\boldsymbol{\sigma})_{s_1',s_1} \equiv \boldsymbol{\sigma}_1$, etc. There is no contribution of order Q^0, but there are several contributions of order Q^1 (namely, at N^{2}LO) coming from the diagrams depicted in figure 1:

$$\begin{aligned}
V^{(a),\mathrm{REL}}_{\alpha_1'\alpha_2',\alpha_1\alpha_2} &= \left(\frac{g_A h^1_\pi}{\sqrt{2}F_\pi}\right)\left(\frac{1}{4M^2}\right)(\boldsymbol{\tau}_1 \times \boldsymbol{\tau}_2)_z \, \frac{1}{k^2+m_\pi^2} \times \\
&\quad \Big\{-4iK^2\boldsymbol{k}\cdot(\boldsymbol{\sigma}_1+\boldsymbol{\sigma}_2) + \boldsymbol{k}\cdot\boldsymbol{\sigma}_1\,(\boldsymbol{k}\times\boldsymbol{K})\cdot\boldsymbol{\sigma}_2 + \boldsymbol{k}\cdot\boldsymbol{\sigma}_2\,(\boldsymbol{k}\times\boldsymbol{K})\cdot\boldsymbol{\sigma}_1\Big\} \,,
\end{aligned} \tag{4}$$

$$V^{(d)}_{\alpha_1'\alpha_2',\alpha_1\alpha_2} = -\frac{g_A h^1_\pi}{\sqrt{2}F_\pi}\frac{1}{4\pi^2 F_\pi^2}(\boldsymbol{\tau}_1 \times \boldsymbol{\tau}_2)_z i\boldsymbol{k}\cdot(\boldsymbol{\sigma}_1+\boldsymbol{\sigma}_2)\,L(k)\,, \tag{5}$$

$$\begin{aligned}
V^{(f+g+g')}_{\alpha_1'\alpha_2',\alpha_1\alpha_2} &= -\frac{g_A h^1_\pi}{\sqrt{2}F_\pi}\frac{g_A^2}{4\pi^2 F_\pi^2}\Big[4(\boldsymbol{\tau}_1+\boldsymbol{\tau}_2)_z\, i\boldsymbol{k}\cdot(\boldsymbol{\sigma}_1\times\boldsymbol{\sigma}_2)\,L(k) \\
&\quad +(\boldsymbol{\tau}_1 \times \boldsymbol{\tau}_2)_z i\boldsymbol{k}\cdot(\boldsymbol{\sigma}_1+\boldsymbol{\sigma}_2)\Big(H(k)-3L(k)\Big)\Big] \,,
\end{aligned} \tag{6}$$

$$\begin{aligned}
V^{(CT)}_{\alpha_1'\alpha_2',\alpha_1\alpha_2} &= \frac{C_1}{\Lambda_\chi m_\pi^2}\boldsymbol{K}\cdot(\boldsymbol{\sigma}_1-\boldsymbol{\sigma}_2) + \frac{C_2}{\Lambda_\chi m_\pi^2} i\boldsymbol{k}\cdot(\boldsymbol{\sigma}_1\times\boldsymbol{\sigma}_2) \\
&+ \frac{C_3}{\Lambda_\chi m_\pi^2}(\boldsymbol{\tau}_1 \times \boldsymbol{\tau}_2)_z i\boldsymbol{k}\cdot(\boldsymbol{\sigma}_1+\boldsymbol{\sigma}_2) + \frac{C_4}{\Lambda_\chi m_\pi^2}(\boldsymbol{\tau}_1+\boldsymbol{\tau}_2)_z\boldsymbol{K}\cdot(\boldsymbol{\sigma}_1-\boldsymbol{\sigma}_2) \\
&+ \frac{C_5}{\Lambda_\chi m_\pi^2}\Big(3\tau_{1z}\tau_{2z} - (\boldsymbol{\tau}_1\cdot\boldsymbol{\tau}_2)\Big)\boldsymbol{K}\cdot(\boldsymbol{\sigma}_1-\boldsymbol{\sigma}_2)\,,
\end{aligned} \tag{7}$$

where M is the nucleon mass, $\Lambda_\chi = 4\pi f_\pi$, $\boldsymbol{K} = (\boldsymbol{p}_1 + \boldsymbol{p}_1')/2$, and

$$s = \sqrt{k^2+4m_\pi^2}\,, \qquad L(k) = \frac{1}{2}\frac{s}{k}\ln\left(\frac{s+k}{s-k}\right)\,, \qquad H(k) = \frac{4m_\pi^2}{s^2}L(k)\,. \tag{8}$$

The term $V^{(a),\mathrm{REL}}$ is a relativistic correction coming from diagram (a), $V^{(d)}$ and $V^{(f+g+g')}$ are TPE contributions, and $V^{(CT)}$ is the contribution coming from the contact interactions. The contributions of diagrams (b) and (c) vanish after integrating over the loop variable, while those of diagrams (e) and (h) (vertex corrections) can be reabsorbed by a redefinition of the coupling constant h^1_π. More details on the derivation of the potential will be reported elsewhere [31].

In order to transform this potential to r-space, we have to multiply the expressions reported above by a cutoff, which is be chosen to be

$$f_\Lambda(k) = \exp\Big(-(k/\Lambda)^4\Big)\,, \tag{9}$$

where $\Lambda = 500 \div 700$ MeV is a cutoff parameter. With such a choice (a function which depends on k only), the resulting potentials are local. The potential contains 6 unknown parameters, the pion-nucleon PV coupling constant h^1_π and 5 LEC's multiplying the contact interactions. Of course, the potential also depends on the cutoff parameter Λ.

As already mentioned, TPE diagrams have been calculated earlier by Zhu *et al.* [12] and Desplanques *et al.* [15]. We find that the expressions reported above are in agreement with those of Ref. [15], but for the contribution deriving from diagram (d), which in Ref. [15] contains a factor g_A^3 instead of g_A. This is probably due to the fact that in calculation of Ref. [15] the authors have Z-type TPE diagrams in place of the "triangle" diagram (d).

In order to estimate the magnitude of the various LEC's, we compare the chiral potential with the DDH model [2], for which there are a number of studies available in the literature. The DDH model contains a OPE contribution identical to that given in Eq. (3). The authors of Ref. [2] proposed a "reasonable range" of h^1_π values, and also a "best choice" value:

$$\text{best choice}: \ h^1_\pi = 4.56 \times 10^{-7} , \qquad \text{reasonable range}: \ 0 \leq h^1_\pi \leq 11.4 \times 10^{-7} . \tag{10}$$

Furthermore, by assuming that the contact interactions represent exchanges of heavy mesons, we can also estimate the magnitude of the LEC's C_i, $i = 1, \ldots, 5$. In practice, one compares the expression of $V^{(CT)}$ given in Eq. (7) with the DDH components coming from exchanges of ρ- and ω-mesons in the approximation $|\boldsymbol{k}| \ll m_\rho$, m_ω. Using the values for the PV ρ and ω coupling constants entering the DDH model, as estimated in Ref. [30] from an analysis of the longitudinal asymmetry in $\vec{p}p$ scattering, we find

$$C_1 \approx 0.3\times 10^{-7} , \ C_2 \approx 13\times 10^{-7} , \ C_3 \approx -0.4\times 10^{-7} , \ C_4 \approx -0.8\times 10^{-7} , \ C_5 \approx 0.7\times 10^{-7} , \tag{11}$$

suggesting that, while C_2 is rather large, the remaining LEC's are of order 1×10^{-7}.

3. Results

In this section, we study several PV observables using our PV potential. We start off by analyzing the longitudinal asymmetry in $\vec{p}$-p scattering, and then study the neutron spin rotation in $\vec{n}$-d scattering and the longitudinal asymmetry in the $^3\text{He}(\vec{n}, p)^3\text{H}$ reaction.

3.1. The $\vec{p}$-p longitudinal asymmetry

The longitudinal asymmetry A_z^{pp} in the collision $\vec{p}$-p is defined as

$$A_z^{pp}(\theta, E) = \frac{\sigma_{+1/2}(\theta, E) - \sigma_{-1/2}(\theta, E)}{\sigma_{+1/2}(\theta, E) + \sigma_{-1/2}(\theta, E)} , \tag{12}$$

where $\sigma_m(\theta, E)$ is the differential cross section for an incident proton beam of energy E (in the laboratory system) and longitudinal polarization $m = \pm 1/2$, while θ is the scattering angle. The observable actually measured in the experiment is defined as (the "average" longitudinal asymmetry)

$$\overline{A}_z^{pp}(E) = \frac{\int_{\theta_1 \leq \theta \leq \frac{\pi}{2}} d\hat{r} \ A_z^{pp}(\theta, k)\sigma(\theta, E)}{\int_{\theta_1 \leq \theta \leq \frac{\pi}{2}} d\hat{r} \ \sigma(\theta, E)} , \qquad \theta_1 \approx 15^0 , \tag{13}$$

where

$$\sigma(\theta, E) = \frac{1}{2} \left(\sigma_{+1/2}(\theta, E) + \sigma_{-1/2}(\theta, E) \right) . \tag{14}$$

To compute this observable, we need the matrix elements of the isospin operators entering the PV potential between $|pp\rangle \equiv |T = 1, T_z = +1\rangle$ states:

$$\langle T = 1, T_z = +1 | (\boldsymbol{\tau}_1 \times \boldsymbol{\tau}_2)_z | T = 1, T_z = +1 \rangle = 0 , \tag{15}$$

$$\langle T = 1, T_z = +1 | (\boldsymbol{\tau}_1 + \boldsymbol{\tau}_2)_z | T = 1, T_z = +1 \rangle = 2 , \tag{16}$$

$$\langle T = 1, T_z = +1 | 3\boldsymbol{\tau}_{1z}\boldsymbol{\tau}_{2z} - \boldsymbol{\tau}_1 \cdot \boldsymbol{\tau}_2 | T = 1, T_z = +1 \rangle = 2 . \tag{17}$$

Table 1. Coefficients $a_i(E)$ at the three energies of the experimental data, for three choices of the cutoff parameter Λ. The calculations have been performed using the PC I-N3LO NN potential model [32] and the PV potential described in the text.

E [MeV]	a_0	a_1	a_2
	$\Lambda = 500$ MeV		
13.6	0.2567	−0.0861	−0.2227
45	0.5667	−0.1832	−0.4517
221	−0.2443	−0.0864	0.1771
	$\Lambda = 600$ MeV		
13.6	0.2130	−0.0704	−0.1991
45	0.5010	−0.1522	−0.4210
221	−0.2079	−0.0525	0.1640
	$\Lambda = 700$ MeV		
13.6	0.1618	−0.0616	−0.1705
45	0.3995	−0.1331	−0.3669
221	−0.1236	−0.0287	0.1203

Thus, the contributions of the OPE terms $V^{(a),\rm LO}$ and $V^{(a),\rm REL}$ vanish and the constant h^1_π can contribute only through the potential terms associated with TPE diagrams. Among the contact terms, that multiplying C_3 does not contribute, while the terms that multiply C_1, C_4 and C_5 differ only for the isospin operator. Taking into account the matrix elements in Eqs. (15)–(17), the longitudinal asymmetry can be expressed as

$$\overline{A}_z^{pp}(E) = a_0(E)h^1_\pi + a_1(E)C'_1 + a_2(E)C_2 \ , \tag{18}$$

where

$$C'_1 = C_1 + 2C_4 + 2C_5 \ , \tag{19}$$

and $a_0(E)$, $a_1(E)$ e $a_2(E)$ are numerical coefficients independent from the values of the LEC's (however, they depend on Λ). The observable $\overline{A}_z^{pp}(E)$ has been measured by several groups. In this contribution we consider only the experiments performed at $E \equiv E_{\rm lab} = 13.6$ MeV [18], 45 MeV [19], and 221 MeV [20], where $\overline{A}_z^{pp}$ has been obtained with a reasonable accuracy:

$$\begin{aligned} \overline{A}_z^{pp}(13.6{\rm MeV}) &= (-0.97 \pm 0.20) \times 10^{-7} \ , \\ \overline{A}_z^{pp}(45{\rm MeV}) &= (-1.53 \pm 0.21) \times 10^{-7} \ , \\ \overline{A}_z^{pp}(221{\rm MeV}) &= (+0.84 \pm 0.34) \times 10^{-7} \ . \end{aligned} \tag{20}$$

The numerical values of the coefficients $a_0(E)$, $a_1(E)$ and $a_2(E)$ for these three energies, calculated using the chiral PC NN potential derived by Entem & Machleidt [32] at next-to-next-to-next-to-leading order (I-N3LO model), and our chiral PV potential, are reported in table 1, for three values of the cutoff parameter Λ.

In principle, one could fix the three unknown parameters h^1_π, C'_1, and C_2 by requiring that the longitudinal asymmetry in Eq. (18) reproduces the experimental values in Eq. (20). However, as can be seen from table 1, the values of a_i at low energy scale as $\sqrt{E}$, since the longitudinal asymmetry is dominated by the contribution of S-waves—that is, the experimental data at $E = 13.6$ MeV and 45 MeV do not provide independent constraints—and therefore, in practice, the number of independent equations reduces to two.

It is therefore necessary to fix the value of one of the constant to determine the remaining two. In the following, we assume that the h^1_π value is in the "reasonable range" given in Eq. (10),

Table 2. Coefficients C_1' and C_2 for different values of h_π^1 and Λ determined to reproduce the experimental values of $\overline{A}_z^{pp}$ at 45 and 221 MeV.

Λ [MeV]	C_1'	C_2	C_1'	C_2	C_1'	C_2
	$h_\pi^1 = 4.56 \times 10^{-7}$		$h_\pi^1 = 0$		$h_\pi^1 = 11.4 \times 10^{-7}$	
500	−2.15516	9.98171	−1.51765	4.00256	−3.11142	18.9504
600	−2.69957	10.03513	−2.18203	4.42237	−3.47588	18.4543
700	−4.22214	10.66532	−4.68110	5.86730	−3.53372	17.8623

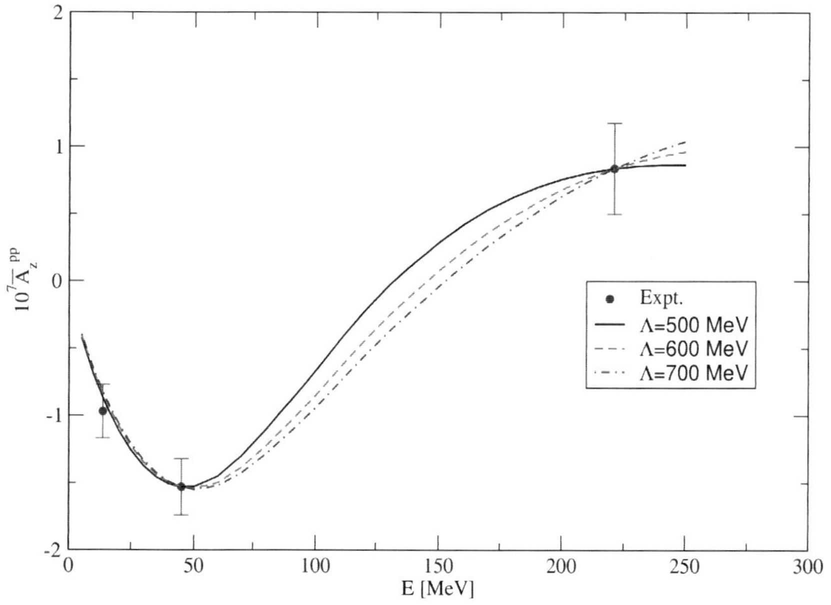

Figure 2. (Color online) Proton-proton longitudinal asymmetry as a function of the laboratory energy E for three values of the cutoff Λ. The calculation is performed using the EFT potential with $h_\pi^1 = 4.56 \times 10^{-7}$ and the values of constants C_1' and C_2 listed in table 2.

see the discussion at the end of the previous section. In particular we perform the calculations for three values of the coupling constant h_π^1:

(i) $h_\pi^1 = 4.56 \times 10^{-7}$ ("best choice")
(ii) $h_\pi^1 = 0$ (minimum value of the "reasonable range")
(iii) $h_\pi^1 = 11.4 \times 10^{-7}$ (maximum value of the "reasonable range")

The values of C_1' and C_2, corresponding to the three choices of h_π^1 and determined so as to reproduce the experimental asymmetries at 45 and 221 MeV, are reported in table 2. Inspection of the table suggests that the dependence on Λ is reasonable: the values of the LEC's are close to 1, as it would naively be expected in the present EFT at order Q/Λ_χ. We also note that the obtained values for C_1' e C_2 corresponding to the case $h_\pi^1 = 4.56 \times 10^{-7}$ are in reasonable agreement with those reported in Eq. (11).

Finally, the longitudinal asymmetries calculated for $h_\pi^1 = 4.56 \times 10^{-7}$ and three choices of Λ are reported, as functions of energy, in figure 2. The experimental asymmetry at $E = 13.6$ MeV is automatically reproduced, even though the corresponding experimental datum was not used for determining the LEC's. This confirms the fact that this measurement does not provide

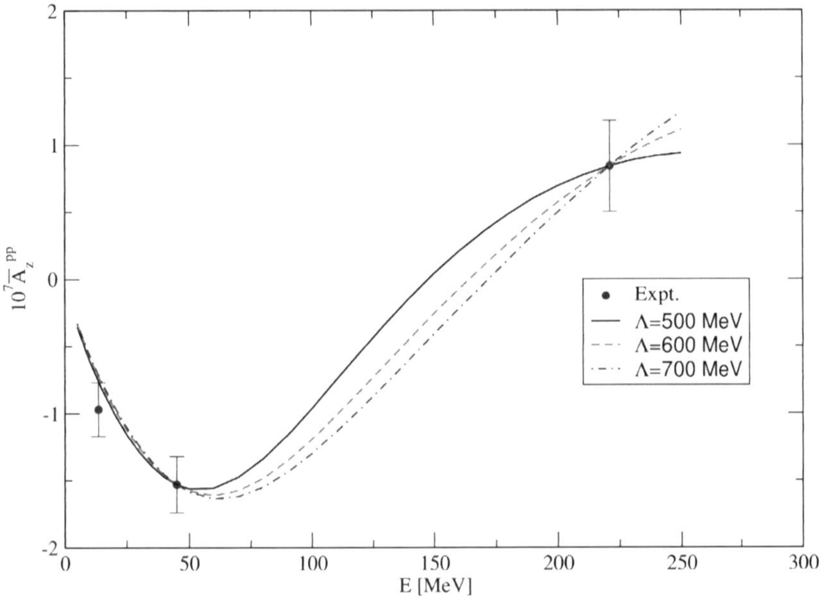

Figure 3. (Color online) Same as figure 2, but for $h_\pi^1 = 0$.

any additional information with respect to those carried out at 45 and 221 MeV. For different energies than those used to determine the constants C_1' and C_2, the asymmetries calculated with the various values of Λ are slightly different. This "spread" in practice gives an indication of the "theoretical uncertainty" inherent in the EFT approach. We note however that the spread is very small, in particular, it is less than the experimental uncertainty. A similar conclusion is also obtained for the other values of h_π^1, as can be seen from figures 3 and 4. Note that for $E \approx 150$ MeV, $\overline{A}_z^{pp}$ shows some sensitivity to the different values of h_π^1.

3.2. $\vec{n}$-d spin rotation

A beam of trasversally polarized neutrons passing through a slab of matter exhibits a spin rotation due to PV interactions with the nuclei in the medium. The observable of interest is the spin rotation angle ϕ about the neutron direction of motion. In first order perturbation theory in the weak interactions, the neutron spin rotation per unit-length of matter traversed, $\mathrm{d}\phi/\mathrm{d}z$, in $\vec{n}$-d scattering is given by [33, 34]

$$\frac{1}{\rho}\frac{\mathrm{d}\phi}{\mathrm{d}z} = \frac{1}{3\,v_{\rm rel}} \mathrm{Re} \sum_{m_n m_d} \epsilon_{m_n} \, {}^{(-)}\langle p\hat{\mathbf{z}}; m_n, m_d \mid V^{\rm PV} \mid p\hat{\mathbf{z}}; m_n, m_d\rangle^{(+)} \,, \tag{21}$$

where ρ is the density of deuterons (a parameter under control of the experimenter), $V^{\rm PV}$ denotes the PV nuclear potential, $|p\hat{\mathbf{z}}; m_n, m_d\rangle^{(\mp)}$ are the n-d scattering states with incoming-wave $(-)$ and outgoing-wave $(+)$ boundary conditions and relative momentum $\mathbf{p} = p\hat{\mathbf{z}}$ taken along the spin-quantization axis, *i.e.* the $\hat{\mathbf{z}}$-axis, and $v_{\rm rel} = p/\mu$ is the magnitude of the relative velocity, μ being the n-d reduced mass. The expression above is averaged over the spin projections $m_d = \pm 1, 0$ of the deuteron, however, the phase factor $\epsilon_{m_n} \equiv (-)^{1/2 - m_n}$ is ± 1 depending on whether the neutron has $m_n = \pm 1/2$.

The $\vec{n}$-d ingoing and outgoing wave functions have been calculated with the HH method [35]. The PC part of the potential is the I-N3LO NN potential [32] with the inclusion of the three-nucleon (3N) force derived from a chiral EFT at N^{2}LO [36], with the constants fixed in Ref. [37]

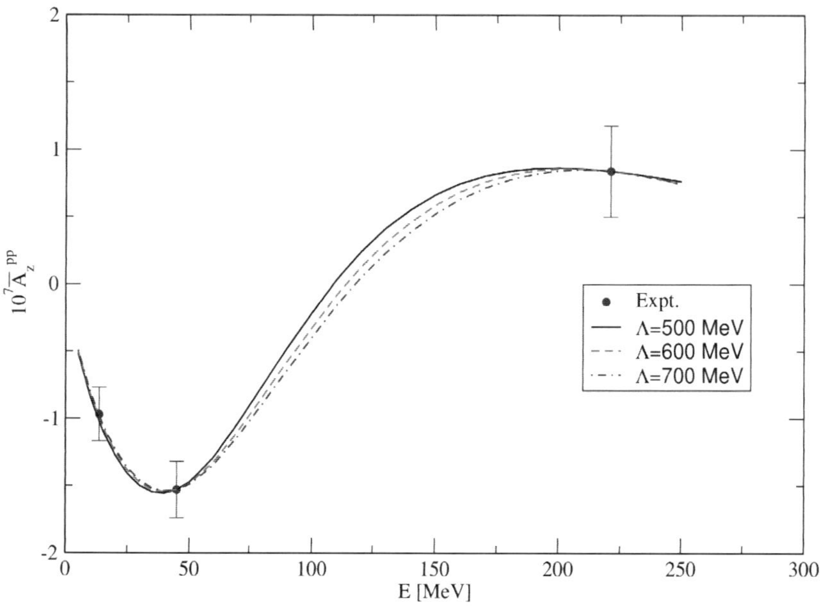

Figure 4. (Color online) Same as figure 2, but for $h_\pi^1 = 11.4 \times 10^{-7}$.

(N-N2LO model). The PV potential used in Eq. (21) is the chiral potential discussed in the previous section. We present the calculated angles in table 3, for the usual choices of h_π^1 and Λ values, and the corresponding values of C_1' and C_2 reported in table 2, by selecting the case $C_1 = C_1'$, $C_{3,4,5} = 0$. Several comments are in order. 1) This observable is dominated by the contribution of the LO OPE potential, and therefore is directly proportional to the value of h_π^1. Indeed, a measurement of this observable could provide an independent verification of the value of h_π^1, extracted from a measurement of the photon asymmetry in the radiative capture $^1\mathrm{H}(\vec{n},\gamma)^2\mathrm{H}$, presently considered the best experiment for extracting this LEC. 2) The contributions of the other terms (due to TPE or contact interactions) are rather small. For $h_\pi^1 = 4.56 \times 10^{-7}$ and $h_\pi^1 = 11.4 \times 10^{-7}$, the contribution of these medium- and short-range terms is about 5%. Clearly, for values of h_π^1 close to zero the observable would be very small, and the tiny contribution would come entirely from contact interactions. 3) For the same reason, the observable is not very sensitive to other choices for the LEC's C_i, $i = 1, 3, 4, 5$. A sensitivity study of ϕ varying these LEC's (but keeping fixed the combination $C_1' = C_1 + 2C_4 + 2C_5$ to the value determined by the *pp* longitudinal asymmetry as discussed above) has been performed. In this study we have chosen several combinations of $C_i = \pm 1$, $i = 3, 4, 5$ and we have found that the spin rotation angle varies less than 5%.

3.3. The $^3He(\vec{n},p)^3H$ longitudinal asymmetry

For ultracold neutrons, the longitudinal asymmetry $A_z^{n^3\mathrm{He}}$ for the reaction $^3\mathrm{He}(\vec{n},p)^3\mathrm{H}$ (defined in analogy with Eq. (12)), is given by $A_z^{n^3\mathrm{He}} = a_z \cos\theta$ [38], where θ is the angle between the outgoing proton momentum and the neutron beam direction. The coefficient a_z can be expressed in terms of products of T-matrix elements involving three PC and three PV transitions (see Ref. [38] for more details). Such T-matrix elements are calculated with the HH method [35], using the same strong interaction Hamiltonian model as for the $n-d$ calculation, namely the I-N3LO [32] NN potential in combination with the N-N2LO 3N force model. The HH calculation is a challenging one, for two reasons. The first is the coupled-channel nature of the scattering

Table 3. Spin rotation for unit-length of matter traversed $d\phi/dz$ for $\vec{n}$-d scattering in units of 10^{-7} rad cm^{-1} at zero energy assuming a liquid deuterium density of $\rho = 0.4 \times 10^{23}$ atoms cm^{-3}. The calculations are performed using the I-N3LO NN plus the N-N2LO 3N potentials for the PC interaction, and the chiral PV potential model discussed in this paper for various choices of the pion-nucleon coupling constant h^1_π and the cutoff parameter Λ. The corresponding values of the LEC's C_i, $i = 1, \ldots, 5$ are discussed in the text. In the columns labeled "OPE/LO" we have reported the spin rotation angle calculated by retaining in the PV potential only the LO OPE contribution, while in the columns labeled "FULL" we have included all terms.

	$h^1_\pi = 4.56 \times 10^{-7}$		$h^1_\pi = 0$		$h^1_\pi = 11.4 \times 10^{-7}$	
	OPE/LO	FULL	OPE/LO	FULL	OPE/LO	FULL
500	0.1028	0.0987	0.0000	+0.0012	0.2570	0.2448
600	0.1028	0.1022	0.0000	+0.0023	0.2569	0.2520
700	0.1019	0.1055	0.0000	+0.0061	0.2547	0.2547

Table 4. Same as table 3, but for the coefficient a_z (in units of 10^{-7}) describing the ^{3}He$(\vec{n}, p)^3$H longitudinal asymmetry (preliminary results).

	$h^1_\pi = 4.56 \times 10^{-7}$		$h^1_\pi = 0$		$h^1_\pi = 1.2 \times 10^{-7}$	
	OPE/LO	FULL	OPE/LO	FULL	OPE/LO	FULL
500	−0.551	−0.544	0.000	+0.044	−1.377	−1.425
600	−0.554	−0.578	0.000	+0.034	−1.385	−1.497
700	−0.546	−0.584	0.000	+0.009	−1.366	−1.473

problem: even at vanishing energies for the incident neutron, the elastic n-^{3}He and charge-exchange p-^{3}H channels are both open. The second is the presence of $J^\pi = 0^\pm$ resonant states between the p-^{3}H and n-^{3}He thresholds, which slows down the convergence of the expansion, and requires a large number of HH basis functions in order to achieve numerically stable results. Further discussion of this aspect of the calculations will be presented elsewhere [31]. Here we will present preliminary results for a_z, obtained using not fully converged calculations.

In table 4, we present the results of this preliminary calculation of a_z using the PV chiral potential, with the choice $C_1 = C_1'$ and $C_{3,4,5} = 0$. We observe that $A_z^{n^3\mathrm{He}}$ too is dominated by the contribution of the LO OPE potential. Naively, one expects that the most important contribution would come from the isoscalar operators. In fact, at this energy, the reaction proceeds mainly through the close 0^+ and 0^- resonances, which are considered to have total isospin $T = 0$ [39]. Thus, the isoscalar operators in the PV potential should give the dominant contribution. However, the Coulomb interaction in the final state induces sizable isospin mixing configurations and, since the LO OPE term is the longest range term, it ultimately ends up giving the most important contribution (with the obvious exception where h^1_π is close to zero). We also observe a rather large cancellation between the contributions coming from the TPE and the different contact interaction terms, while the contribution of $V^{(a),\mathrm{REL}}$ is always very tiny. For the case $h^1_\pi = 0$, a_z is given mainly by the contribution of the isoscalar operator multiplying the LEC C_2.

We now discuss the dependence of a_z on the LEC's $C_{3,4,5}$. Due to the aforementioned cancellation among the various contributions of the contact potential, we find in this case greater sensitivity to these LEC's, of the order of 20%. Therefore, the measurement of this observable could be very useful to extract them. We also observe that a_z is not very sensitive to Λ, suggesting that contributions beyond N^{2}LO to the PV potential may not be significant.

4. Conclusions and perspectives

In this contribution, we have provided a preliminary account of an ongoing study of PV effects in few-nucleon systems, based on chiral EFT. First, we have presented a re-derivation of the PV potential at N^2LO, in order to clarify ambiguities present in previous studies. Second, we have carried out calculations of a number of PV observables in A = 2–4, based on chiral potentials for both the strong- and weak-interaction sector. We should note that there are slight inconsistencies in the strong-interaction potentials, adopted in the present study, which however could be easily removed. For example, the 3N force at N^2LO includes the parameters c_1, c_3, and c_4 which also enter the NN force. In the N-N2LO model, the values of these parameters have been taken from the NN force model derived in Ref. [40], rather than from the I-N3LO model by Entem and Machleidt.

At order Q our PV potential contains 6 low-energy constants. One can envisage, at least in principle, a suite of experiments involving A = 2–5 systems which would constrain, in fact over-constrain, these six LEC's. The PV pion-nucleon coupling constant h^1_π should be accurately determined in the near future by the NPDGAMMA experiment [23], aimed at the measurement of the photon asymmetry in the ${}^1\mathrm{H}(\vec{n},\gamma)^2\mathrm{H}$ radiative capture. However, for electromagnetic processes, one has to include in the Lagrangian additional LEC's needed to fix the strengths of PV two-body current operators of pion range [12], and this could complicate the extraction of h^1_π. Alternatively, as discussed in this contribution, one could use the measurements of spin-rotation angles in $\vec{n}$-p and $\vec{n}$-d scattering. Such experiments are planned to be performed in a few years [25, 26, 28, 27]. In this contribution, the sensitivity of the $\vec{n}$-d spin rotation angle to h^1_π has been discussed in detail.

Once h^1_π is known, the other LEC's could be determined as follows. From the existing measurements of the $\vec{p}$-p longitudinal asymmetry we can fix C_2 and the combination $C_1' = C_1 + 2C_4 + 2C_5$. We have also found that the the longitudinal asymmetry in the charge-exchange reaction ${}^3\mathrm{He}(\vec{n},p)^3\mathrm{H}$ at ultracold neutron energies is sensitive to the $C_{3,4,5}$ LEC's, and an experiment to measure it is being planned at the SNS facility at Oak Ridge [29]. Additional information could come from the old measurement of the $\vec{p}$-α longitudinal analyzing power at 46 MeV [21] and from the study of the neutron spin rotation in $\vec{n}$-α scattering, which is in progress [27]. However, an accurate theoretical analysis of these two reactions (which involve five nucleons) is still lacking, and this will certainly be the subject of future theoretical efforts.

Other PV observables which could be interesting to study both experimentally and theoretically are the photon asymmetries in ${}^2\mathrm{H}(\vec{n},\gamma)^3\mathrm{H}$ and ${}^3\mathrm{He}(\vec{n},\gamma)^4\mathrm{He}$ radiative captures. The normal PC captures are strongly suppressed: the experimental values for the corresponding cross sections [41, 42] are, respectively, almost 3 and 4 orders of magnitude smaller than measured in ${}^1\mathrm{H}(\vec{n},\gamma)^2\mathrm{H}$. One would naively expect relatively large PV asymmetries in these cases, possibly orders of magnitude larger than in the A=2 system. Clearly, accurate theoretical estimates for them could be useful in motivating our experimental colleagues to carry out these extremely challenging measurements. Work to construct PV electromagnetic transition operators using the same EFT described here is in progress (see, for example, Ref. [43] for the application of chiral EFT to derive PC electromagnetic transition operators).

References

[1] Ramsey-Musolf M J and Page S A 2006 *Ann. Rev. Nucl. Part. Sci.* **56** 1 [arXiv:hep-ph/0601127]
[2] Desplanques B, Donoghue J F and Holstein B R 1980 *Ann. Phys. (N.Y.)* **124** 449
[3] Weinberg S 1990 *Phys. Lett. B* **251** 288; 1991 *Nucl. Phys. B* **363** 3; 1992 *Phys. Lett. B* **295** 114
[4] Bernard V, Kaiser N and Meissner U G 1995 *Int. J. Mod. Phys. E* **4** 193–346
[5] Epelbaum E 2006 *Prog. Part. Nucl. Phys.* **57** 654
[6] Epelbaum E, Hammer H W and Meissner U G 2009 *Rev. Mod. Phys.* **81** 1773
[7] Machleidt R and Entem D R 2011 *Phys. Rep.* **503** 1

[8] Kaplan D B and Savage M J 1993 *Nucl. Phys. A* **556** 653; 1994 Erratum *Nucl. Phys. A* **570** 833; 1994 Erratum *Nucl. Phys. A* **580** 679
[9] Haxton W C 2008 arXiv:0802.2984
[10] Wasem J 2011 arXiv:1108.1151
[11] Zhu S L, Puglia S J, Holstein B R and Ramsey-Musolf M J 2000 *Phys. Rev. D* **62** 033008
[12] Zhu S L, Maekawa C M, Holstein B R, Ramsey-Musolf M J and van Kolck U 2005 *Nucl. Phys. A* **748** 435
[13] Danilov G S 1965 *Phys. Lett.* **18** 40
[14] Hyun C H, Ando S and Desplanques B 2007 *Phys. Lett. B* **651** 257
[15] Desplanques B, Hyun C H, Ando S and Liu C P 2008 *Phys. Rev. C* **77** 064002
[16] Hyun C H, Desplanques B, Ando S and Liu C P 2008 *Mod. Phys. Lett. A* **23** 2293
[17] Girlanda L 2008 *Phys. Rev. C* **77** 067001
[18] Eversheim P D *et al.* 1991 *Phys. Lett. B* **256** 11
[19] Kistryn S *et al.* 1987 *Phys. Rev. Lett.* **58** 1616
[20] Berdoz A R *et al.* 2003 *Phys. Rev. C* **68** 034004
[21] Lang J *et al.*, 1985 *Phys. Rev. Lett.* **54** 170
[22] Cavaignac J F, Vignon B and Wilson R 1977 *Phys. Lett. B* **67** 148
[23] Gericke M T *et al.* 2009 *Nucl. Inst. Meth. A* **611** 239
[24] Knyaz'kov V A *et al.* 1994 *Nucl. Phys. A* **417** 209
[25] Snow W M 2009 *Nucl. Inst. Meth. A* **611** 248
[26] Bass C D *et al.* 2009 *Nucl. Inst. Meth. A* **612** 69
[27] Snow W M *et al.* 2011 *Phys. Rev. C* **83** 022501(R)
[28] Markoff D M (private communication)
[29] Bowman C D *et al.* 2007 *A Measurement of the Parity Violating Proton Asymmetry in the Capture of Polarized Cold Neutrons on* ^3He, proposal submitted to the SNS FNPB PRAC T, 2007-11-25
[30] Carlson J, Schiavilla R, Brown V R and Gibson B F 2002 *Phys. Rev. C* **65** 035502
[31] Viviani M, Baroni A, Girlanda L, Kievsky A, Marcucci L E and Schiavilla R (in preparation)
[32] Entem D R and Machleidt R 2003 *Phys. Rev. C* **68** 041001
[33] Schiavilla R, Carlson J and Paris M 2004 *Phys. Rev. C* **70** 044007
[34] Fermi E 1950 *Nuclear Physics* (Chicago: The University of Chicago Press)
[35] Kievsky A, Rosati S, Viviani M, Marcucci L E and Girlanda L 2008 *J. Phys. G: Nucl. Part. Phys.* **35** 063101
[36] Epelbaum E, Nogga A, Glöckle W, Kamada H, Meissner U G and Wital H, *Phys. Rev. C* **66** 064001
[37] Navrátil P 2007 *Few-Body Syst.* **41** 117
[38] Viviani M, Schiavilla R, Girlanda L, Kievsky A and Marcucci L E 2010 *Phys. Rev. C* **82** 044001
[39] Tilley D R, Weller H R and Hale G M 1992 *Nucl. Phys. A* **451** 1
[40] Epelbaum E, Glöckle W and Meissner U G 2005 *Nucl. Phys. A* **747** 362
[41] Jurney E T, Bendt P J and Browne J C 1982 *Phys. Rev. C* **25** 2810; Konijnenberg M W, Abrahams K, Kopecky J, Stecher-Rasmussen F, Wervelman R and Koch J H 1988 *Phys. Lett. B* **205** 215
[42] Wolfs F L H, Freedman S J, Nelson J E, Dewey M S and Greene G L 1989 *Phys. Rev. Lett.* **63** 2721; Wervelman R, Abrahams K, Postma H, Booten J G L and Van Hees A G M 1991 *Nucl. Phys. A* **526** 265
[43] Pastore S, Girlanda L, Schiavilla R, Viviani M and Wiringa R B 2009 *Phys. Rev. C* **80** 034004; Pastore S, Girlanda L, Schiavilla R and Viviani M 2011 *Phys. Rev. C* **84** 024001

Towards a more refined model of three-nucleon interaction

L Girlanda[1,2], A Kievsky[3] and M Viviani[3]

[1] Department of Physics, University of Salento I-73100 Lecce, Italy
[2] INFN, Sezione di Lecce, I-73100 Lecce, Italy
[3] INFN, Sezione di Pisa, I-56127 Pisa, Italy

E-mail: girlanda@le.infn.it

Abstract. We derive the minimal form of the two-derivative three-nucleon contact interaction by imposing all constraints from discrete symmetries and Fierz identities. In order to comply with the requirements of Poincaré covariance, a basis of operators depending on relative momenta is used. The resulting interaction depends on 10 unknown low-energy constants and leads to a three-nucleon potential which we give in local form in coordinate space.

1. Introduction

Understanding the nuclear interaction from first principles remains a challenging task. At the two-nucleon ($2N$) level, phenomenological potentials have reached a remarkable degree of accuracy, providing an excellent fit to the very large body of experimental data. Comparable accuracy is attained by a new class of potentials, so called chiral potentials, which are developed in the framework of an effective field theory, incorporating all symmetry properties of the underlying theory of the strong interactions (see Refs. [1, 2, 3, 4, 5] for reviews).

The effective theory is formulated in terms of nucleons and pions, whose interactions are constrained by the chiral symmetry of QCD. The effect of heavier degrees of freedom is encoded in the value of (an infinite set of) coupling constants, the low-energy constants (LECs). The theory is organized as a systematic low-energy expansion, in powers of small momenta or quark masses (or rather, pion masses), so that at each order of the chiral expansion, only a finite number of LECs is involved. The occurrence of new LECs, as one proceeds in the expansion, limits the predictive power of the theory, which is lower and lower as the scale of "new physics" (mass of particles not explicitly included in the theory) becomes closer and closer. Nevertheless, it can be said that the effective theory approach represents a solid link between nuclear physics and QCD. It allows, for instance, to justify the hierarchy of nuclear forces: n-nucleon forces are twice suppressed with respect to the $(n-1)$-nucleon forces.

Parallel to these developments there has been tremendous progress in the numerical techniques for the determination of bound and scattering states of few-nucleon systems. The latter then constitute a privileged theoretical laboratory to test different models of the nuclear interaction. Presently the low-energy expansion for the $2N$ potential has been carried up to the next-to-next-to-next-to leading order (N3LO), yielding a χ^2 per datum very close to one.

The status of the three-nucleon ($3N$) interaction is not as satisfactory. While its presence is necessary if one wants to reproduce the 3- and 4-nucleon binding energies, all currently available

Published under licence by IOP Publishing Ltd

Table 1. Leading (no-derivative) $3N$ contact operators satisfying rotational, isospin and time-reversal invariance. Subscripts refer to the nucleon bilinear on which the operators act. For instance, $O_5^{(0)} = N^\dagger\sigma^i NN^\dagger\sigma^i\tau^a NN^\dagger\tau^a N$.

i	$O_i^{(0)}$
1	$\mathbf{1}$
2	$\boldsymbol{\sigma}_1 \cdot \boldsymbol{\sigma}_2$
3	$\boldsymbol{\tau}_1 \cdot \boldsymbol{\tau}_2$
4	$\boldsymbol{\sigma}_1 \cdot \boldsymbol{\sigma}_2 \boldsymbol{\tau}_1 \cdot \boldsymbol{\tau}_2$
5	$\boldsymbol{\sigma}_1 \cdot \boldsymbol{\sigma}_2 \boldsymbol{\tau}_2 \cdot \boldsymbol{\tau}_3$
6	$\boldsymbol{\sigma}_1 \times \boldsymbol{\sigma}_2 \cdot \boldsymbol{\sigma}_3 \boldsymbol{\tau}_1 \times \boldsymbol{\tau}_2 \cdot \boldsymbol{\tau}_3$

$3N$ potentials fail to simultaneously describe also the neutron-deuteron doublet scattering length [6]. As a matter of fact, several low-energy data in the nucleon-deuteron system remain still unexplained (e.g. the so-called A_y puzzle). This is not really surprising, since the adopted $3N$ interaction models only contain a few free parameters, contrary to what happens for the $2N$ interaction, which is parametrized by more than 20 adjustable parameters. For example, the chiral $3N$ force at N2LO only contains 2 adjustable LECs [7, 8], and its extension to N3LO does not involve any new LEC [9, 10, 11].

In this context, starting with the observation that the discrepancies between theory and experiment in the $3N$ system arise at very low energy, we propose to refine that part of the interaction which stems from purely nucleonic vertices. Indeed, at very low energies, even the pions can be integrated out of the theory (their effect being subsumed in the values of the LECs), giving rise to the "pionless" effective theory, and the interaction among nucleons is due to contact vertices. It should be noticed that purely contact vertices are also part of the "pionful" version of the effective theory: a matching procedure could in principle be used to find the relationship between the values of the LECs in both theories. A $3N$ contact term is already part of the N2LO chiral ("pionful") $3N$ interaction. Invariance under parity dictates that subleading $3N$ contact terms contain two (covariant) derivatives of nucleon fields; they would therefore contribute, in the pionful version of the effective theory, at N4LO: they are not considered in any of the currently available models. Our strategy to identify a minimal set of such operators is based on the use of Fierz-like relations, which are consequences of the anticommuting nature of fermion fields.

2. Preliminaries: the leading $3N$ contact interaction

Already at the leading order the most general Lagrangian density satisfying rotational, isospin and time-reversal invariance contains redundant operators. Indeed, it is written as

$$\mathcal{L}_{3N}^{(0)} = -\sum_{i=1}^{6} E_i O_i^{(0)}, \tag{1}$$

where the operators $O_i^{(0)}$ are defined in table 1. By simultaneously changing two nucleon fields and the corresponding spin-isospin indices one can derive relations among the different operators, analogously to what was done in Ref. [12] for the parity-violating NN interaction. The properties under exchange of spin indices are expressed by the following relations

$$\begin{aligned}
(\mathbf{1})[\mathbf{1}] &= \tfrac{1}{2}(\mathbf{1}][\mathbf{1}) + \tfrac{1}{2}(\boldsymbol{\sigma}]\cdot[\boldsymbol{\sigma}) \\
(\sigma^i)[\mathbf{1}] &= \tfrac{1}{2}(\sigma^i][\mathbf{1}) + \tfrac{1}{2}(\mathbf{1}][\sigma^i) - \tfrac{i}{2}\epsilon^{ijk}(\sigma^j][\sigma^k) \\
(\sigma^i)[\sigma^j] &= \tfrac{1}{2}\left\{\delta^{ij}(\mathbf{1}][\mathbf{1}) - \delta^{ij}(\boldsymbol{\sigma}]\cdot[\boldsymbol{\sigma}) + (\sigma^i][\sigma^j) + (\sigma^j][\sigma^i) + i\epsilon^{ijk}(\sigma^k][\mathbf{1}) - i\epsilon^{ijk}(\mathbf{1}][\sigma^k)\right\},
\end{aligned} \tag{2}$$

where **1** is the identity operator in the one-particle spin space and (,) and [,] denote spin indices of the enclosed operator. Analogous relations hold in the one-particle isospin space for the identity operator and τ matrices. If we interchange the spin-isospin indices of the first two bilinears we get the following relations

$$\begin{aligned} O_1^{(0)} &= -\tfrac{1}{4}\left(O_1^{(0)} + O_2^{(0)} + O_3^{(0)} + O_4^{(0)}\right) \\ O_2^{(0)} &= -\tfrac{1}{4}\left(3O_1^{(0)} - O_2^{(0)} + 3O_3^{(0)} - O_4^{(0)}\right) \\ O_3^{(0)} &= -\tfrac{1}{4}\left(3O_1^{(0)} + 3O_2^{(0)} - O_3^{(0)} - O_4^{(0)}\right) \\ O_4^{(0)} &= -\tfrac{1}{4}\left(9O_1^{(0)} - 3O_2^{(0)} - 3O_3^{(0)} + O_4^{(0)}\right) \\ O_5^{(0)} &= -\tfrac{1}{2}\left(3O_3^{(0)} - O_5^{(0)}\right) \\ O_6^{(0)} &= 2\left(O_4^{(0)} - O_5^{(0)}\right), \end{aligned} \tag{3}$$

while the relations deriving from the remaining permutations are linearly dependent on the ones above. As a consequence of these relations, all six operators are proportional to each other, so that only one operator is left, in agreement with Ref. [8]. It is important to realize that the equivalence among the six operators is in general lost in the presence of a cutoff: in such a case the three nucleons do not occupy exactly the same point in space. However, in the framework of the effective theory, different choices of the operator are equivalent up to higher order corrections.

3. Constraining the subleading contact interaction

The subleading $3N$ contact interaction involves operators with 2 spatial derivatives, as dictated by parity symmetry. Momentum conservation implies that the only possible space-structures have the general form

$$\begin{aligned} X^+_{A,ij} &= (N^\dagger \overleftrightarrow{\nabla}_i N)(N^\dagger \overleftrightarrow{\nabla}_j N)(N^\dagger N) \\ X^+_{B,ij} &= \nabla_i(N^\dagger N)\nabla_j(N^\dagger N)(N^\dagger N) \\ X^-_{C,ij} &= i\nabla_i(N^\dagger N)(N^\dagger \overleftrightarrow{\nabla}_j N)(N^\dagger N) \\ X^+_{D,ij} &= (N^\dagger \overleftrightarrow{\nabla}_i \overleftrightarrow{\nabla}_j N)(N^\dagger N)(N^\dagger N). \end{aligned} \tag{4}$$

In the above expressions, and the following ones, the superscripts refer to the time-reversal properties, once appropriate factors of i have been included in order to ensure hermiticity. We are only interested in the isospin symmetric limit, therefore the relevant structures are

$$T^+ = \mathbf{1}, \quad \boldsymbol{\tau}_1\cdot\boldsymbol{\tau}_2, \quad \boldsymbol{\tau}_1\cdot\boldsymbol{\tau}_3, \quad \boldsymbol{\tau}_2\cdot\boldsymbol{\tau}_3, \quad T^- = \boldsymbol{\tau}_1\times\boldsymbol{\tau}_2\cdot\boldsymbol{\tau}_3, \tag{5}$$

where the subscripts of the Pauli matrices refer to the nucleon bilinears they belong to. Even (odd) combinations of $X \otimes T$ structures under time-reversal are to be associated with spin structures containing even (odd) numbers of σ matrices. Finally, the spin-space indices have to be contracted with Kronecker δ's of Levi-Civita tensors ϵ's. An overcomplete list of 146 operators has been presented in Ref. [13], constructed according to the above criteria. Here we present an alternative list of operators constructed in terms of relative momenta, defined as

$$\nabla_a = \nabla_2 - \nabla_1, \quad \nabla_b = \nabla_3 - \frac{1}{2}(\nabla_1 + \nabla_2). \tag{6}$$

The use of such variables allows us to automatically comply with the requirements of Poincaré covariance, which in the context of the non-relativistic low-energy effective theory reduce to Galileo covariance. Indeed the use of the above combinations of momenta will entail a **P**-independent interaction, **P** being the total momentum of the three particles. Notice however that the full Poincaré covariance can be consistently implemented order by order in the effective theory, as done in Ref.[14]: as a result one would obtain the so called fixed terms, i.e. two derivative terms with fixed coupling constants. We will not take this kind of terms into account in the present work. The gradients in Eq. (6) can be either $\overrightarrow{\nabla}$, acting of the whole nucleon bilinear and depending on the single-nucleon momentum transfer $\mathbf{k}_i = \mathbf{p}'_i - \mathbf{p}_i$, or $\overleftrightarrow{\nabla}$, depending on the single-nucleon momentum combination $\mathbf{Q}_= \mathbf{p}'_i + \mathbf{p}_i$. However not all the combinations need to be taken into account, as one can interchange the labels of the three particles to express some combinations in terms of others, for instance

$$\overrightarrow{\nabla}_a \overleftrightarrow{\nabla}_b f(1,2,3) \;\rightarrow\; \overrightarrow{\nabla}_a \overleftrightarrow{\nabla}_a [f(1,3,2) - f(1,2,3)] + 2 \overleftrightarrow{\nabla}_a \overrightarrow{\nabla}_b f(1,3,2), \tag{7}$$

$$\begin{aligned} \overrightarrow{\nabla}_b \overrightarrow{\nabla}_b f(1,2,3) \;\rightarrow\; & -\frac{1}{4}\Big\{\overrightarrow{\nabla}_a \overrightarrow{\nabla}_a f(1,2,3) - 4 \overrightarrow{\nabla}_a \overrightarrow{\nabla}_a f(1,3,2) \\ & +4 \overrightarrow{\nabla}_a \overrightarrow{\nabla}_b [f(1,2,3) + f(1,3,2)]\Big\} \end{aligned} \tag{8}$$

$$\begin{aligned} \overrightarrow{\nabla}_b \overleftrightarrow{\nabla}_b f(1,2,3) \;\rightarrow\; & -\frac{1}{4}\Big\{\overrightarrow{\nabla}_a \overleftrightarrow{\nabla}_a f(1,2,3) - 4 \overrightarrow{\nabla}_a \overleftrightarrow{\nabla}_a f(1,3,2) \\ & +4 \overleftrightarrow{\nabla}_a \overrightarrow{\nabla}_b [f(1,2,3) + f(1,3,2)]\Big\} \end{aligned} \tag{9}$$

$$\begin{aligned} \overleftrightarrow{\nabla}_b \overleftrightarrow{\nabla}_b f(1,2,3) \;\rightarrow\; & -\frac{1}{4}\Big\{\overleftrightarrow{\nabla}_a \overleftrightarrow{\nabla}_a f(1,2,3) - 4 \overleftrightarrow{\nabla}_a \overleftrightarrow{\nabla}_a f(1,3,2) \\ & +4 \overleftrightarrow{\nabla}_a \overleftrightarrow{\nabla}_b [f(1,2,3) + f(1,3,2)]\Big\} \end{aligned} \tag{10}$$

where f denotes an arbitrary spin-isospin function of the three particles. Using these and similar identities, we are left with the operators listed in table 2. The list is overcomplete, since the operators are related to each other through Fierz rearrangements. Such relations leave only 10 independent operators, in agreement with the findings of Ref. [13]. In order to write the minimal subleading contact Lagrangian, it is convenient to go back to the operator basis used in Ref. [13],

$$\begin{aligned} \mathcal{L}^{(2)}_{3N} = \; & E'_1 \overrightarrow{\nabla}(N^\dagger N) \cdot \overrightarrow{\nabla}(N^\dagger N) N^\dagger N + E'_2 \overrightarrow{\nabla}(N^\dagger \tau^a N) \cdot \overrightarrow{\nabla}(N^\dagger \tau^a N) N^\dagger N \\ & + E'_3 \overrightarrow{\nabla}(N^\dagger \tau^a N) \cdot \overrightarrow{\nabla}(N^\dagger N) N^\dagger \tau^a N + E'_4 \overrightarrow{\nabla} \cdot (N^\dagger \vec{\sigma} N) \overrightarrow{\nabla} \cdot (N^\dagger \vec{\sigma} N) N^\dagger N \\ & + E'_5 \overrightarrow{\nabla} \cdot (N^\dagger \vec{\sigma} \tau^a N) \overrightarrow{\nabla} \cdot (N^\dagger \vec{\sigma} \tau^a N) N^\dagger N + E'_6 \overrightarrow{\nabla}_i (N^\dagger \sigma^j \tau^a N) \overrightarrow{\nabla}_j (N^\dagger \sigma^i \tau^a N) N^\dagger N \\ & + E'_7 \overrightarrow{\nabla}_i (N^\dagger \sigma^j N) \overrightarrow{\nabla}_i (N^\dagger \sigma^j N) N^\dagger N + E'_8 \overrightarrow{\nabla}_i (N^\dagger \sigma^j \tau^a N) \overrightarrow{\nabla}_i (N^\dagger \sigma^j \tau^a N) N^\dagger N \\ & + E'_9 \overrightarrow{\nabla} \cdot (N^\dagger \vec{\sigma} N) \overrightarrow{\nabla}(N^\dagger N) \cdot N^\dagger \vec{\sigma} N + E'_{10} \epsilon^{abc} \overrightarrow{\nabla}(N^\dagger \tau^a N) \times \overrightarrow{\nabla}(N^\dagger \tau^b N) \cdot N^\dagger \vec{\sigma} \tau^c N. \end{aligned} \tag{11}$$

4. Subleading $3N$ contact potential

Taking the matrix element of the interaction between $3N$ states we can obtain the $3N$ potential. If we define $\mathbf{k}_i = \mathbf{p}_i - \mathbf{p}'_i$ and $\mathbf{Q}_i = \mathbf{p}_i + \mathbf{p}'_i$, where $\mathbf{p}_i$ and $\mathbf{p}'_i$ denote the initial and final momenta of nucleon i, the potential in momentum space reads

$$\begin{aligned} V = \sum_{i \neq j \neq k} \Big[& -E_1 \mathbf{k}_i^2 - E_2 \mathbf{k}_i^2 \boldsymbol{\tau}_i \cdot \boldsymbol{\tau}_j - E_3 \mathbf{k}_i^2 \boldsymbol{\sigma}_i \cdot \boldsymbol{\sigma}_j - E_4 \mathbf{k}_i^2 \boldsymbol{\sigma}_i \cdot \boldsymbol{\sigma}_j \boldsymbol{\tau}_i \cdot \boldsymbol{\tau}_j \\ & - E_5 \left(3\mathbf{k}_i \cdot \boldsymbol{\sigma}_i \mathbf{k}_i \cdot \boldsymbol{\sigma}_j - \mathbf{k}_i^2\right) - E_6 \left(3\mathbf{k}_i \cdot \boldsymbol{\sigma}_i \mathbf{k}_i \cdot \boldsymbol{\sigma}_j - \mathbf{k}_i^2\right)) \boldsymbol{\tau}_i \cdot \boldsymbol{\tau}_j \\ & + \frac{i}{2} E_7 \mathbf{k}_i \times (\mathbf{Q}_i - \mathbf{Q}_j) \cdot (\boldsymbol{\sigma}_i + \boldsymbol{\sigma}_j) + \frac{i}{2} E_8 \mathbf{k}_i \times (\mathbf{Q}_i - \mathbf{Q}_j) \cdot (\boldsymbol{\sigma}_i + \boldsymbol{\sigma}_j) \boldsymbol{\tau}_j \cdot \boldsymbol{\tau}_k \\ & - E_9 \mathbf{k}_i \cdot \boldsymbol{\sigma}_i \mathbf{k}_j \cdot \boldsymbol{\sigma}_j - E_{10} \mathbf{k}_i \cdot \boldsymbol{\sigma}_i \mathbf{k}_j \cdot \boldsymbol{\sigma}_j \boldsymbol{\tau}_i \cdot \boldsymbol{\tau}_j \Big], \end{aligned} \tag{12}$$

Table 2. Complete list of 2-derivative three-nucleon contact operators compatible with rotational, isospin, parity and time-reversal invariance. Subscripts refer to the nucleon bilinear on which the operators act.

O_{1-3}	$\overleftrightarrow{\nabla}_a \cdot \overleftrightarrow{\nabla}_a[\mathbf{1}, \boldsymbol{\tau}_1 \cdot \boldsymbol{\tau}_2, \boldsymbol{\tau}_1 \cdot \boldsymbol{\tau}_3]$
O_{4-6}	$\overleftrightarrow{\nabla}_a \cdot \overleftrightarrow{\nabla}_a \vec{\sigma}_1 \cdot \vec{\sigma}_2[\mathbf{1}, \boldsymbol{\tau}_1 \cdot \boldsymbol{\tau}_2, \boldsymbol{\tau}_1 \cdot \boldsymbol{\tau}_3]$
O_{7-10}	$\overleftrightarrow{\nabla}_a \cdot \overleftrightarrow{\nabla}_a \vec{\sigma}_2 \cdot \vec{\sigma}_3[\mathbf{1}, \boldsymbol{\tau}_1 \cdot \boldsymbol{\tau}_2, \boldsymbol{\tau}_1 \cdot \boldsymbol{\tau}_3, \boldsymbol{\tau}_2 \cdot \boldsymbol{\tau}_3]]$
O_{11-13}	$\overleftrightarrow{\nabla}_a \cdot \vec{\sigma}_1 \overleftrightarrow{\nabla}_a \cdot \vec{\sigma}_2[\mathbf{1}, \boldsymbol{\tau}_1 \cdot \boldsymbol{\tau}_2, \boldsymbol{\tau}_1 \cdot \boldsymbol{\tau}_3]$
O_{14-17}	$\overleftrightarrow{\nabla}_a \cdot \vec{\sigma}_1 \overleftrightarrow{\nabla}_a \cdot \vec{\sigma}_3[\mathbf{1}, \boldsymbol{\tau}_1 \cdot \boldsymbol{\tau}_2, \boldsymbol{\tau}_1 \cdot \boldsymbol{\tau}_3, \boldsymbol{\tau}_2 \cdot \boldsymbol{\tau}_3]$
O_{18}	$\overleftrightarrow{\nabla}_a \cdot \overleftrightarrow{\nabla}_a \vec{\sigma}_1 \times \vec{\sigma}_2 \cdot \vec{\sigma}_3[\boldsymbol{\tau}_1 \times \boldsymbol{\tau}_2 \cdot \boldsymbol{\tau}_3]$
O_{19}	$\overleftrightarrow{\nabla}_a \cdot \vec{\sigma}_1 \overleftrightarrow{\nabla}_a \times \vec{\sigma}_2 \cdot \vec{\sigma}_3[\boldsymbol{\tau}_1 \times \boldsymbol{\tau}_2 \cdot \boldsymbol{\tau}_3]$
O_{20}	$\overleftrightarrow{\nabla}_a \cdot \vec{\sigma}_3 \overleftrightarrow{\nabla}_a \times \vec{\sigma}_1 \cdot \vec{\sigma}_2[\boldsymbol{\tau}_1 \times \boldsymbol{\tau}_2 \cdot \boldsymbol{\tau}_3]$
O_{21-23}	$\overrightarrow{\nabla}_a \cdot \overrightarrow{\nabla}_a[\mathbf{1}, \boldsymbol{\tau}_1 \cdot \boldsymbol{\tau}_2, \boldsymbol{\tau}_1 \cdot \boldsymbol{\tau}_3]$
O_{24-26}	$\overrightarrow{\nabla}_a \cdot \overrightarrow{\nabla}_a \vec{\sigma}_1 \cdot \vec{\sigma}_2[\mathbf{1}, \boldsymbol{\tau}_1 \cdot \boldsymbol{\tau}_2, \boldsymbol{\tau}_1 \cdot \boldsymbol{\tau}_3]$
O_{27-30}	$\overrightarrow{\nabla}_a \cdot \overrightarrow{\nabla}_a \vec{\sigma}_2 \cdot \vec{\sigma}_3[\mathbf{1}, \boldsymbol{\tau}_1 \cdot \boldsymbol{\tau}_2, \boldsymbol{\tau}_1 \cdot \boldsymbol{\tau}_3, \boldsymbol{\tau}_2 \cdot \boldsymbol{\tau}_3]]$
O_{31-33}	$\overrightarrow{\nabla}_a \cdot \vec{\sigma}_1 \overrightarrow{\nabla}_a \cdot \vec{\sigma}_2[\mathbf{1}, \boldsymbol{\tau}_1 \cdot \boldsymbol{\tau}_2, \boldsymbol{\tau}_1 \cdot \boldsymbol{\tau}_3]$
O_{34-37}	$\overrightarrow{\nabla}_a \cdot \vec{\sigma}_1 \overrightarrow{\nabla}_a \cdot \vec{\sigma}_3[\mathbf{1}, \boldsymbol{\tau}_1 \cdot \boldsymbol{\tau}_2, \boldsymbol{\tau}_1 \cdot \boldsymbol{\tau}_3, \boldsymbol{\tau}_2 \cdot \boldsymbol{\tau}_3]$
O_{38}	$\overrightarrow{\nabla}_a \cdot \overrightarrow{\nabla}_a \vec{\sigma}_1 \times \vec{\sigma}_2 \cdot \vec{\sigma}_3[\boldsymbol{\tau}_1 \times \boldsymbol{\tau}_2 \cdot \boldsymbol{\tau}_3]$
O_{39}	$\overrightarrow{\nabla}_a \cdot \vec{\sigma}_1 \overrightarrow{\nabla}_a \times \vec{\sigma}_2 \cdot \vec{\sigma}_3[\boldsymbol{\tau}_1 \times \boldsymbol{\tau}_2 \cdot \boldsymbol{\tau}_3]$
O_{40}	$\overrightarrow{\nabla}_a \cdot \vec{\sigma}_3 \overrightarrow{\nabla}_a \times \vec{\sigma}_1 \cdot \vec{\sigma}_2[\boldsymbol{\tau}_1 \times \boldsymbol{\tau}_2 \cdot \boldsymbol{\tau}_3]$
O_{41-44}	$(\overleftrightarrow{\nabla}_a \cdot \vec{\sigma}_1 \overleftrightarrow{\nabla}_b \cdot \vec{\sigma}_3 - \overleftrightarrow{\nabla}_a \cdot \vec{\sigma}_3 \overleftrightarrow{\nabla}_b \cdot \vec{\sigma}_1)[\mathbf{1}, \boldsymbol{\tau}_1 \cdot \boldsymbol{\tau}_2, \boldsymbol{\tau}_1 \cdot \boldsymbol{\tau}_3, \boldsymbol{\tau}_2 \cdot \boldsymbol{\tau}_3]$
O_{45-48}	$(\overrightarrow{\nabla}_a \cdot \vec{\sigma}_1 \overrightarrow{\nabla}_b \cdot \vec{\sigma}_3 - \overrightarrow{\nabla}_a \cdot \vec{\sigma}_3 \overrightarrow{\nabla}_b \cdot \vec{\sigma}_1)[\mathbf{1}, \boldsymbol{\tau}_1 \cdot \boldsymbol{\tau}_2, \boldsymbol{\tau}_1 \cdot \boldsymbol{\tau}_3, \boldsymbol{\tau}_2 \cdot \boldsymbol{\tau}_3]$
O_{49}	$i \overleftrightarrow{\nabla}_a \cdot \vec{\sigma}_1 \overrightarrow{\nabla}_a \cdot \vec{\sigma}_2[\boldsymbol{\tau}_1 \times \boldsymbol{\tau}_2 \cdot \boldsymbol{\tau}_3]$
O_{50}	$i \overleftrightarrow{\nabla}_a \cdot \vec{\sigma}_1 \overrightarrow{\nabla}_a \cdot \vec{\sigma}_3[\boldsymbol{\tau}_1 \times \boldsymbol{\tau}_2 \cdot \boldsymbol{\tau}_3]$
O_{51}	$i \overleftrightarrow{\nabla}_a \cdot \vec{\sigma}_3 \overrightarrow{\nabla}_a \cdot \vec{\sigma}_1[\boldsymbol{\tau}_1 \times \boldsymbol{\tau}_2 \cdot \boldsymbol{\tau}_3]$
O_{52}	$i \overleftrightarrow{\nabla}_a \cdot \overrightarrow{\nabla}_a \vec{\sigma}_1 \cdot \vec{\sigma}_3[\boldsymbol{\tau}_1 \times \boldsymbol{\tau}_2 \cdot \boldsymbol{\tau}_3]$
O_{53-56}	$i \overleftrightarrow{\nabla}_a \times \overrightarrow{\nabla}_a \cdot \vec{\sigma}_1[\mathbf{1}, \boldsymbol{\tau}_1 \cdot \boldsymbol{\tau}_2, \boldsymbol{\tau}_1 \cdot \boldsymbol{\tau}_3, \boldsymbol{\tau}_2 \cdot \boldsymbol{\tau}_3]$
O_{57-60}	$i \overleftrightarrow{\nabla}_a \times \overrightarrow{\nabla}_a \cdot \vec{\sigma}_3[\mathbf{1}, \boldsymbol{\tau}_1 \cdot \boldsymbol{\tau}_2, \boldsymbol{\tau}_1 \cdot \boldsymbol{\tau}_3, \boldsymbol{\tau}_2 \cdot \boldsymbol{\tau}_3]$
O_{61}	$i \overleftrightarrow{\nabla}_a \cdot \overrightarrow{\nabla}_a \vec{\sigma}_1 \times \vec{\sigma}_2 \cdot \vec{\sigma}_3[\boldsymbol{\tau}_1 \cdot \boldsymbol{\tau}_3]$
O_{62-65}	$i \overleftrightarrow{\nabla}_a \cdot \vec{\sigma}_1 \overrightarrow{\nabla}_a \times \vec{\sigma}_2 \cdot \vec{\sigma}_3[\mathbf{1}, \boldsymbol{\tau}_1 \cdot \boldsymbol{\tau}_2, \boldsymbol{\tau}_1 \cdot \boldsymbol{\tau}_3, \boldsymbol{\tau}_2 \cdot \boldsymbol{\tau}_3]$
O_{66-67}	$i \overleftrightarrow{\nabla}_a \cdot \vec{\sigma}_3 \overrightarrow{\nabla}_a \times \vec{\sigma}_1 \cdot \vec{\sigma}_2[\boldsymbol{\tau}_1 \cdot \boldsymbol{\tau}_3, \boldsymbol{\tau}_2 \cdot \boldsymbol{\tau}_3]$
O_{68-71}	$i \overrightarrow{\nabla}_a \cdot \vec{\sigma}_1 \overleftrightarrow{\nabla}_a \times \vec{\sigma}_2 \cdot \vec{\sigma}_3[\mathbf{1}, \boldsymbol{\tau}_1 \cdot \boldsymbol{\tau}_2, \boldsymbol{\tau}_1 \cdot \boldsymbol{\tau}_3, \boldsymbol{\tau}_2 \cdot \boldsymbol{\tau}_3]$
O_{72-73}	$i \overrightarrow{\nabla}_a \cdot \vec{\sigma}_3 \overleftrightarrow{\nabla}_a \times \vec{\sigma}_1 \cdot \vec{\sigma}_2[\boldsymbol{\tau}_1 \cdot \boldsymbol{\tau}_3, \boldsymbol{\tau}_2 \cdot \boldsymbol{\tau}_3]$
O_{74-77}	$i \overleftrightarrow{\nabla}_a \times \overrightarrow{\nabla}_a \cdot \vec{\sigma}_1 \vec{\sigma}_2 \cdot \vec{\sigma}_3[\mathbf{1}, \boldsymbol{\tau}_1 \cdot \boldsymbol{\tau}_2, \boldsymbol{\tau}_1 \cdot \boldsymbol{\tau}_3, \boldsymbol{\tau}_2 \cdot \boldsymbol{\tau}_3]$
O_{78-80}	$i \overleftrightarrow{\nabla}_a \times \overrightarrow{\nabla}_a \cdot \vec{\sigma}_3 \vec{\sigma}_1 \cdot \vec{\sigma}_2[\mathbf{1}, \boldsymbol{\tau}_1 \cdot \boldsymbol{\tau}_2, \boldsymbol{\tau}_1 \cdot \boldsymbol{\tau}_3]$

continues...

Table 3. Continuation of table 2.

O_{81}	$i\overleftrightarrow{\nabla}_a\cdot\overrightarrow{\nabla}_b[\boldsymbol{\tau}_1\times\boldsymbol{\tau}_2\cdot\boldsymbol{\tau}_3]$
O_{82}	$i\overleftrightarrow{\nabla}_a\cdot\overrightarrow{\nabla}_b\,\vec{\sigma}_1\cdot\vec{\sigma}_2[\boldsymbol{\tau}_1\times\boldsymbol{\tau}_2\cdot\boldsymbol{\tau}_3]$
O_{83}	$i\overleftrightarrow{\nabla}_a\cdot\vec{\sigma}_1\,\overrightarrow{\nabla}_b\cdot\vec{\sigma}_2[\boldsymbol{\tau}_1\times\boldsymbol{\tau}_2\cdot\boldsymbol{\tau}_3]$
O_{84}	$i\overleftrightarrow{\nabla}_a\cdot\vec{\sigma}_1\,\overrightarrow{\nabla}_b\cdot\vec{\sigma}_3[\boldsymbol{\tau}_1\times\boldsymbol{\tau}_2\cdot\boldsymbol{\tau}_3]$
O_{85}	$i\overleftrightarrow{\nabla}_a\cdot\vec{\sigma}_3\,\overrightarrow{\nabla}_b\cdot\vec{\sigma}_1[\boldsymbol{\tau}_1\times\boldsymbol{\tau}_2\cdot\boldsymbol{\tau}_3]$
O_{86}	$i\overleftrightarrow{\nabla}_a\cdot\overrightarrow{\nabla}_b\,\vec{\sigma}_1\cdot\vec{\sigma}_3[\boldsymbol{\tau}_1\times\boldsymbol{\tau}_2\cdot\boldsymbol{\tau}_3]$
O_{87-90}	$i\overleftrightarrow{\nabla}_a\times\overrightarrow{\nabla}_b\cdot\vec{\sigma}_1[\mathbf{1},\boldsymbol{\tau}_1\cdot\boldsymbol{\tau}_2,\boldsymbol{\tau}_1\cdot\boldsymbol{\tau}_3,\boldsymbol{\tau}_2\cdot\boldsymbol{\tau}_3]$
O_{91-92}	$i\overleftrightarrow{\nabla}_a\times\overrightarrow{\nabla}_b\cdot\vec{\sigma}_3[\boldsymbol{\tau}_1\cdot\boldsymbol{\tau}_3,\boldsymbol{\tau}_2\cdot\boldsymbol{\tau}_3]$
O_{93-95}	$i\overleftrightarrow{\nabla}_a\cdot\overrightarrow{\nabla}_b\,\vec{\sigma}_1\times\vec{\sigma}_2\cdot\vec{\sigma}_3[\mathbf{1},\boldsymbol{\tau}_1\cdot\boldsymbol{\tau}_2,\boldsymbol{\tau}_1\cdot\boldsymbol{\tau}_3]$
O_{96-99}	$i\overleftrightarrow{\nabla}_a\cdot\vec{\sigma}_1\,\overrightarrow{\nabla}_b\times\vec{\sigma}_2\cdot\vec{\sigma}_3[\mathbf{1},\boldsymbol{\tau}_1\cdot\boldsymbol{\tau}_2,\boldsymbol{\tau}_1\cdot\boldsymbol{\tau}_3,\boldsymbol{\tau}_2\cdot\boldsymbol{\tau}_3]$
$O_{100-103}$	$i\overleftrightarrow{\nabla}_a\cdot\vec{\sigma}_3\,\overrightarrow{\nabla}_b\times\vec{\sigma}_1\cdot\vec{\sigma}_2[\mathbf{1},\boldsymbol{\tau}_1\cdot\boldsymbol{\tau}_2,\boldsymbol{\tau}_1\cdot\boldsymbol{\tau}_3,\boldsymbol{\tau}_2\cdot\boldsymbol{\tau}_3]$
$O_{104-107}$	$i\overrightarrow{\nabla}_b\cdot\vec{\sigma}_1\,\overleftrightarrow{\nabla}_a\times\vec{\sigma}_2\cdot\vec{\sigma}_3[\mathbf{1},\boldsymbol{\tau}_1\cdot\boldsymbol{\tau}_2,\boldsymbol{\tau}_1\cdot\boldsymbol{\tau}_3,\boldsymbol{\tau}_2\cdot\boldsymbol{\tau}_3]$
$O_{108-111}$	$i\overrightarrow{\nabla}_b\cdot\vec{\sigma}_3\,\overleftrightarrow{\nabla}_a\times\vec{\sigma}_1\cdot\vec{\sigma}_2[\mathbf{1},\boldsymbol{\tau}_1\cdot\boldsymbol{\tau}_2,\boldsymbol{\tau}_1\cdot\boldsymbol{\tau}_3,\boldsymbol{\tau}_2\cdot\boldsymbol{\tau}_3]$
$O_{112-115}$	$i\overleftrightarrow{\nabla}_a\times\overrightarrow{\nabla}_b\cdot\vec{\sigma}_1\,\vec{\sigma}_2\cdot\vec{\sigma}_3[\mathbf{1},\boldsymbol{\tau}_1\cdot\boldsymbol{\tau}_2,\boldsymbol{\tau}_1\cdot\boldsymbol{\tau}_3,\boldsymbol{\tau}_2\cdot\boldsymbol{\tau}_3]$
O_{116}	$i\overleftrightarrow{\nabla}_a\times\overrightarrow{\nabla}_b\cdot\vec{\sigma}_3\,\vec{\sigma}_1\cdot\vec{\sigma}_2[\boldsymbol{\tau}_1\cdot\boldsymbol{\tau}_3]$

with E_{1-10} combinations of the low-energy constants E_i'. The potential in coordinate space is obtained as a Fourier transform. In order to do this we need to regularize the potential by a momentum cutoff. Using a cutoff which only depends on the momentum transfer $\mathbf{k}_i$, e.g. $F(\mathbf{k}_j^2;\Lambda)F(\mathbf{k}_k^2;\Lambda)$, the coordinate-space potential assumes a local form,

$$\begin{aligned} V = \sum_{i\neq j\neq k} \quad & (E_1 + E_2\boldsymbol{\tau}_i\cdot\boldsymbol{\tau}_j + E_3\boldsymbol{\sigma}_i\cdot\boldsymbol{\sigma}_j + E_4\boldsymbol{\tau}_i\cdot\boldsymbol{\tau}_j\boldsymbol{\sigma}_i\cdot\boldsymbol{\sigma}_j)\left[Z_0''(r_{ij}) + 2\frac{Z_0'(r_{ij})}{r_{ij}}\right] Z_0(r_{ik}) \\ & +(E_5 + E_6\boldsymbol{\tau}_i\cdot\boldsymbol{\tau}_j)S_{ij}\left[Z_0''(r_{ij}) - \frac{Z_0'(r_{ij})}{r_{ij}}\right] Z_0(r_{ik}) \\ & +(E_7 + E_8\boldsymbol{\tau}_i\cdot\boldsymbol{\tau}_k)(\mathbf{L}\cdot\mathbf{S})_{ij}\frac{Z_0'(r_{ij})}{r_{ij}}Z_0(r_{ik}) \\ & +(E_9 + E_{10}\boldsymbol{\tau}_j\cdot\boldsymbol{\tau}_k)\boldsymbol{\sigma}_j\cdot\hat{\mathbf{r}}_{ij}\boldsymbol{\sigma}_k\cdot\hat{\mathbf{r}}_{ik}Z_0'(r_{ij})Z_0'(r_{ik}) \end{aligned} \tag{13}$$

where S_{ij} and $(\mathbf{L}\cdot\mathbf{S})_{ij}$ are respectively the tensor and spin-orbit operators for particles i and j and the function $Z_0(r)$ (also depending on the cutoff Λ) is defined as

$$Z_0(r) = \int \frac{d\mathbf{p}}{(2\pi)^3} e^{i\mathbf{p}\cdot\mathbf{r}} F(\mathbf{p}^2;\Lambda) \tag{14}$$

With the above choice of basis, most terms in the potential can be viewed as an ordinary interaction of particles ij with a further dependence on the coordinate of the third particle. In particular the terms multiplying E_7 and E_8 are of a spin-orbit character. Such kind of terms was suggested in Ref. [15] to solve the so-called A_y puzzle (see also Ref. [16, 17]). Notice also that some of the spin-isospin structures implied by the leading contact interaction, which were equivalent up to cutoff effects, are resolved at the two-derivative level.

5. Concluding remarks

As we have outlined, the $3N$ contact interaction consists of a leading contribution, at the N2LO of the chiral expansion, and a subleading one, arising at the N4LO. Parity requires that no

$3N$ contact interactions appear at N3LO or N5LO. On the other hand, the leading $4N$ contact interaction arises at N5LO, therefore we are lead to consider such terms at the same level of accuracy, at least in the framework of pionless EFT. It turns out [13] that there is only one independent $4N$ contact operator. Using cutoff functions depending only on momentum transfer $\mathbf{k}_i$, as done in the $3N$ case, e.g. choosing $F(\mathbf{k}_j^2;\Lambda)F(\mathbf{k}_k^2;\Lambda)F(\mathbf{k}_l^2;\Lambda)$, the 4-body contact potential is local in coordinate space,

$$V_{4N} = F \sum_{i\neq j\neq k\neq l} Z_0(r_{ij})Z_0(r_{ik})Z_0(r_{il}), \tag{15}$$

where F is the associated low-energy constant. Therefore we can decouple, to a certain extent, the $3N$ sector from the $4N$ sector, in the course of fixing all low-energy constants from data: indeed, one can adjust the subleading $3N$ potential in the $A=3$ systems without worrying about the consequences for the α particle binding energy, which could always be reproduced by adjusting the $4N$ contact term.

It should be stressed that the terms considered here start to contribute at N4LO and are therefore beyond the accuracy of the presently developed nuclear interactions. In particular, a complete EFT calculation should also address the problem of the NN interaction at the same order. Notice that the four-derivatives $2N$ operators are already part of the N3LO interaction while the six-derivative ones would start to contribute at N5LO, and have not been considered in the literature so far. Nevertheless, we can expect that, despite representing only part of the N4LO interaction, the terms derived here could play an important role in the accurate description of the three-nucleon systems, since they are completely unconstrained by symmetries. This is at variance with the $3N$ N3LO interaction, which contains no free parameters [9, 10, 11]. Moreover, the same terms will also appear in the pionless version of the effective theory as next-to-leading $3N$ interaction and leading $4N$ interaction. After a sensitivity study of the 10 subleading terms of the $3N$ potentials, the corresponding low-energy constants could be determined from a fit of several polarization observables in $N-d$ scattering at low energies. There are well established discrepancies between theoretical predictions and experimental data in some of the polarization observables as for example the vector analyzing powers A_y and iT_{11} and the vector analyzing power T_{21} in elastic scattering [18, 19, 20], and in several unpolarized breakup cross sections [21, 20]. We expect, based on the results of Ref. [15], that some of the 10 operators of the $3N$ potential will have sufficient sensitivity to fix these discrepancies with reasonable values of their low-energy constants. Correspondingly, the $4N$ contact term could be used to reproduce the ^{4}He binding energy, and the predictions of the derived potentials could be tested in the description of the four-body scattering states. In fact, also in this case there exist large and still unexplained discrepancies between theory and experiment [22, 23, 24, 25]. Studies along these lines are in progress.

References

[1] Bernard V, Kaiser N and Meissner U -G 1995 *Int. J. Mod. Phys.* **E4** 193
[2] van Kolck U 1999 *Prog. Part. Nucl. Phys.* **43** 337
[3] Bedaque P F and van Kolck U 2002 *Ann. Rev. Nucl. Part. Sci.* **52** 339
[4] Epelbaum E 2006 *Prog. Part. Nucl. Phys.* **57** 654
[5] Epelbaum E, Hammer H -W and Meissner U -G 2009 *Rev. Mod. Phys.* **81** 1773
[6] Kievsky A, Viviani M, Girlanda L and Marcucci L E 2010 *Phys. Rev. C* **81** 044003
[7] van Kolck U 1994 *Phys. Rev. C* **49** 2932
[8] Epelbaum E, Nogga A, Gloeckle W, Kamada H, Meissner U -G and Witala H 2002 *Phys. Rev. C* **66** 064001
[9] Epelbaum E 2008 *Few Body Syst.* **43** 57
[10] Bernard V, Epelbaum E, Krebs H and Meissner U -G 2008 *Phys. Rev. C* **77** 064004
[11] Bernard V, Epelbaum E, Krebs H, Meissner U -G 2011 [arXiv:1108.3816 [nucl-th]]
[12] Girlanda L 2008 *Phys. Rev. C* **77** 067001

[13] Girlanda L, Kievsky A and Viviani M 2011 *Phys. Rev. C* **84** 014001
[14] Girlanda L, Pastore S, Schiavilla R and Viviani M 2010 *Phys. Rev. C* **81** 034005
[15] Kievsky A 1999 *Phys. Rev. C* **60** 034001
[16] Canton L and Schadow W 2000 *Phys. Rev. C* **62** 044005
[17] Canton L and Schadow W 2001 *Phys. Rev. C* **64** 031001
[18] Koike Y and Haidenbauer J 1987 *Nucl. Phys. A* **463** 365
[19] Kievsky A, Viviani M and Rosati S 2001 *Phys. Rev. C* **64** 024002
[20] Sagara K 2010 *Few Body Syst.* **48** 59
[21] Gloeckle W, Witala H, Huber D *et al.* 1996 *Phys. Rep.* **274** 107
[22] Fisher B M, Brune C R, Karwowski H J *et al.* 2006 *Phys. Rev. C* **74** 034001
[23] Deltuva A and Fonseca A C 2007 *Phys. Rev. C* **75** 014005
[24] Deltuva A and Fonseca A C 2007 *Phys. Rev. Lett.* **98** 162502
[25] Deltuva A and Fonseca A C 2007 *Phys. Rev. C* **76** 021001

Journal of Physics: Conference Series **336** (2011) 012005 doi:10.1088/1742-6596/336/1/012005

The scattering matrix from bound state solutions

A Kievsky[1]

[1] Istituto Nazionale di Fisica Nucleare, Sezione di Pisa, Largo B. Pontecorvo 3, I-56127, Pisa, Italy

E-mail: kievsky@pi.infn.it

Abstract. Two integral relations, that have been recently derived from the Kohn variational principle (KVP), are used to describe scattering states. In usual applications the variational scattering wave function requires the explicit form of its asymptotic behavior. This is not the case when the integral relations are applied since, due to their short range nature, the only condition is that the scattering wave function Ψ be the solution of $(H-E)\Psi = 0$ in the internal region. In order to show the applicability of the method, two examples are analyzed as the computation of phase-shifts from bound state type wave functions in the $A = 2$ using a model potential. As a last example we discuss the use of the integral relations in the $A = 3$ system using a realistic nucleon-nucleon potential.

1. Introduction

Much of the information of the nucleon-nucleon (NN) interaction is obtained from the study of $p-p$ and $n-p$ scattering. Specific interaction models reproduce the available NN data with a χ^2 per datum close to one. The application of these models to study the nuclear dynamics in systems with $A > 2$ is limited from our capability of solving the corresponding Schrödinger equation. For example, the fact that the spectrum of each hydrogen and helium isotope has only one bound state in the mass region $A = 2-4$ and the $A = 5$ nucleus does not exist, limits the applicability of bound state methods in the study of these nuclei to a few states. In Ref. [1] a detailed study of the three-nucleon bound states has been done, whereas a similar analysis in the case of ^{4}He can be found in Refs. [2, 3]. In recent years much of the study in the three-nucleon system has been done in the three-nucleon continuum (see Refs. [4, 5] and references therein). Results in the four-body system have been obtained so far in the energy region below the breakup in three particles [6, 7].

Well established methods to treat both, bound and scattering states, are the solution of the Faddeev equations ($A = 3$) or Faddeev-Yakubovsky equations ($A = 4$) in configuration or momentum space and the Hyperspherical Harmonic (HH) expansion in conjunction with the Kohn variational principle (KVP). These methods have proven to be of great accuracy. They have been tested through different benchmarks [8, 9]. On the other hand, many other methods are presently used to describe bound states: for example the Green Function Montecarlo (GFMC) and No Core Shell Model (NCSM) methods have been used in nuclei up to $A = 10$ and $A = 12$ respectively [10, 11]. Attempts to use these methods in the description of scattering states recently appeared [12, 13].

The possibility of employing bound state techniques to describe scattering states has always attracted particular attention [14]. Also recently continuum-discretized states obtained from

Published under licence by IOP Publishing Ltd

the stochastic variational method have been used to study $\alpha + n$ scattering [15]. In those two approaches the tangent of the phase-shift results a quotient of two numbers. In the former the numerator and denominator are obtained from two integral relations after projecting the Schrödinger equation, whereas in the latter the numerator results from an integral relation derived by means of the Green's function formalism and the denominator from the normalization of the continuum-discretized state.

Recently two integral relations have been derived from the KVP [16]. It has been shown that starting from the KVP, the tangent of the phase-shift can be expressed in a form of a quotient where both, the numerator and the denominator, are given as two integral relations. This is similar to what was proposed in Ref. [14], however the variational character of the quotient and its strict relation with the KVP were not recognized. In fact, it is this characteristic that makes possible many different and interesting applications of the integral relations. Accordingly, in the present study we would like to discuss some specific examples. We will show that the integral relations can be used to compute phase-shifts from bound-state-like functions. We start our analysis from the simplest case, the $A = 2$ system, using a model potential. Then, using a semirealistic interaction, $n - d$ as well as $p - d$ scattering are considered. This is of particular interest since $p-d$ scattering has been a subject of intense investigations. A second application of the integral relations regards the possibility of determining $p - d$ phase-shifts from a calculation in which the Coulomb potential has been screened. Finally, as a third application, we will discuss the use of the integral relations in which more than one channel is open. All these examples serve to demonstrate the general validity of the KVP formulated in terms of integral relations. Due to their short-range nature, they are determined by the wave function in the interaction region and not from its explicit asymptotic behavior. This means that each wave function Ψ verifying $(H - E)\Psi = 0$ in the interaction region can be used to determine the corresponding scattering amplitude even if its asymptotic behavior is not the physical one.

2. Integral relations from the Kohn variational principle

Following Ref. [16] we give a brief derivation of the integral relations. Let us first consider a two-body system interacting through a short-range potential $V(r)$ at the center of mass energy E in a relative angular momentum state $l = 0$. The solution of the Schrödinger equation in configuration space (m is twice the reduced mass),

$$(H - E)\Psi(\boldsymbol{r}) = (-\frac{\hbar^2}{m}\nabla^2 + V - E)\Psi(\boldsymbol{r}) = 0 \ , \tag{1}$$

can be obtained after specifying the corresponding boundary conditions. For $E > 0$, with $k^2 = E/(\hbar^2/m)$ and assuming a short-range potential V, $\Psi(\boldsymbol{r}) = \phi(r)/\sqrt{4\pi}$ and

$$\phi(r \to \infty) \longrightarrow \sqrt{k}\left[A\frac{\sin(kr)}{kr} + B\frac{\cos(kr)}{kr}\right] \ . \tag{2}$$

from which one gets $\Psi \to AF + BG$, where

$$\begin{aligned} F &= \sqrt{\frac{k}{4\pi}}\frac{\sin(kr)}{kr} \\ G &= \sqrt{\frac{k}{4\pi}}\frac{\cos(kr)}{kr} \ . \end{aligned} \tag{3}$$

Using the normalization condition

$$\frac{m}{\hbar^2}\left[< F|H - E|\tilde{G} > - < \tilde{G}|H - E|F >\right] = 1 \ . \tag{4}$$

and assuming that Ψ is an exact solution of Eq.(1), the wave function verifies the following integral relations:

$$\begin{aligned} -\frac{m}{\hbar^2} &< \Psi|H-E|F >= B, \\ \frac{m}{\hbar^2} &< \Psi|H-E|\tilde{G} >= A, \\ \tan\delta &= \frac{B}{A} \ . \end{aligned} \tag{5}$$

In the above expressions we have introduced the regularized function $\tilde{G} = f_{reg}G$ with the property $|\tilde{G}(r=0)| < \infty$ and $\tilde{G} = G$ outside the interaction region. A possible choice is

$$\tilde{G} = \sqrt{\frac{k}{4\pi}}\frac{\cos(kr)}{kr}(1-\mathrm{e}^{-\gamma r}) \ , \tag{6}$$

with γ being a non linear parameter which will be discussed below. Values verifying $\gamma > 1/r_0$, with r_0 the range of the potential, could be appropriate.

In practical cases the solution of the Schrödinger equation is obtained numerically. Then, $\tan\delta$ is extracted from $\phi(r)$ analyzing its behavior outside the range of the potential. The equivalence between the extracted value and that one obtained from the integral relations defines the accuracy of the numerical computation. A relative difference of the order of 10^{-7} of the two values is usually achieved using standard numerical techniques to solve the differential equation and to compute the two one-dimensional integrals. To be noticed the short range character of the integral relations. This means that the phase-shift is determined by the internal structure of the wave function.

In the following we demonstrate that the relation $\tan\delta = B/A$, which is an exact relation when the exact wave function Ψ is used in Eq (5), can be considered accurate up to second order when a trial wave function is used, as it has a strict connection with the Kohn variational principle. The connection of the integral relations with the KVP is straightforward. Defining a trial wave function Ψ_t as

$$\Psi_t = \Psi_c + AF + B\,\tilde{G} \ , \tag{7}$$

with $\Psi_c \to 0$ as $r \to \infty$, the condition $\Psi_t \to AF + B\,G$ as $r \to \infty$ is fulfilled. The KVP states that the second order estimate for $\tan\delta$ is

$$[\tan\delta]^{2^{nd}} = \tan\delta - \frac{m}{\hbar^2} < (1/A)\Psi_t|H-E|(1/A)\Psi_t > \ . \tag{8}$$

The above functional is stationary with respect to variations on Ψ_c and $\tan\delta$. As shown in Ref. [16] it can be given as

$$\begin{aligned} B^{2^{nd}} &= -\frac{m}{\hbar^2} <\Psi_t|H-E|F>, \\ A &= \frac{m}{\hbar^2} <\Psi_t|H-E|\tilde{G}>, \\ [\tan\delta]^{2^{nd}} &= B^{2^{nd}}/A \ . \end{aligned} \tag{9}$$

These equations extend the validity of the integral relations, given in Eq.(5) for the exact wave functions, to trial wave functions. Moreover these relations are valid for generic values of the number of particles. In the case in which more than one channel is open the coefficients A and B of the above equations correspond to matrices

$$\left.\begin{aligned} B_{ij}^{2^{nd}} &= -\tfrac{m}{\hbar^2} < \Psi_i|H-E|F_j > \\ A_{ij} &= \tfrac{m}{\hbar^2} < \Psi_i|H-E|\tilde{G}_j > \end{aligned}\right\} R^{2^{nd}} = A^{-1}B^{2^{nd}} . \tag{10}$$

with $R^{2^{nd}}$ the second order estimate of the scattering matrix whose eigenvalues are the phase shifts and the indeces (i, j) indicate the different asymptotic configurations accessible at the specific energy under consideration.

To be noticed that $F, \tilde{G}$ are solutions of the Schrödinger equation in the asymptotic region, therefore $(H - E)F \to 0$ and $(H - E)\tilde{G} \to 0$ as the distance between the particles increases. As a consequence the decomposition of Ψ_t in the three terms of Eq. (7) can be considered formal since, due to the short-range character of the relation integrals, it is sufficient for the trial wave function to be a solution of $(H - E)\Psi_t = 0$ in the interaction region, without an explicit indication of its asymptotic behavior. This fact, together with the variational character of the relations allows for a number of applications to be discussed in the next section.

3. Application of the integral relations

We first discuss an application of the integral relations of Eq. (9) to a two-body system. To make contact with the results given in Refs. [16, 17], we use a central, s-wave gaussian potential

$$V(r) = -V_0 \exp(-r^2/r_0^2) \ , \tag{11}$$

with $V_0 = -51.5$ MeV, $r_0 = 1.6$ fm and $\hbar^2/m = 41.4696$ MeV fm^2. This potential has a shallow $L = 0$ bound state with energy $E_{2B} = -0.397743$ MeV.

We introduce the orthogonal basis

$$\phi_m = \mathcal{L}_m^{(2)}(z) \exp -(z/2) \ , \tag{12}$$

with $\mathcal{L}_m$ a (normalized) Laguerre polynomial and $z = \beta r$, where β is a nonlinear parameter, to expand the wave function of the system

$$\Psi_0 = \sum_{m=0}^{M-1} a_m^0 \phi_m \ . \tag{13}$$

We solve the eigenvalue problem of H for different dimensions M of the basis. The variational principle states that

$$E_0 = \langle \Psi_0 | H | \Psi_0 \rangle \geq E_{2B} \ , \tag{14}$$

with the equality valid when $M \to \infty$. The nonlinear parameter β can be fixed to improve the convergence properties of the basis. In fact, for each value of M there is a value of β that minimizes the energy. Increasing M, the minimum of the energy becomes less and less dependent on β resulting in a plateau. Increasing further the dimension of the basis, the extension of the plateau increases as well, without any appreciable improvement in the eigenvalue, indicating that the convergence has been reached to a certain accuracy. At each step Ψ_0 represents a first order estimate of the exact bound state wave function. Since, in our example, the system has only one bound state, with appropriate values of M and β, the diagonalization of H results in one negative eigenvalue E_0 and $M-1$ positive eigenvalues E_j ($j = 1, ..., M-1$). The corresponding wave functions

$$\Psi_j = \sum_{m=0}^{M-1} a_m^j \phi_m \quad j = 1, ..., M-1 \ , \tag{15}$$

are approximate solutions of $(H - E_j)\Psi_j = 0$ in the interaction region. As $r \to \infty$ they go to zero exponentially and therefore they do not represent physical scattering states. The eigenvectors can be used to compute the integral relations of Eq. (9) and to calculate the second order estimate of the phase-shifts δ_j at the specific energies E_j. This analysis is shown in table 1 in which the non linear parameter β of the Laguerre basis has been chosen to be 1.2 fm^{-1}. In the first row of the table the ground state energy is given for different values of the number M of Laguerre polynomials. The stability of E_0 at the level of 1 keV is achieved already with $M = 20$. For a given value of M, E_j, with $j = 1, 2, 3$, are the first three positive eigenvalues. The eigenvectors corresponding to positive energies approximate the scattering states at these specific energies. Since the lowest scattering state appears at zero energy, none of the positive eigenvalues can reach this value for any finite values of M. We observe that the eigenvalues diminish as M increases. Defining $k_j^2 = \frac{m}{\hbar^2} E_j$, the second order estimate for the phase shift at each energy and at each value of M is obtained as

$$\begin{aligned}
&-\frac{m}{\hbar^2} < \Psi_j|H - E|F_j >= B_j \qquad \text{with} \qquad F_j = \sqrt{\frac{k_j}{4\pi}} \frac{\sin(k_j r)}{k_j r} \\
&\frac{m}{\hbar^2} < \Psi_j|H - E|\tilde{G}_j >= A_j \quad \text{with} \qquad \tilde{G}_j = f_{reg} \sqrt{\frac{k_j}{4\pi}} \frac{\cos(k_j r)}{k_j r} \\
&[\tan \delta_j]^{2^{nd}} = \frac{B_j}{A_j}.
\end{aligned} \tag{16}$$

On the other hand, as we are considering the $A = 2$ system, at each specific energy value E_j the phase shift $\tan \delta_j$ can be obtained by solving the Schrödinger equation numerically. The two values, $[\tan \delta_j]^{2^{nd}}$ and $\tan \delta_j$, are given in the table 1 at the corresponding energies as a function of M. We observe that, as M increases, the relative difference between the variational estimate and the exact value reduces, for example at $M = 40$ it is about 10^{-6}. In fact, as M increases, each eigenvector gives a better representation of the exact wave function in the internal region and the second order estimates, $[\tan \delta_j]^{2^{nd}}$ approach the exact result.

Table 1. The two-nucleon bound state E_0 and the first three positive eigenvalues E_j ($j = 1, 3$), as a function of the number of Laguerre polynomials M. The second order estimates, $[\tan \delta_j]^{2^{nd}}$, obtained applying the integral relations are given in each case and compared to the exact results, $\tan \delta_j$.

M	10	20	30	40
E_0	-0.395079	-0.397740	-0.397743	-0.397743
E_1	0.536349	0.116356	0.048091	0.026008
$[\tan \delta_1]^{2^{nd}}$	-1.507280	-0.622242	-0.392005	-0.286479
$\tan \delta_1$	-1.522377	-0.621938	-0.392021	-0.286480
E_2	1.984580	0.449655	0.190019	0.103503
$[\tan \delta_2]^{2^{nd}}$	-5.919685	-1.353736	-0.812313	-0.584389
$\tan \delta_2$	-5.703495	-1.354691	-0.812270	-0.584388
E_3	4.512635	0.994433	0.423117	0.231645
$[\tan \delta_3]^{2^{nd}}$	13.998124	-2.451174	-1.302799	-0.908128
$\tan \delta_3$	12.684474	-2.448343	-1.302887	-0.908131

In the following application the study of the integral relations is extended to describe a 2 + 1 collision in the three-body system, below the breakup threshold into three particles. Following Refs. [18, 19, 20, 21] a general three-nucleon bound state can be expand in terms of the hyperspherical harmonic basis as

$$\Psi_n = \sum_{\alpha,K,m} A^n_{\alpha,K,m} |\alpha, K, m>, \tag{17}$$

where the index n indicates the level of the state with energy E_n and $|\alpha, K, m>$ is a totally antisymmetric element of the expansion basis for the $A = 3$ system. The linear coefficients $A^n_{\alpha,K,m}$ of the wave function and the energy of the state are obtained by solving the following generalized eigenvalue problem

$$\sum_{\alpha',K',m'} A^n_{\alpha',K',m'} < \alpha, K, m|H - E_n|\alpha', K', m' >= 0 \,. \tag{18}$$

Considering a realistic nucleon-nucleon interaction as the AV14 potential and focussing in the $J\pi = 1/2^+$ state, the lowest solution of the eigenvalue problem ($n = 0$) is the triton bound state. A certain number of negative eigenvalues verifying $E_n > E_d$ (with E_d the deuteron energy) also appear. Defining the positive energy $E^0_n = E_n - E_d$, the corresponding eigenvectors Ψ_n approximately describe a scattering process at the center-of-mass energy E^0_n, though asymptotically they go to zero. The lowest eigenvalue E_0 shows a high stability with the dimension of the basis conversely, the succesive eigenvalues E_n with $n > 0$ show an appreciable variation with the size of the basis. On the other hand the $J = 1/2+$ has two degenerate states in the continuum spectrum corresponding to the two possible couplings of the total spin S and orbital angular momentum L, $(L, S) = (0, 1/2)$ or $(L, S) = (2, 3/2)$, to $J = 1/2$. Varying the size of the basis, maintaining the lower bound state stable, it is possible to obtain solutions having equal eigenvalues at different position in the spectrum. Selecting two of these solutions and calling them Ψ^1 and Ψ^2, these two states are solutions of $(H - E_n)\Psi^{1,2}_n = 0$ in the internal region and, since they are square integrable states, they go to zero asymptotically. However the integral relations of Eq.(9) depend on the internal part of the wave function and therefore it would be possible to use Ψ^1_n and Ψ^2_n as trial wave functions. In this case the second order estimate of the scattering matrix is

$$\begin{aligned} A_{ij} &= \frac{m}{\hbar^2} < \Psi^i_n|H - E_n|\Omega^1_j >, \\ B_{ij} &= -\frac{m}{\hbar^2} < \Psi^i_n|H - E_n|\Omega^0_j >, \\ R &= A^{-1}[B], \end{aligned} \tag{19}$$

where i, j indicate either the two solutions, $\Psi^{1,2}_n$, and the two possible values of the set of quantum numbers (L, S, J) in $J = 1/2^\pm$. The regular and irregular asymptotic functions Ω^0_j and Ω^1_j are defined as

$$\Omega^\lambda_{LSJ}(\mathbf{x}_i, \mathbf{y}_i) = \sum_{l_\alpha=0,2} w_{l_\alpha}(x_i)\mathcal{R}^\lambda_L(y_i) \left\{ \left[[Y_{l_\alpha}(\hat{x}_i)s^{jk}_\alpha]_1 s^i \right]_S Y_L(\hat{y}_i) \right\}_{JJ_z} [t^{jk}_\alpha t^i]_{TT_z} \,, \tag{20}$$

where $w_{l_\alpha}(x_i)$ is the $l_\alpha = 0, 2$ deuteron wave function, $s^{jk}_\alpha = 1, t^{jk}_\alpha = 0$, and L is the relative angular momentum of the deuteron and the incident nucleon. The superscript λ indicates the regular ($\lambda \equiv 0$) or the irregular ($\lambda \equiv 1$) solution of the Schrödinger equation in the asymptotic region. It should be noticed that the irregular solution has been opportunely regularized at the origin

$$\mathcal{R}^1_L(y) = (1 - \mathrm{e}^{-\gamma r_{Nd}})^{L+1} G_L(y), \tag{21}$$

where $r_{Nd} = (\sqrt{3}/2)\, y$ is the nucleon-deuteron separation and the parameter γ is fixed requiring that $\mathcal{R}^1_L(y) \equiv G_L(y)$ asymptotically.

Results of phase-shifts and mixing parameters for the $n - d$ system, calculated using the AV14 NN potential, are presented for the $J^\pi = 1/2^+$ state in figure 1 at the three selected energies $E_{lab} = 1, 2, 3$ MeV. The stability of the results with γ, the regularization parameter introduced in Eq.(21), is chosen as a convergence criterion. This criterion has been discussed in Refs. [16, 22] and essentially it establishes the quality of Ψ_n^i as solution of $(H - E_n)\Psi_n^i = 0$. In fact, if Ψ_n^i is a good solution, the integrals of Eq.(19) are largely independent of γ. The results are compared to the benchmark of Ref. [8] given in the figures as a red line. The results of the application of Eq.(19) are shown as filled circles corresponding to values of γ varying from 0.25 fm^{-1} to 1.25 fm^{-1}. We can observed a good stability on this interval and, furthermore, the results are in very good agreement with those of Ref [8].

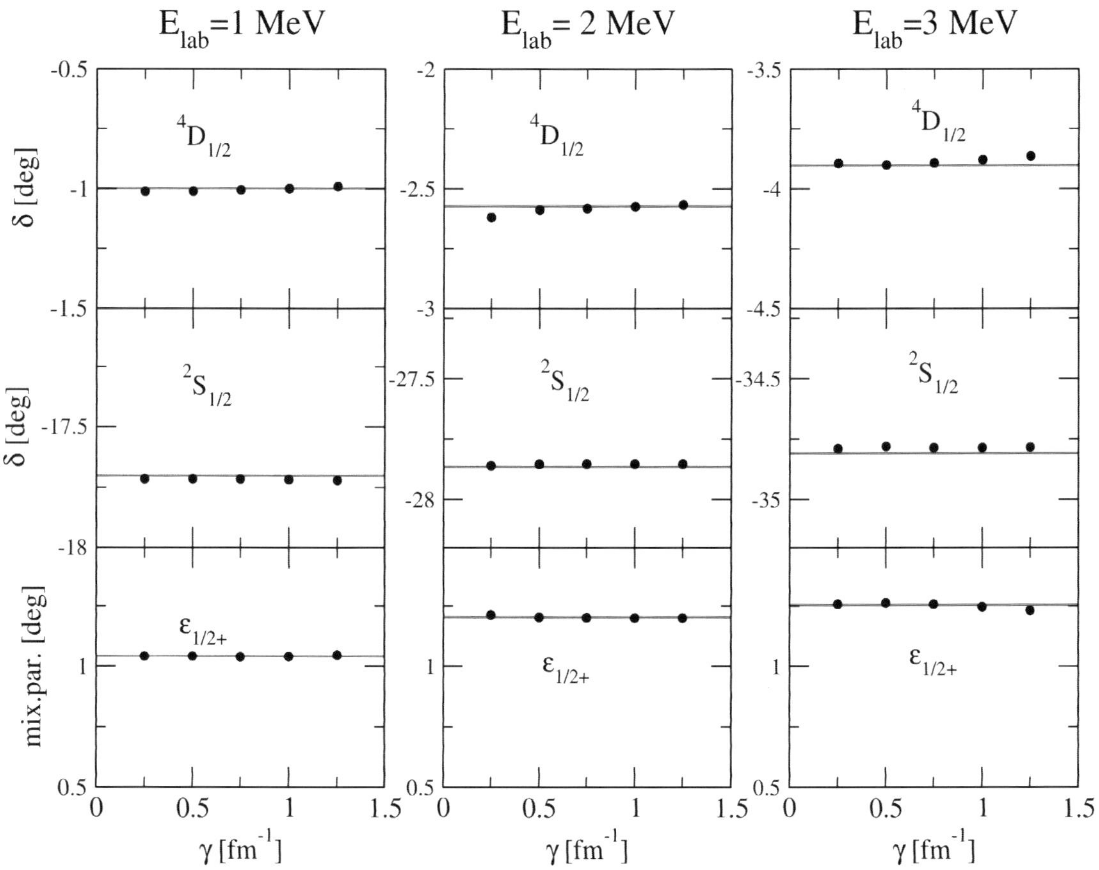

Figure 1. (Color online) The $n-d$ $J^\pi = 1/2^+$ phase-shifts and mixing parameters as a function of the regularization parameter γ at the three indicated energies. The red line corresponds to the results of Ref. [8].

4. Conclusions

In this work the elastic scattering matrix has been determined using bound-state-like wave functions. To this end two integral relations derived from the KVP have been used. Initially, these integral relations were derived in Ref. [16] in order to extract phase shifts from the solutions calculated using the hyperspherical adiabatic expansion in the three-nucleon system. Here applications to elastic scattering of a nucleon on a deuteron ($A = 3$) below the breakup threshold, using realistic nucleon-nucleon potentials has been discussed. In particular two solutions at the same energy have to be determined corresponding to the different possible asymptotic configurations of the system. The obtained results are in close agreement with those presented in the $A = 3$ benchmark of Ref. [8]. A more detailed analysis of the use of the integral relations in the $A = 3, 4$ systems is in progress [23].

References

[1] Nogga A *et al.* 2003 *Phys. Rev. C* **67** 034004
[2] Nogga A, Kamada H, Glöckle W and Barrett B R 2002 *Phys. Rev. C* **65** 054003
[3] Viviani M, Kievsky A and Rosati S 2005 *Phys. Rev. C* **71** 024006
[4] Glöckle W., Witała H, Hüber D, Kamada H and Golak J 1994 *Phys. Rep.* **274** 107
[5] Kievsky A, Viviani M, and Rosati S 2001 *Phys. Rev. C* **64** 024002
[6] Fisher B M *et al.* 2006 *Phys. Rev. C* **74** 034001
[7] Deltuva A and Fonseca A C 2007 *Phys. Rev. C* **75** 014005
[8] Kievsky A *et al.* 1998 *Phys. Rev. C* **58** 3085
[9] Lazauskas R *et al.* 2005 *Phys. Rev. C* **71** 064003
[10] Pieper S C, Varga K and Wiringa R B 2002 *Phys. Rev. C* **66** 044310
[11] Navrátil P, Gueorguiev V G, Vary J P, Ormand W E and Nogga A 2007 *Phys. Rev. Lett.* **99** 042501
[12] Nollett K N, Pieper S C, Wiringa R B, Carlson J and Hale G M 2007 *Phys. Rev. Lett.* **99** 022502
[13] Quaglioni S and Navrátil P 2009 *Phys. Rev. C* **79** 044606
[14] Harris F E 1967 *Phys. Rev. Lett.* **19** 173
[15] Suzuki Y, Horiuchi W and Arai K 2009 *Nucl. Phys. A* **823** 1
[16] Barletta P, Romero-Redondo C, Kievsky A, Viviani M, and Garrido E 2009 *Phys. Rev. Lett.* **103** 090402
[17] Barletta P and Kievsky A 2009 *Few-Body Syst.* **45** 25
[18] Kievsky A, Viviani M and Rosati S 1994 *Nucl. Phys. A* **577** 511
[19] Kievsky A, Viviani M and Rosati S 1993 *Nucl. Phys. A* **551** 241
[20] Kievsky A 1997 *Nucl. Phys. A* **624** 125
[21] Kievsky A, Rosati S, Viviani M, Marcucci L E and Girlanda L 2008 *J. Phys. G: Nucl. Part. Phys.* **35** 063101
[22] Kievsky A, Viviani M, Barletta P, Romero-Redondo C and Garrido E 2010 *Phys. Rev. C* **81** 034002
[23] Kievsky A, Viviani M and Marcucci L E 2011 Theoretical description of three- and four-nucleon scattering states using bound-state-like wave functions *arXiv:1109.3976*

Journal of Physics: Conference Series **336** (2011) 012006 doi:10.1088/1742-6596/336/1/012006

Few-nucleon bound states using the unsymmetrized HH expansion

M Gattobigio[1], A Kievsky[2] and M Viviani[2]

[1] INLN, Université de Nice-Sophia Antipolis, CNRS, 1361 route des Lucioles, 06560 Valbonne, France

[2] Istituto Nazionale di Fisica Nucleare, Sezione di Pisa, Largo B. Pontecorvo 3, I-56127, Pisa, Italy

E-mail: Mario.Gattobigio@inln.cnrs.fr

Abstract. The hyperspherical harmonic basis is used to describe bound states in an A–body system up to six bodies. In this basis the kinetic energy matrix is diagonal whereas the potential matrix is presented in a compact form, well suited for numerical implementation. The basis is neither symmetrized nor antisymmetrized, as required in the case of identical particles. However, after the diagonalization of the Hamiltonian matrix, the eigenvectors reflect the symmetries present in it, and the identification of the physical states is possible, as it will be shown in specific cases. As an example we solve the case of $A = 3, 4, 5, 6$ particles interacting through a short-range central interaction with and without the inclusion of the Coulomb potential.

1. Introduction

The *ab initio* description of light nuclear systems, starting from the nucleon-nucleon (NN) interaction, requires well established methods to solve the Schrödinger equation. Among them, the Green's function Monte Carlo (GFMC) method has been extensively used to describe light nuclei up to $A = 10$ and the no-core shell model (NCSM) up to $A = 12$ [1, 2]. In the $A \leq 4$ systems, well established methods for treating both bound and scattering states exist as the Faddeev equations ($A = 3$) and the Faddeev-Yakubovsky equations ($A = 4$) in configuration or momentum space, and the Hyperspherical Harmonic (HH) expansion. All these methods have proven to be of great accuracy and they have been tested using different benchmarks [3, 4, 5].

The HH method provides a systematic way of constructing a complete basis for the expansion of the A-particle wave function. In the specific case of applications to nuclear physics, the wave function has to be antisymmetric and, therefore, the HH basis has been managed to produce basis states having well defined properties under particle permutations. However, it is possible to use the HH basis without a previous symmetrization procedure [6]. It has been observed that the eigenvectors of the Hamiltonian matrix reflects the symmetries present in it, even if it has been constructed using the non-symmetrized basis. The only requirement was to include all the HH basis elements having the same grand angular quantum number K. It is a property of the HH basis that basis elements having well defined behavior under particle permutation can be constructed as a linear combination of HH elements having the same value of K. Accordingly, the diagonalization procedure generates eigenvectors with a well defined permutation symmetry that can be organized in accordance with the irreducible representations of the permutation group of A objects, S_A. Identifying those eigenvectors with the desired symmetry, the corresponding

Published under licence by IOP Publishing Ltd

energies can be considered variational estimates. In particular, in Ref. [6], it was possible to identify a subset of eigenvectors and eigenvalues corresponding exactly to those that would be obtained performing the preliminary symmetrization of the states. It should be noticed that the simplicity of using the HH basis without a preliminary antisymmetrization step has to be counterbalanced with the large dimension of the matrices to be diagonalized. However, at present, different techniques are available to treat (at least partially) this problem.

As mentioned, one of the main problem in using the HH basis is its large degeneracy, resulting in very large matrices. On the other hand, the potential energy matrix, expressed as a sum of pairwise interactions, cannot connect arbitrary basis elements differing in some specific quantum numbers. This means that in some representation each pairwise-interaction term is represented by a sparse matrix. The matrix representation of the potential $V(1,2)$, constructed in terms of basis elements in which the quantum numbers of particles $(1,2)$ are well defined, is sparse in $A \geq 3$ systems. Its matrix elements connecting basis elements with different quantum numbers labelling states which do not involve particles $(1,2)$ are zero. A problem arises when the matrix elements of the generic term $V(i,j)$, defining the interaction between particles (i,j), has to be calculated using basis elements in which the quantum numbers of particles (i,j) are not well defined. One operative way to solve this problem consists in rotating the basis to a system of coordinates in which particles (i,j) have well defined quantum numbers. This makes the matrix $V(i,j)$ sparse. Conversely, the rotation matrix is not sparse but it can be expressed as a product of sparse matrices, each one representing a rotation which involves a permutation of particles of successive numbering. After these manipulations the potential energy matrix results in a sum of products of sparse matrices suitable for numerical implementations.

2. The unsymmetrized HH basis

Following Refs. [6, 7], we present a brief overview of the properties of the HH basis and its implementation without generating basis elements with well defined permutational symmetry. We start with the following definition of the Jacobi coordinates for an equal-mass A-body system with Cartesian coordinates $\mathbf{r}_1 \ldots \mathbf{r}_A$

$$\mathbf{x}_{N-j+1} = \sqrt{\frac{2j}{j+1}}\,(\mathbf{r}_{j+1} - \mathbf{X}_j)\,, \qquad j = 1, \ldots, N\,, \tag{1}$$

where $N = A - 1$ and $\mathbf{X}_j = \sum_{i=1}^{j} \mathbf{r}_i / j$. For a given set of Jacobi coordinates $\mathbf{x}_1, \ldots, \mathbf{x}_N$, we can introduce the hyperradius ρ

$$\rho = \left(\sum_{i=1}^{N} \mathbf{x}_i^2\right)^{1/2} = \left(2\sum_{i=1}^{A} (\mathbf{r}_i - \mathbf{X})^2\right)^{1/2} = \left(\frac{2}{A}\sum_{j>i}^{A} (\mathbf{r}_j - \mathbf{r}_i)^2\right)^{1/2}, \tag{2}$$

with $X = \sum_{i=1}^{A} \mathbf{r}_i / A$ the center of mass, and the hyperangular coordinates Ω_N

$$\Omega_N = (\hat{\mathbf{x}}_1, \ldots, \hat{\mathbf{x}}_N, \phi_2, \ldots, \phi_N)\,, \tag{3}$$

with the hyperangles ϕ_i defined via

$$\begin{aligned}
x_N &= \rho\cos\phi_N \\
x_{N-1} &= \rho\sin\phi_N\cos\phi_{N-1} \\
&\;\vdots \\
x_i &= \rho\sin\phi_N\cdots\sin\phi_{i+1}\cos\phi_i \\
&\;\vdots \\
x_2 &= \rho\sin\phi_N\cdots\sin\phi_3\cos\phi_2 \\
x_1 &= \rho\sin\phi_N\cdots\sin\phi_3\sin\phi_2\,.
\end{aligned} \tag{4}$$

The explicit expression for the HH functions, having well defined values of LM, is

$$\mathcal{Y}^{LM}_{[K]}(\Omega_N) = \left[\prod_{j=2}^{N} \mathcal{P}^{\alpha_{l_j},\alpha_{K_{j-1}}}_{K_j}(\phi_j)\right]\left[Y_{l_1}(\hat{\boldsymbol{x}}_1)\otimes Y_{l_2}(\hat{\boldsymbol{x}}_2)|_{L_2}\ldots\otimes Y_{l_{N-1}}(\hat{\boldsymbol{x}}_{N-1})|_{L_{N-1}}\otimes Y_{l_N}(\hat{\boldsymbol{x}}_N)\right]_{LM}, \tag{5}$$

with the indicated coupling scheme. The hyperspherical polynomial is

$$\mathcal{P}^{\alpha_{l_j},\alpha_{K_{j-1}}}_{K_j}(\phi_j) = \mathcal{N}^{\alpha_{l_j},\alpha_{K_j}}_{n_j}(\cos\phi_j)^{l_j}(\sin\phi_j)^{K_{j-1}}P^{\alpha_{K_{j-1}},\alpha_{l_j}}_{n_j}(\cos 2\phi_j)\,. \tag{6}$$

The set of quantum numbers $[K]$ includes the $n_2\ldots n_N$ indices of the Jacobi polynomials, the $l_1\ldots l_N$ angular momenta of the particles and the intermediate couplings $L_2\ldots L_{N-1}$. The K_j quantum numbers are defined as

$$K_j = \sum_{i=1}^{j}(l_i + 2n_i)\,, \qquad n_1 = 0\,, \qquad K \equiv K_N\,, \tag{7}$$

$K \equiv K_N$ is known as the grand angular momentum, and $\mathcal{N}^{\alpha\beta}_n$ is a normalization factor. For the definition of the α_a, where a can be either an angular momentum l_j or a quantum number K_j, one needs to introduce the hyperspherical-binary-tree structure [8]. For example the tree of figure 1 corresponds to the choice of hyperangles given by Eq. (4), in which the coefficients specialize to $\alpha_{K_j} = K_j + 3j/2 - 1$ and $\alpha_{l_j} = l_j + 1/2$.

Hyperspherical functions constructed on top of different hyperspherical-coordinate definitions can be related using the $\mathcal{T}$-coefficients [9, 10]. Schematically, these coefficients relate the following tree structures

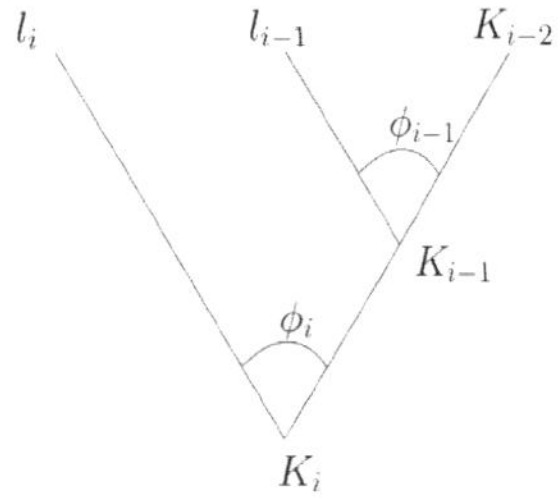

$$= \sum_{\tilde{n}_{i-1}=0}^{N_i} \mathcal{T}^{\alpha_{K_{i-2}}\alpha_{l_{i-1}}\alpha_{l_i}}_{n_{i-1}\tilde{n}_{i-1}K_i}$$

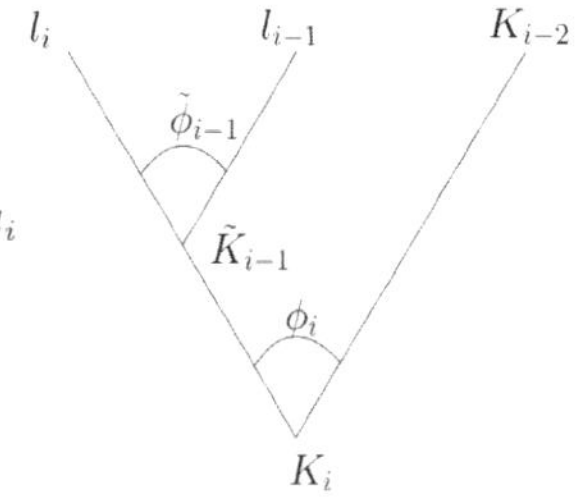

, (8)

where $K_i = K_{i-1} + l_i + 2n_i = \tilde{K}_{i-1} + l_i + 2\tilde{n}_i$. The explicit definition of the coefficients is given in Ref. [6]. Let us call $\mathcal{Y}^{LM}_{[K]}(\Omega^i_N)$ the HH basis element constructed in terms of a set of Jacobi

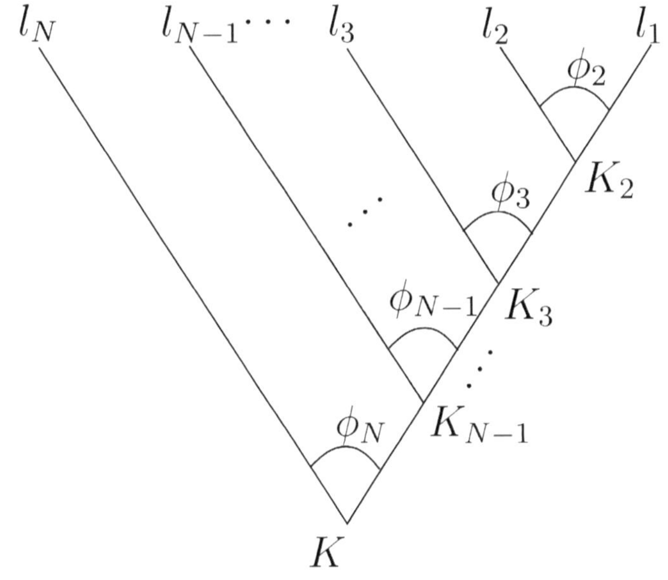

Figure 1. Hyperspherical tree corresponding to Eq. (4).

coordinates in which the i-th and $i+1$-th Jacobi vectors result from the transposition between particles $j, j+1$

$$\begin{aligned} \mathbf{x}'_i &= -\frac{1}{j}\,\mathbf{x}_i + \frac{\sqrt{(j+1)^2 - 2(j+1)}}{j}\,\mathbf{x}_{i+1} \\ \mathbf{x}'_{i+1} &= \frac{\sqrt{(j+1)^2 - 2(j+1)}}{j}\,\mathbf{x}_i + \frac{1}{j}\,\mathbf{x}_{i+1}\,, \end{aligned} \tag{9}$$

with all the other vectors equal to the original ones (transposed basis). The coefficients

$$\mathcal{A}^{i,LM}_{[K][K']} = \int d\Omega_N [\mathcal{Y}^{LM}_{[K]}(\Omega_N)]^* \mathcal{Y}^{LM}_{[K']}(\Omega^i_N)\,, \tag{10}$$

are the matrix elements of a matrix $\mathcal{A}^{LM}_i$ that allows to express the transposed HH basis elements in terms of the reference basis. The coefficients $\mathcal{A}^{i,LM}_{[K][K']}$ form a very sparse matrix and they can be calculated analytically using the $\mathcal{T}$- coupling coefficients and the Raynal-Revai matrix elements. A generic rotation between the reference HH basis and a basis in which the last Jacobi vector is defined as $\mathbf{x}'_N = \mathbf{r}_j - \mathbf{r}_i$ can be constructed as successive products of the $\mathcal{A}^{k,LM}_{[K][K']}$ coefficients. Defining $\mathcal{Y}^{LM}_{[K]}(\Omega^{ij}_N)$ the HH basis element constructed in terms of a set of Jacobi coordinates in which the N-th Jacobi vector is defined $\mathbf{x}'_N = \mathbf{r}_j - \mathbf{r}_i$, this coefficient can be given in the following form

$$\mathcal{B}^{ij,LM}_{[K][K']} = \int d\Omega [\mathcal{Y}^{LM}_{[K]}(\Omega_N)]^* \mathcal{Y}^{LM}_{[K]}(\Omega^{ij}_N) = \left[\mathcal{A}^{LM}_{i_1} \cdots \mathcal{A}^{LM}_{i_n}\right]_{[K][K']}\,. \tag{11}$$

The particular values of the indices $i_1, \ldots, i_n$, labelling the matrices $\mathcal{A}^{LM}_{i_1}, \ldots, \mathcal{A}^{LM}_{i_n}$, depend on the pair (i,j). The matrix

$$\mathcal{B}^{LM}_{ij} = \mathcal{A}^{LM}_{i_1} \cdots \mathcal{A}^{LM}_{i_n}\,, \tag{12}$$

is written as a product of the sparse matrices $\mathcal{A}_i^{LM}$'s.

We consider now the potential energy of an A-body system constructed in terms of two-body interactions

$$V = \sum_{i<j} V(i,j) \quad . \tag{13}$$

In terms of the HH basis it results

$$\sum_{ij} V_{ij}(\rho) = \sum_{ij} [\mathcal{B}_{ij}^{LM}]^t \, V_{12}(\rho) \, \mathcal{B}_{ij}^{LM} \, , \tag{14}$$

where the matrix elements of the matrix $V_{12}(\rho)$ are defined as

$$\begin{aligned} V^{(1,2)}_{[K][K']}(\rho) &= \langle \mathcal{Y}^{LM}_{[K]}(\Omega_N) | V(1,2) | \mathcal{Y}^{LM}_{[K']}(\Omega_N) \rangle = \\ \delta_{l_1,l_1'} &\cdots \delta_{l_N,l_N'} \delta_{L_2,L_2'} \cdots \delta_{L_N,L_N'} \delta_{K_2,K_2'} \cdots \delta_{K_N,K_N'} \\ \times &\int d\phi_N (\cos\phi_N \sin\phi_N)^2 \, \mathcal{P}^{\alpha_{l_N},\alpha_{K_{N-1}}}_{K_N}(\phi_N) \, V(\rho\cos\phi_N) \, \mathcal{P}^{\alpha_{l_N},\alpha_{K_{N-1}}}_{K_N'}(\phi_N) \, . \end{aligned} \tag{15}$$

Each term of the sum in Eq.(14) results in a product of sparse matrices, a property which allows an efficient implementation of matrix-vector product.

3. Results

In this section we present results for $A = 3 - 6$ systems obtained by a direct diagonalization of the Hamiltonian of the system. The corresponding Hamiltonian matrix is obtained using the following orthonormal basis

$$\langle \rho\,\Omega \,|\, m\,[K] \rangle = \left(\beta^{(\alpha+1)/2} \sqrt{\frac{m!}{(\alpha+m)!}} \, L^{(\alpha)}_m(\beta\rho) \, \mathrm{e}^{-\beta\rho/2} \right) \mathcal{Y}^{LM}_{[K]}(\Omega_N) \, , \tag{16}$$

where $L^{(\alpha)}_m(\beta\rho)$ is a Laguerre polynomial with $\alpha = 3N - 1$ and β a variational non-linear parameter. The matrix elements of the Hamiltonian are obtained after integrations in the ρ, Ω spaces. They depend on the indices m, m' and $[K], [K']$ as follows

$$\begin{aligned} \langle m'\,[K'] | H | \, m\,[K] \rangle = &-\frac{\hbar^2 \beta^2}{m} (T^{(1)}_{m'm} - K(K+3N-2) T^{(2)}_{m'm}) \delta_{[K'][K]} \\ &+ \sum_{ij} \left[\sum_{[K''][K''']} \mathcal{B}^{ij,LM}_{[K][K'']} \mathcal{B}^{ij,LM}_{[K'''][K']} V^{m,m'}_{[K''][K''']} \right] . \end{aligned} \tag{17}$$

The matrices $T^{(1)}$ and $T^{(2)}$ have an analytical form and are given in Ref. [6]. The matrix elements $V^{m,m'}_{[K][K']}$ are obtained after integrating the matrix $V_{12}(\rho)$ in ρ-space (we will call the corresponding matrix V_{12}). Introducing the diagonal matrix D such that $\langle [K'] \,|\, D \,| [K] \rangle = \delta_{[K],[K']} K(K + 3N - 2)$, and the identity matrix I in K-space, we can rewrite the Hamiltonian schematically as

$$H = -\frac{\hbar^2 \beta^2}{m} ({}^{(1)}T \otimes I + {}^{(2)}T \otimes D) + \sum_{ij} [\mathcal{B}^{LM}_{ij}]^t \, V_{12} \, \mathcal{B}^{LM}_{ij} \, , \tag{18}$$

in which the tensor product character of the kinetic energy is explicitly given.

As a central potential we have chosen the Volkov potential

$$V(r) = V_R \, e^{-r^2/R_1^2} + V_A \, e^{-r^2/R_2^2} , \quad (19)$$

with $V_R = 144.86$ MeV, $R_1 = 0.82$ fm, $V_A = -83.34$ MeV, and $R_2 = 1.6$ fm. The nucleons are considered to have the same mass chosen to be equal to the reference mass m and corresponding to $\hbar^2/m = 41.47$ MeV fm^2. With this parametrization of the potential, the two-nucleon system has a binding energy $E_{2N} = 0.54592$ MeV and a scattering length $a_{2N} = 10.082$ fm. This potential has been used several times in the literature making its use very useful to compare different methods [11, 12, 13, 14]. The use of central potentials in general produces too much binding, in particular the $A = 5$ system results bounded. Conversely, the use of the s-wave version of the potential produces a spectrum much closer to the experimental situation. This is a direct consequence of the weakness of the nuclear interaction in p-waves. Accordingly, we analyze both versions of the potential, the central Volkov potential and the s-wave projected potential. The results are obtained after a direct diagonalization of the Hamiltonian matrix of Eq.(17) including $m_{max} + 1$ Laguerre polynomials with a fix value of β, and all HH states corresponding to maximum value of the grand angular momentum K_{max}. The scale parameter β can be used as a non-linear parameter to study the convergence in the index $m = 0, 1, \ldots, m_{max}$, with m_{max} the maximum value considered. In the present analysis the convergence will be studied with respect to the index K_{max}, therefore, the number of Laguerre polynomials at each step, $m_{max} + 1$, will be sufficiently large to guarantee independence from β of the physical eigenvalues and eigenvectors. We found that $m_{max} + 1 \approx 20$ Laguerre polynomials (with proper values of β) were sufficient for an accuracy of 0.1% in the calculated eigenvalues.

In table 1 we present the pattern of convergence for selected bound states in $A = 3, 4, 5, 6$ systems using the Volkov potential acting in all partial waves or acting only in s-wave. For $A = 3, 4$ the ground state of the system corresponds to the states $(L, S, T)J^\pi = (0, 1/2, 1/2)1/2^+$ ($A = 3$) and $(L, S, T)J^\pi = (0, 0, 0)0^+$ ($A = 4$). In the case of systems with $A > 4$ the spatially-symmetric state cannot be antisymmetrized using the corresponding spin-isospin functions. In the $A = 5$ system the deepest two levels having positive parity and $L = 0$ correspond to a completely symmetric state (the irreducible representation of S_5 [**5**]). The third level belongs to the irreducible representation of S_5 [**4 1**]; it can be antisymmetrized using the $A = 5$ spin-isospin functions with $S = 1/2, T = 1/2$, therefore it represents the lowest level of the $(L, S, T)J^\pi = (0, 1/2, 1/2)1/2^+$ state of five nucleons. The convergence of this state in terms of K_{max} is given in the fourth column of the table 1. It can be observed that the [**4 1**] state is not bounded, in agreement with the fact that the $A = 5$ nucleus does not exist. Its energy results to be above the threshold of 30.42 MeV describing an ^{4}He nucleus plus a fifth nucleon far away (here the Coulomb interaction has not been included). In the case of the $A = 6$ system we concentrate the analysis in the $(L, S, T)J^\pi = (0, 0, 1)0^+$ and $(0, 1, 0)1^+$ states which, using a central potential, and disregarding the Coulomb interaction, are degenerate. Including the Coulomb interaction between two protons, the first state has the quantum numbers of ^{6}He. A direct diagonalization of the six body Hamiltonian using the non-symmetrized HH basis, with the Volkov potential, produces a spectrum in which the first two levels belongs to the [**6**] irreducible representation of S_6. The third level belongs to the [**5 1**] representation whereas the fourth level belongs to the [**4 2**] representation, and it is the first one that can be symmetrized using the $A = 6$ spin-isospin functions with $S = 0, T = 1$ or $S = 1, T = 0$. The convergence pattern of this level in terms of K_{max} is shown in the last column of table 1. Similar to the $A = 5$ case, the Volkov potential acting in all waves predicts large binding energies. Using the s-wave potential a much more reasonable value of ≈ 34 MeV is obtained for this level.

For $A = 6$, when the Coulomb interaction between two protons is considered, the degeneracy of the [**4 2**] level (of dimension 9) is broken and four different states appear. It is possible to identify the physical state looking at the symmetry of the four neutrons. One of the states

Table 1. Ground state of the $A = 3, 4$ systems (in MeV) and selected states of the $A = 5, 6$ systems (see text) using the all-waves (upper part) and s-wave (lower part) Volkov potential as a function of the maximum grand angular quantum number K_{max}.

			all-waves			
K_{max}	$A=3$	$A=4$	K_{max}	$A=5$	K_{max}	$A=6$
0	7.7075	28.580	4	22.043	2	61.142
10	8.4157	30.278	8	25.568	6	63.377
20	8.4623	30.416	12	27.043	10	65.354
30	8.4647	30.418	16	27.862	14	66.201
40	8.4649	30.418	20	28.371	18	66.437
			24	28.719	22	66.491
SVM [12]	8.46	30.42		43.00		66.25
Ref. [11]	8.462	30.406		42.383		
			s-wave			
K_{max}	$A=3$	$A=4$	K_{max}	$A=5$	K_{max}	$A=6$
0	7.7075	28.580	3	24.317	2	24.793
10	8.3812	30.116	7	26.923	6	30.723
20	8.4283	30.250	11	27.971	10	32.244
30	8.4307	30.252	15	28.521	14	33.075
40	8.4309	30.252	19	28.823	18	33.561
			23	29.005	22	33.814

belongs to the $[\mathbf{4}]$ representation, two belong to the $[\mathbf{3\,1}]$ representation and the last one belongs to the $[\mathbf{2}^2]$ representation of S_4. This last state is the only one that can be antisymmetrized using the spin functions of four neutrons having $S_n = 0$. Moreover, the proton state is spatially symmetric and therefore can be antisymmetrized with the spin function $S_p = 0$ making a total $S = 0$ state. This can be seen as

$$[\mathbf{4\,2}] \to [\mathbf{1}^2] \otimes [\mathbf{3\,1}] + [\mathbf{2}] \otimes [\mathbf{4}] + [\mathbf{2}] \otimes [\mathbf{3\,1}] + [\mathbf{2}] \otimes [\mathbf{2}^2], \tag{20}$$

where only $[\mathbf{2}] \otimes [\mathbf{2}^2]$ is physical and describes the ground state of ^{6}He. When the Coulomb interaction is extended to three particles, the symmetry breaking is $S_6 \to S_3 \otimes S_3$, and the split reads

$$[\mathbf{4\,2}] \to [\mathbf{2\,1}] \otimes [\mathbf{3}] + [\mathbf{2\,1}] \otimes [\mathbf{2\,1}] + [\mathbf{3}] \otimes [\mathbf{3}] + [\mathbf{3}] \otimes [\mathbf{2\,1}], \tag{21}$$

with the only physical state, describing ^{6}Li, being $[\mathbf{2\,1}] \otimes [\mathbf{2\,1}]$. In table 2 we show the results for the $L = 0$ state.

4. Conclusions

We have shown the power and the flexibility of the unsymmetrized HH approach. In particular we were able to include basis states up to $K = 22$ for a six-body system (corresponding to a basis set of 38·798·760 elements using 11 Laguerre polynomials). For these large matrices it is possible to restrict the search of the eigenvectors to those having a particular symmetry using a symmetry-adapted Lanczos method [15]. Essentially, starting with a vector having the desired symmetry, after each iteration of the matrix-vector product, the new vector is projected onto the sub-space of the selected symmetry. Following Ref. [15], an intermediate purification step is also implemented. This method has the characteristic of finding eigenvalues corresponding to eigenvectors of one particular symmetry simplifying the search procedure and the identification of the eigenvectors. The extension of this method to treat realistic NN potentials is in progress.

Table 2. Binding energies calculated with Volkov's potential in s-wave, for $A = 6$ particles and $L = 0$, with and without Coulomb interaction, using 11 Laguerre's polynomials.

K_{max}	N_{HH}	$E_{[^6\text{He}]}$ (MeV) $[\mathbf{2}] \otimes [\mathbf{2}^2]$	$E_{[^6\text{Li}]}$ (MeV) $[\mathbf{2\,1}] \otimes [\mathbf{2\,1}]$
2	15	24.064	22.974
4	120	28.016	26.988
6	680	29.935	28.947
8	3045	30.851	29.889
10	11427	31.446	30.496
12	37310	31.908	30.964
14	108810	32.275	31.334
16	288990	32.558	31.620
18	709410	32.762	31.827
20	1628328	32.912	31.980
22	3527160	33.016	32.087

References

[1] Pieper S C, Varga K and Wiringa R B 2002 *Phys. Rev. C* **66** 044310
[2] Navrátil P, Gueorguiev V G, Vary J P, Ormand W E and Nogga A 2007 *Phys. Rev. Lett.* **99** 042501
[3] Kievsky A *et al.* 1998 *Phys. Rev. C* **58** 3085
[4] Lazauskas R *et al.* 2005 *Phys. Rev. C* **71** 034004
[5] Kamada H *et al.* 2001 *Phys. Rev. C* **64** 044001
[6] Gattobigio M, Kievsky A and Viviani M 2011 *Phys. Rev. C* **83** 024001
[7] Gattobigio M, Kievsky A, Viviani M and Barletta P 2009 *Phys. Rev. A* **79** 032513
[8] Vilenkin N Ya, Kuznetsov G I and Smorodinskii Ya A 1966 *Sov. J. Nucl. Phys.* **2** 645
[9] Kil'dyushov M S 1972 *Yad. Fiz.* **15** 197 [1972 *Sov. J. Nucl. Phys.* **15** 113]
[10] Kil'dyushov M S 1972 *Yad. Fiz.* **16** 217 [1973 *Sov. J. Nucl. Phys.* **16** 117]
[11] Barnea N 1999 *Phys. Rev. A* **59** 1135
[12] Varga K and Suzuki Y 1995 *Phys. Rev. C* **52** 2885
[13] Viviani M, Kievsky A and Rosati S 2005 *Phys. Rev. C* **71** 024006
[14] Timofeyuk N K 2002 *Phys. Rev. C* **65** 064306
[15] Wang X-G and Carrington T Jr 2001 *J. Chem. Phys.* **114** 1473

Journal of Physics: Conference Series **336** (2011) 012007 doi:10.1088/1742-6596/336/1/012007

The inclusion of tensor terms in the nonrelativistic nuclear Energy Density Functionals

Gianluca Colò

Dipartimento di Fisica, Università degli Studi di Milano and INFN, Sezione di Milano, via Celoria 16, 20133 Milano (Italy)

E-mail: colo@mi.infn.it

Abstract. This contribution is intended to briefly review recent attempts to include tensor terms in effective nuclear Energy Density Functionals (EDFs). We stress that effective tensor forces should not be conflused with the well-known *bare* tensor force. Consistently with the basic idea of EDFs, collective states are shown to provide better constraints for the tensor parameters, as compared with single-particle states. In fact, basic limitations of the description of the single-particle strength in atomic nuclei provided by EDFs, are highlighted in the last part of this contribution.

1. Introduction

Modern EDFs constitute the basic tool for microscopic nuclear structure calculations if a systematic exploration of wide regions of the nuclear chart is envisaged, and if also excited states at relatively large excitation energies are to be considered. There exist several kinds, and many parameterizations, for nuclear EDFs. In the nonrelativistic description, one often starts from effective Hamiltonians based either on zero-range Skyrme interactions or finite-range Gogny interactions. Recently, local functionals derived from the Skyrme Hamiltonians are generalized by including systematically all terms consistent with symmetries (up to a given order in the derivative operators). Relativistic Mean Field (RMF), or covariant, functionals have also been devoted much attention: in this case the EDF is derived from a Lagrangian that includes nucleons treated as Dirac particles and effective mesons (plus the associated couplings).

The basic features of EDFs for nuclei are reviewed, e.g., in Ref. [1]. They are, as a rule, fitted by using empirical properties of uniform nuclear matter together with the masses and charge radii of selected reference nuclei. They can describe in a reasonable way the global trends of the ground-state properties of nuclei (binding energies, radii, and deformations). They can be used to describe superheavy nuclei, drip-line isotopes, or even exotic forms of nuclear matter like that found inside specific layers of neutron stars (at least, at densities that are not much larger, or smaller, than usual terrestrial nuclear densities).

Starting from an EDF, the ground-state properties are obtained my minimizing the total energy. The scheme is in principle based on the Density Functional Theory (DFT) originally introduced by Hohenberg and Kohn. In practice, if one starts from an Hamiltonian $H_{\rm eff} = T + V_{\rm eff}$ (where T is the kinetic energy and $V_{\rm eff}$ is the effective force) and calculates the total energy as

$$E[\rho] = \langle \Phi | H_{\rm eff} | \Phi \rangle, \tag{1}$$

Published under licence by IOP Publishing Ltd

where $|\Phi\rangle$ is a general Slater determinant, formally the minimization of the total energy amounts to solving the Hartree-Fock (HF) equations. However, if the interaction is adequately designed and the parameters are well fitted, the model, although formally is a mean field model, can effectively include a large class of many-body correlations.

The time-dependent extension of the static DFT describe nuclear oscillations around the ground-state, namely collective states like the low-lying density modes or the giant resonances. Starting from an effective Hamiltonian $H_{\rm eff}$, time-dependent DFT looks the same as time-dependent HF (in close analogy with the discussion of the previous paragraph). It is well known from textbooks that the small amplitude limit of time-dependent HF is the so-called Random Phase Approximation (RPA) theory. Self-consistent RPA on top of Skyrme-HF, or Gogny-HF, or RMF calculations for the nuclear ground-state, have been performed for several years.

In the last five years or so, much interest has concerned the relevance of effective tensor terms within, e.g., the Skyrme Hamiltonian. In fact, only central terms (and one or two spin-orbit terms) are included in most of the Skyrme parameter sets which have been widely used so far in the literature. Many groups have recently performed HF calculations by including tensor terms on top of the usual Skyrme central terms (see Refs. [2, 3, 4, 5, 6, 7, 8, 9]). In Ref. [10], Skyrme results the for single-particle HF gaps are compared with results obtained within the framework of HF with the Gogny force (including the tensor force introduced in [11]), as well as with relativistic HF (RHF) results [12, 13, 14]. A similar yet less complete comparison had been performed in Ref. [15]. It has to be stressed that in the relativistic case the tensor interaction is associated with pion and ρ-meson exchange and cannot be accomodated in the Hartree, or RMF, description.

As far as RPA calculations are concerned, tensor terms have been introduced in the Skyrme framework, both for normal states [16, 17] and charge-exchange states [18, 19, 20, 21]. We will argue below that these calculations seem to lead to more clear ways to constrain the tensor force parameters, as compared with those in which HF single-particle states are considered. Gogny-RPA calculations including tensor correlations have been recently reported in Ref. [22].

The outline of the present contribution is the following. After recalling the basic features of tensor terms added on top of Skyrme zero-range forces in Sec. 2, we discuss the main results concerning the properties of single-particle states and collective vibrations in Secs. 3 and 4, respectively. Some intrinisc limitations of the DFT description of single-particle states are eventually discussed in Sec. 5, before presenting some short conclusion in Sec. 6.

2. Tensor force: formalism

The zero-range form of the effective nucleon-nucleon force has been proposed originally by T.H.R. Skyrme [23, 24]. We do not repeat here many well-known features of the models based on the use of the Skyrme interaction. Since its first application in HF calculations until now, it has been proven to be quite useful for the understanding of many nuclear properties. It should be never forgotten, however, that this force is meant as an effective force for HF or RPA calculations.

The tensor terms that can be added on top of the central ones have been written already in the original papers, and read

$$\begin{aligned} V^T \; = \; & \frac{T}{2}\left\{\left[(\sigma_1\cdot\mathbf{k}')\,(\sigma_2\cdot\mathbf{k}') - \frac{1}{3}\,(\sigma_1\cdot\sigma_2)\,\mathbf{k}'^2\right]\delta(\mathbf{r}_1-\mathbf{r}_2)\right. \\ & \left. + \,\delta(\mathbf{r}_1-\mathbf{r}_2)\left[(\sigma_1\cdot\mathbf{k})(\sigma_2\cdot\mathbf{k}) - \frac{1}{3}\,(\sigma_1\cdot\sigma_2)\,\mathbf{k}^2\right]\right\} \\ & + \frac{U}{2}\left\{(\sigma_1\cdot\mathbf{k}')\,\delta(\mathbf{r}_1-\mathbf{r}_2)\,(\sigma_2\cdot\mathbf{k}) + (\sigma_2\cdot\mathbf{k}')\,\delta(\mathbf{r}_1-\mathbf{r}_2)(\sigma_1\cdot\mathbf{k})\right. \\ & \left. -\frac{2}{3}\left[(\sigma_1\cdot\sigma_2)\mathbf{k}'\cdot\delta(\mathbf{r}_1-\mathbf{r}_2)\mathbf{k}\right]\right\}. \end{aligned} \qquad (2)$$

In the above expression, the operator $\mathbf{k}$ is defined as $(\nabla_1 - \nabla_2)/2i$, whereas $\mathbf{k}'$ is its complex conjugate and it reads $-(\nabla'_1 - \nabla'_2)/2i$. Operators without (with) prime symbols are meant to act at right (left). The coupling constants T and U denote the strengths of the triplet-even and triplet-odd tensor interactions, respectively.

These terms provide additional contributions to both the energy density $\mathcal{H}$ (the total energy is defined as $\int d^3r\ \mathcal{H}(\vec{r})$), and to the spin-orbit part of the HF average potential. In fact, these additional contributions have the same structure as other contributions coming from the exchange part of the central Skyrme terms. We will keep this distinction clear in what follows. However, from the viewpoint of pure DFT they cannot be distinguished.

The central exchange and tensor contributions to the energy density $\mathcal{H}$ are given by

$$\mathcal{H} = \frac{1}{2}\alpha(J_n^2 + J_p^2) + \beta J_n J_p, \tag{3}$$

where J_n and J_p are the so-called spin-orbit densities for neutrons and protons, respectively. Their definition in spherical systems is (see, for instance, Ref. [25])

$$J_q(r) = \frac{1}{4\pi r^3}\sum_i (2j_i+1)\left[j_i(j_i+1) - l_i(l_i+1) - \frac{3}{4}\right] R_i^2. \tag{4}$$

Here (and in what follows) q=0(1) labels neutrons (protons). The sum over $i \equiv n, l, j$ runs over all occupied states having the given q, and R_i are the radial wavefunctions. The HF spin-orbit potential is given by

$$U_{\text{s.o.}}^{(q)} = \frac{W_0}{2r}\left(2\frac{d\rho_q}{dr} + \frac{d\rho_{q'}}{dr}\right) + \left(\alpha\frac{J_q}{r} + \beta\frac{J_{q'}}{r}\right). \tag{5}$$

The first term comes from the Skyrme two-body spin-orbit interaction and depends on the usual densities ρ_q, whereas the second term includes both the central exchange and the tensor contributions. These central exchange and tensor contributions enter the above expressions as

$$\alpha = \alpha_C + \alpha_T, \tag{6}$$

$$\beta = \beta_C + \beta_T, \tag{7}$$

with

$$\alpha_C = \frac{1}{8}(t_1 - t_2) - \frac{1}{8}(t_1x_1 - t_2x_2), \tag{8}$$

$$\beta_C = -\frac{1}{8}(t_1x_1 + t_2x_2), \tag{9}$$

$$\alpha_T = \frac{5}{12}U, \tag{10}$$

$$\beta_T = \frac{5}{24}(T + U). \tag{11}$$

The first thing to be noted is that J_q is almost negligible in the $\mathbf{l}\cdot\mathbf{s}$ nuclei, that is, in those systems in which both spin-orbit partners ($j_< = l - 1/2$ and $j_> = l + 1/2$) are either occupied or unoccupied. In these cases, the tensor force gives no contribution either to the total energy or to the spin-orbit potential. These contributions can be appreciated by looking at the trends of those quantities along isotopic, or isotonic, chains when J_q is varying according to the positive (negative) contributions provided by the occupation of the $j_>$ ($j_<$) orbitals, while the usual densities ρ_q vary more smoothly.

This fact explains while the first attempt [26] to introduce the tensor force on top of the old SIII parameter set [27] was considered unsatisfactory. The results for the single-particle states were of the same quality, or even worse, than the results obtained with the interaction SIII, since few magic nuclei had been studied.

As mentioned in the Introduction, the issue of the tensor force has been considered again, after a long time interval, since its role has been shown to be relevant in the context of the shell-model calculations for exotic nuclei [28]. However, another general word of caution is worth to be mentioned here. In Ref. [28] the bare tensor force due to pion and ρ-meson exchange has been employed. While this force is certainly not small, most of its effects, when fitting EDFs, are absorbed into the central parameters. In this respect, we do not expect to necessarily find consistency between the strength of the bare tensor force, and that of the effective tensor force added on top of Skyrme sets. This latter has to be adjusted by looking at some kind of experimental data.

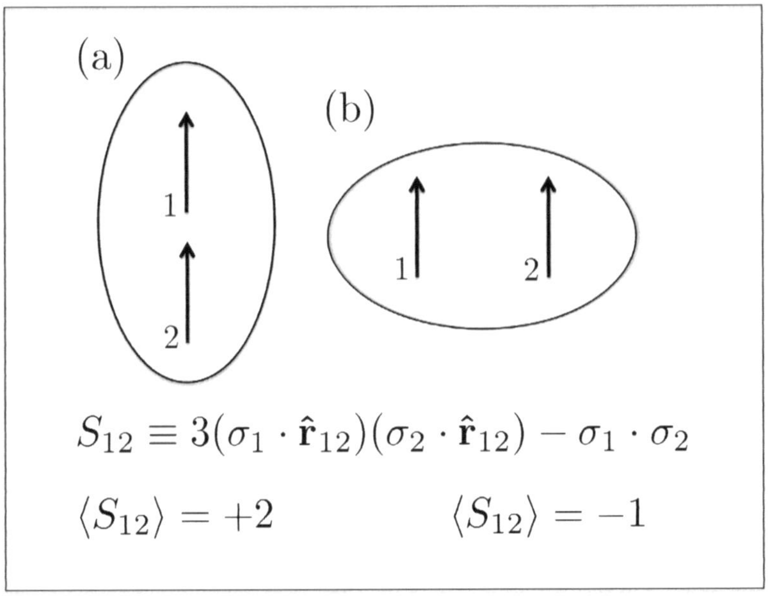

Figure 1. Schematic picture of states having either (a) parallel spins and spatial wavefunction concentrated in space or (b) parallel spins and wavefunctions more spread out in space. See the text for an explanation why case (a) and (b) should respectively correspond to $(j_>, j_<)$ and to $(j_>, j_>)$ (or $(j_<, j_<)$).

3. Effect of the tensor force on single-particle states

As mentioned in the previous Section, in Ref. [28] the proton-neutron tensor force has been advocated as one of the driving mechanisms of the shell structure evolution in exotic nuclei. This force has been shown to produce a strong attraction between neutron and protons with $j_>$ and $j_<$ (for instance, either attraction or repulsion has been shown to take place, respectively, between $f_{7/2}$ neutrons and $d_{3/2}$ or $d_{5/2}$ protons). The intuitive picture for such a mechanism is depicted in the left part (a) of figure 1. The two states $j_>$ and $j_<$ have spatial angular momenta that are opposite to the spins depicted in the figure. In other words, the semiclassical picture is that of two orbits with large relative momentum, and due to the uncertainty principle the wavefunction is spatially confined (as represented schematically by the curve that surrounds the spins). The configuration is deuteron-like and the tensor is known to act attractively in this case. The complementary picture (spatial angular momenta parallel to the spins depicted, small relative momenta of the orbits, and spread wavefunctions) holds for the case (b). In that case, therefore, a repulsive effect is expected.

Although we have stressed that the effective tensor force produced by some kind of fit performed within the Skyrme framework does not necessarily have the strength of the bare tensor force, we expect to see qualitatively the same effects.

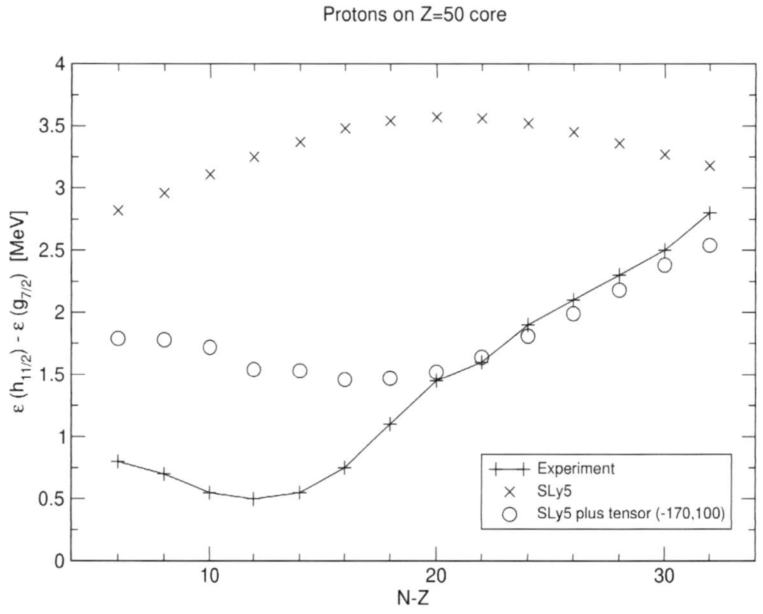

Figure 2. Energy differences between the $1h_{11/2}$ and $1g_{7/2}$ single-proton states along the $Z = 50$ isotopes. The calculations are performed without (crosses) and with (circles) the tensor terms in the spin-orbit potential (5), on top of the set SLy5 [29]. The experimental data (joined by the line) are from Ref. [30].

One of the first studies of this kind is the one of Ref. [4]. The motivation of this study has been the remark that even modern Skyrme sets are unable to produce the correct trend of the single-particle main peaks along isotopic or isotonic chains. From figure 2, it is evident that the calculation with the force SLy5 [29] is not able to reproduce the experimental data, while the insertion of the tensor terms provides a clear improvement. The tensor interaction employed here has $(\alpha_T, \beta_T) = (170, 100)$ MeV fm^5.

The interpretation of this result follows the line of the above argument. In the $Z = 50$ core, only the proton $g_{9/2}$ orbital dominates the proton spin-orbit density J_p. Consequently, with a negative value of α_T, the spin-orbit potential (5) is enlarged in absolute value (note that W_0 is positive and the radial derivatives of the densities are negative), the values of the proton spin-orbit splittings are increased, and the energy difference $\Delta\epsilon \equiv \epsilon(h_{11/2}) - \epsilon(g_{7/2})$ is reduced with respect to the case without tensor. However, the term in α does not give any isospin dependence to the spin-orbit potential for a fixed proton number, at variance with the term in

β. From $N - Z = 6$ to 14, the $g_{7/2}$ neutron orbit is gradually filled and J_n is reduced. Then, the positive value of β_T enlarges in absolute value the spin-orbit potential and increases the spin-orbit splitting, so that the energy difference $\Delta\epsilon$ becomes smaller. From $N - Z = 14$ to 20, the $s_{1/2}$ and $d_{3/2}$ neutron orbits are occupied, and in this region the spin-orbit density is not so much changed as the $s_{1/2}$ orbital does not provide any contribution. Instead, for $N - Z = 20$ to 32, the $h_{11/2}$ orbital is gradually filled. This gives a positive contribution to the spin-orbit potential and the spin-orbit splitting becomes smaller. $\Delta\epsilon$ consequently increases, and this effect is well pronounced in our theoretical results. The magnitude of β determines the slope of the isospin dependence, so that a larger β would give a steeper slope. In any event, it is clear that a positive value of β provides an effect of attraction between $j_>$ and $j_<$ states, and repulsion otherwise, just as in the case of Ref. [28].

It must be added that in more systematic calculations that have been done later, the situation looks more complicated. The negative value of α has been questioned in Ref. [6]. More generally, in that work it has been shown that it is difficult to accomodate in the framework of Skyrme Hamiltonians plus tensor terms, the systematics of single-particle levels in nuclei. This remains true even if one does not add the tensor terms in a perturbative way on top of existing Skyrme sets, but also if one fits on an equal footing central and tensor parameters. The issue remains contradictory even if one restricts to states around the single-particle gap, and tries to learn from the comparision of Skyrme plus tensor, Gogny plus tensor and RHF including pion plus ρ-meson exchange as in Ref. [10]. We will argue below (see Sec. 5) that single-particle states in nuclei are too much affected by correlations that cannot be included in the DFT framework (for instance, particle-vibration coupling). Therefore, although they can give some first hint about the relevance of tensor terms, they cannot be used as a stringent constraint.

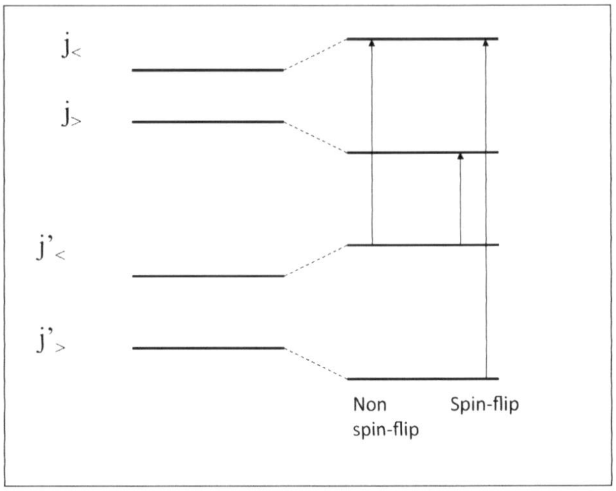

Figure 3. Schematic view of the effect of the tensor force on the mean-field responses to different operators. The levels displayed on the left side evolve to those depicted on the right side due to the effect of the tensor force. The arrows show the transitions that are excited mainly by either non spin-flip (the single arrow in the left part) or spin-flip (the two arrows in the right part) external fields.

4. Effect of the tensor force on collective states

Collective states like the giant resonances are more in the framework of (time-dependent) DFT. One should aim at studying the effects of the tensor force on their properties, and identifying contraints on the tensor force parameters accordingly. This program has been carried out in Refs. [16, 17, 18, 19, 20, 21] in the Skyrme case, and will be reviewed in the present Section.

Naively, one could imagine that the tensor force will affect more the spin-flip rather than the non spin-flip states. At the mean-field level, this can be seen schematically in figure 3. Also, we expect that the residual RPA tensor force will be more relevant for the spin-flip particle-hole (ph) transitions than for the non spin-flip ones. In fact, our systematic investigation of 2^+ and 3^- low-lying density modes, and high-lying giant resonances [16, 17] has concluded that (i) the impact of the tensor force on giant resonances is indeed small, (ii) although the low-lying levels are more affected because they are more sensitive to the detail of the shell structure, they cannot provide a clear and unique signature of how one could fix the value of the tensor force parameters. At the same time, one of the conclusions of Ref. [17] has been that some of the Skyrme forces that belong to the family TIJ introduced in Ref. [6], namely the sets T36, T44, T45 and T46 give the best results for the 2^+ and 3^- states in ^{40}Ca and ^{208}Pb. The same is true for SGII plus tensor terms (added perturbatively). The values for energies and electromagnetic reduced transition probabilities to the ground-state turn out to be accurate at the level of ≈10-20%.

In Refs. [16, 17], the spin-flip M1 excitations have been studied as well. Although in principle they are more sensitive to the introduction of the tensor force, many of the Skyrme sets produce larger values of the transition strength for the isoscalar rather than the isovector M1 operator. This isoscalar dominance is not in agreement with the experimental findings.

Therefore, we have found that spin-flip charge-exchange transitions (where by definition the isoscalar component is absent) are better candidates to study the impact of the tensor force. The first implementation of charge-exchange RPA on top of HF that includes the tensor force consistently, can be found in Ref. [18]. The model has been applied to the case of the Gamow-Teller (GT) response which is associated with the operator

$$\hat{O}_{GT\pm} = \sum_i t^i_{\pm}\sigma^i_1 = \sqrt{4\pi}\sum_i t^i_{\pm}\left[Y_0 \otimes \sigma^i_1\right]_1. \tag{12}$$

The main outcomes of this work are: (i) the peak of the GT$_-$ response in ^{90}Zr and ^{208}Pb is lowered by about 2 MeV when tensor correlations are included, and (ii) at the same time, about 10% of the GT strength is pushed in the high-energy region between 30 and 60 MeV by the tensor correlations. The first conclusion shows the relevance of the GT resonance if one is interested in fixing the tensor force parameters by considering sizeable effects induced by this force. The second conclusion is interesting, and consistent with the previous finding [31] that a large shift of strength towards the higher excitation energy region, far from the GT peak, can be obtained in the full model space of 1 particle-1 hole (1p-1h) plus 2 particle-2 hole (2p-2h) excitations also due to the effect of the tensor force. We find that a precursor of this effect takes place in the more limited 1p-1h space.

In Ref. [19] it has been shown that the downward shift of the GT$_-$ peak is associated with an upward shift of the spin-quadrupole SQ strength associated with the operator

$$\hat{O}_{SQ\pm} = \sum_i t^i_{\pm} r_i^2 \left[Y^i_2 \otimes \sigma^i_1\right]_1. \tag{13}$$

So, the effect induced by the tensor force on the GT resonance can be seen as a "deuteron-like" coupling between modes having L=0 and L=2. Because of the lack of detailed experimental information, however, the position of GT and SQ peaks are not two observables that can constrain the two parameters T and U that characterize a zero-range tensor force.

It has been found in Ref. [20] that a more powerful way to constrain these parameters is provided by the spin-dipole (SD) strength. This latter is associated with the operator

$$\hat{O}^{\lambda}_{\pm} = \sum_i t^i_{\pm} r_i \left[Y^i_1 \otimes \sigma^i_1 \right]_{\lambda} . \tag{14}$$

In the case of ^{208}Pb, experimentally it has been possible to disentangle the strength associated with the three total angular momentum components λ equal to 0^-, 1^- and 2^- [32]. At the same time, from theory it has been found that the residual tensor force introduced in the charge-exchange RPA has a *unique* multipole dependence. Because of this, only specific values for the tensor force parameters T and U can reproduce the experimentally observed sequence of peaks associated with the different values of λ.

This dependence can be understood as follows. The diagonal matrix elements of the residual RPA interaction, in the isovector channel, are given after proper antisymmetrization by

$$V^{(\lambda)}_{ph,\mathrm{AS}} = \left(-\frac{T}{2} + \frac{U}{2} \right) a_{\lambda} \langle \tau_1 \cdot \tau_2 \rangle , \tag{15}$$

where

$$a_{\lambda} = -\frac{5}{12} \left\{ \begin{array}{c} 1 \\ -1/6 \\ 1/50 \end{array} \right\} |\langle p || \hat{O}^{\lambda} || h \rangle|^2 \ \text{for} \ \lambda = \left\{ \begin{array}{c} 0^- \\ 1^- \\ 2^- \end{array} \right\} . \tag{16}$$

So, we can see that diagonal p-h matrix elements in the 0^- case are the largest, and that in the 1^- case they are somewhat smaller. The effect on the 2^- states is expected to be very small. For interactions in which the coupling constant T is positive and U is negative (or positive but small), the tensor correlations are strongly repulsive for 0^- and weakly repulsive for 2^-, whereas for 1^- they are attractive. In this way, one can explain the experimental findings.

The constraints set by the Gamow-Teller and spin-dipole resonances are put together in Ref. [21]. One of the results is illustrated in figure 4. If one wishes to add a tensor force on top of SLy5, the acceptable area for the parameters is roughly triangle-shaped. This area is defined by demanding that the difference δE between theoretical and experimental energies is smaller than 2.5 MeV. U can range from -300 to -50 MeV fm^5, and the minimum value for T is 450 MeV fm^5. The maximum value of T depend on U (the absolute maximum being close to 650 MeV fm^5). In the figure, we also show that some of the TIJ forces display quite reasonable results.

In conclusion, we have shown that it is possible to select Skyrme forces that, after the inclusion of tensor correlations, have a satisfactory behaviour when compared towards experiment in the case of quite different collective modes. T44, T45 and T46 are reasonably accurate for low-lying modes while T43 has passed several tests associated with charge-exchange spin-flip modes. These sets have quite similar values for the parameters.

We believe that the procedures we have undertaken can be useful if one wishes to improve the fine tuning of an effective tensor force. Another issue that has been discussed in Ref. [33] is the stability of functionals with the tensor terms added. We do not discuss this topic here.

5. Limits of the DFT description for the single-particle strength

It is well-known that mean field models do not account well for the single-particle energies, in particular close to the Fermi level. This fact can be considered to be consistent with the idea that these models are an implementation of DFT for atomic nuclei, and single-particle states may not be within the DFT framework.

The issue has been recently addressed in Ref. [34]. In this paper, results for the single-particle states obtained from a recent implementation of the DFT are compared with those coming from the particle-vibration coupling (PVC) approach. PVC is supposed to take care of most of the

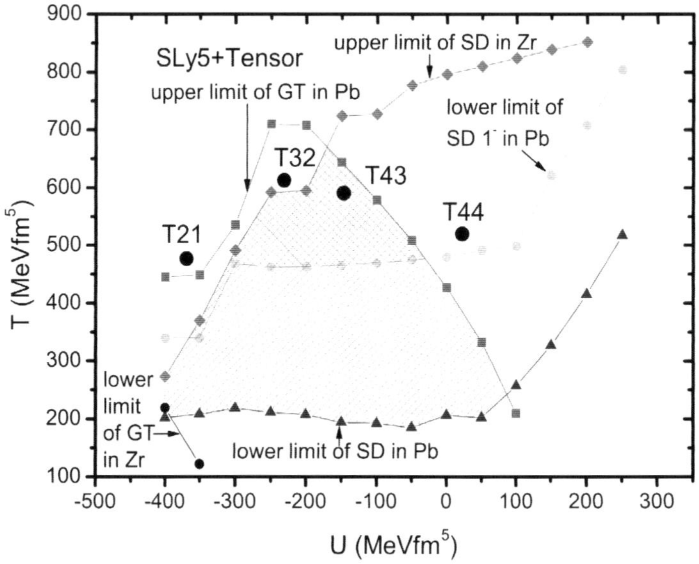

Figure 4. (Color online) The region of T and U values constrained by the criterion $\delta E \leq 2.5$ MeV for the GT and total SD centroid energies in ^{90}Zr and ^{208}Pb, and for the SD 1^- centroid energy in ^{208}Pb. The points correspond to some of the forces TIJ already described in the main text (however, the central part of them is different from that of SLy5, so their position in this plot is only indicative).

dynamical correlations missing in the static DFT models, in keeping with the fact that collective vibrations are the most important low-lying states of spherical nuclei.

Experimentally, the single-particle strength displays also a non-negligible fragmentation, and theory should account for it. The fragmentation lies definitely outside the framework of DFT. This is perhaps the main motivation to try to generalize this framework, in which the basic variable is the static density, to a many-body theory where dynamical correlations are included. A simple version of this, is the PVC model (also called "dynamical" shell model). There are of course other motivations: for instance, the damping width of single-particle and collective states (like the nuclear giant resonances) can be described only by means of theories beyond mean field or present-day DFT.

In the review paper [35], several phenomenological PVC calculations performed in the 80's are compared and discussed. Qualitatively, they all demonstrate the relevance of the PVC to produce the single-particle strength fragmentation, as well as the increase of the level density (that is, the enhancement of the effective mass m^*) close to the Fermi energy which is empirically observed. However, none of them is consistent and it is hard to extract quantitative conclusions due to the many approximations involved. Recently, microscopic PVC calculations have become available, based on the consistent use of either Skyrme interactions [36], or RMF Lagrangians [37, 38].

The purpose of this Section is to make a short summary of the PVC formalism, and to show a few results aimed to demonstrate that single-particle states are affected enough by PVC, so that accurate fitting of the tensor forces based on their properties is not probably a useful strategy.

5.1. Basic elements of PVC theory

Using standard many-body techniques, one can introduce the energy-dependent part $\Sigma(\omega)$ of the self-energy. It enters the well-known Dyson equation,

$$G(\omega) = G_0(\omega) + G_0(\omega)\Sigma(\omega)G(\omega), \tag{17}$$

that allows determining the Green's function G in terms of the unperturbed one G_0. In the present context, G_0 is the mean-field (i.e., Hartree-Fock) Green's function. The $\Sigma(\omega)$ term is associated with the coupling to (mainly density) vibrations.

In the so-called diagonal approximation, as defined in Ref. [39], the quantities appearing in the Dyson equation (17) are written in the basis in which the mean-field is diagonal (that is, the HF basis), and only the diagonal matrix elements of $\Sigma(\omega)$ are considered. If the indices i, p and h label, respectively, a generic single-nucleon state, a particle state above the Fermi energy and a hole state below the Fermi energy, the diagonal matrix element Σ_i reads

$$\Sigma_i(\omega) = \frac{1}{2j_i+1}\left(\sum_{nL,\ p>F} \frac{|\langle i||V||p,nL\rangle|^2}{\omega - \varepsilon_p - \omega_{nL} + i\eta} + \sum_{nL,\ h<F} \frac{|\langle i||V||h,nL\rangle|^2}{\omega - \varepsilon_h + \omega_{nL} - i\eta}\right). \tag{18}$$

The energies ω_{nL} are the energies of the vibrational states (phonons) with angular momentum L, labeled by the index n. A (small) imaginary part η is added in the denominators (it is set at 0.05 MeV in the calculations presented below). The numerators contain the squared modulus of a reduced matrix element called PVC vertex (see Ref. [36] for details).

The solution of the Dyson equation in the diagonal approximation is equivalent to a matrix diagonalization [37, 39]. In several cases, the results obtained by means of full diagonalization do not differ significantly from those obtained by means of perturbation theory (see, e.g., [40]). In this latter case, dressed single-particle energies are calculated by means of the equation

$$\tilde{\varepsilon}_i = \varepsilon_i + \Sigma_i(\omega)|_{\omega=\varepsilon_i}. \tag{19}$$

5.2. Results for a typical nucleus: ^{132}Sn

We show results obtained by starting from the Hartree-Fock (HF) solution associated with the Skyrme parameter set SLy5 [29]. With the same parameter set we perform fully self-consistent RPA calculations in order to obtain the vibrations. The tensor force already employed in Ref. [4] is included in HF but not in the self-consistent RPA. The PVC vertex includes the central momentum-independent part of the Skyrme force. The self-energy is treated in the perturbative approximation, that is, the dressed single-particle energies are obtained by means of Eq. (19).

Results for ^{132}Sn are displayed in figure 5. The different columns, from left to right, refer to calculations at the level of HF without the tensor force, HF including the tensor interaction of Ref. [4] and HF plus PVC, respectively. The main message of this figure is that correlations associated with PVC may produce shifts of the single-particle levels that are comparable, or larger, with respect to the shifts produced by the tensor force.

We have chosen to display the result for this nucleus, since some of his properties have been already discussed in Sec. 3, despite the fact that this calculation neglects the momentum-dependent part of the Skyrme force in the PVC vertex. This approximation is not quantitatively satisfactory, as it has been shown in Ref. [36] to overestimate the shifts associated with PVC. However, the qualitative conclusion that these shifts cannot be neglected when compared with shifts induced by the tensor correlations, remains true.

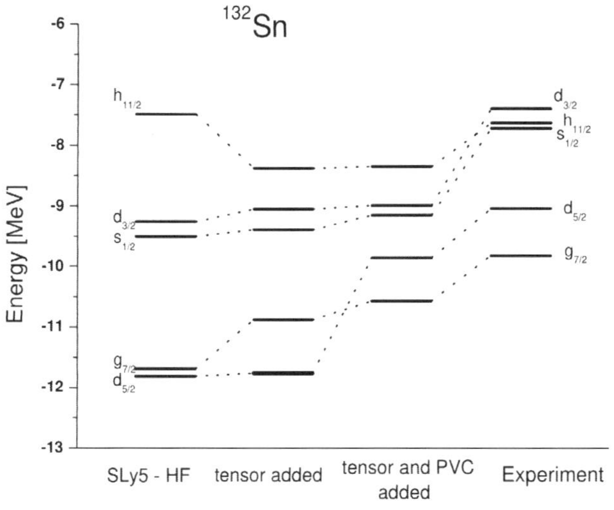

Figure 5. Results obtained for the neutron levels of ^{132}Sn at various levels of approximation: HF with SLy5, HF with SLy5 plus tensor, HF plus PVC (from left to right). The last column at right displays experimental levels.

6. Conclusion

Recently, much attention has been devoted to the inclusion of an effective tensor force in the mean field or DFT framework. After a few years, it is timely to review the main conclusions, at least those that can be drawn by considering tensor terms added to the Skyrme effective interactions.

We have found that the most effective way to constrain the parameters of the tensor force, is the study of collective modes. In particular, charge-exchange spin-flip resonances are clearly affected by the tensor correlations. Since recent experimental data have been provided on the position of the peaks associated with the different components of the spin-dipole resonance, this represents one of the strongest tests for effective Hamiltonians that include the tensor terms. We have shown that these observables, together with the properties of the Gamow-Teller resonance, allow fixing the tensor force parameters to a large extent.

Non charge-exchange modes are also affected by tensor correlations although the effects are more subtle and less clear to disentangle from other effects. This becomes even more clear if one turns the attention towards single-particle properties. We have shown indeed that dynamical correlations associated with particle-vibration coupling can produce similar, or larger, shifts of the single-particle energy levels when compared with tensor correlations.

Acknowledgments

It should be stressed that the results reported in this paper are obtained together with collaborators, whose major contribution should be clearly acknowledged: C L Bai, P F Bortignon, S Fracasso, M Grasso, C Ligang, Z Y Ma, K Moghrabi, H Sagawa, L Sciacchitano, N Van Giai, F R Xu, H Q Zhang, X Z Zhang, W Zou. Also, we would like to thank G P Co' and T Suzuki for useful discussions.

References

[1] Bender M, Heenen P-H, Reinhard P G 2003 *Rev. Mod. Phys.* **75** 121
[2] Brown B A, Duguet T, Otsuka T, Abe D, and Suzuki T 2006 *Phys. Rev. C* **74** 061303(R)
[3] Dobaczewski J 2006 *Proceedings of the Third ANL/MSU/JINA/INT RIA Workshop* ed Duguet T, Esbensen H, Nollett K M and Roberts C D (Singapore, World Scientific)
[4] Colò G, Sagawa H, Fracasso S and Bortignon P F 2007 *Phys. Lett. B* **646** 227
[5] Brink D M and Stancu Fl 2007 *Phys. Rev. C* **75** 064311
[6] Lesinski T, Bender M, Bennaceur K, Duguet T and Meyer J 2007 *Phys. Rev. C* **76** 014312
[7] Grasso M, Ma Z Y, Khan E, Margueron M and Van Giai N 2007 *Phys. Rev. C* **76** 044309
[8] Zou W, Colò G, Ma Z Y, Sagawa H and Bortignon P F 2008 *Phys. Rev. C* **77** 014314
[9] Zalewski M, Dobaczewski J, Satula W and Werner T R 2008 *Phys. Rev. C* **77** 024316
[10] Moreno-Torres M, Grasso M, Liang H, De Donno V, Anguiano M and Van Giai N 2010 *Phys. Rev. C* **81** 064327
[11] Otsuka T, Matsuo T and Abe D 2006 *Phys. Rev. Lett.* **97** 162501
[12] Long W H, Van Giai N and Meng J 2006 *Phys. Lett. B* **640** 150
[13] Long W H, Sagawa H, Van Giai N and Meng J 2007 *Phys. Rev. C* **76** 034314
[14] Long W H, Van Giai N and Meng J 2008 *Europhys. Lett.* **82** 12001
[15] Tarpanov D, Liang H, Van Giai N and Stoyanov C 2008 *Phys. Rev. C* **76** 034314
[16] Cao L, Colò G, Sagawa H, Bortignon P F and Sciacchitano L 2009 *Phys. Rev. C* **80** 064304
[17] Cao L, Sagawa H and Colò G 2011 *Phys. Rev. C* **83** 034324
[18] Bai C L, Sagawa H, Zhang H Q, Zhang X Z, Colò G and Xu F R 2009 *Phys. Lett. B* **675** 28
[19] Bai C L, Zhang H Q, Zhang X Z, Xu F R, Sagawa H and Colò G 2009 *Phys. Rev. C* **79** 041301
[20] Bai C L, Zhang H Q, Sagawa H, Zhang X Z, Colò G and Xu F R 2010 *Phys. Rev. Lett.* **105** 072501
[21] Bai C L, Zhang H Q, Sagawa H, Zhang X Z, Colò G and Xu F R 2011 *Phys. Rev. C* **83** 054316
[22] Anguiano M, Cò G, De Donno V and Lallena A M 2011 *Phys. Rev. C* **83** 064306
[23] Skyrme T H R 1956 *Phil. Mag.* **1** 1043
[24] Skyrme T H R 1959 *Nucl. Phys.* **9** 615
[25] Vautherin D and Brink D M 1972 *Phys. Rev. C* **5** 626
[26] Stancu Fl, Brink D M and Flocard H 1977 *Phys. Lett. B* **68** 108
[27] Beiner M, Flocard H, Van Giai N and Quentin P 1975 *Nucl. Phys. A* **238** 29
[28] Otsuka T, Suzuki T, Fujimoto R, Grawe H and Akaishi Y 2005 *Phys. Rev. Lett.* **95** 232502
[29] Chabanat E, Bonche P, Haensel P, Meyer J and Schaeffer 1997 *Nucl. Phys. A* **627** 710
[30] Schiffer J P, Freeman S J, Caggiano J A, Deibel C, Heinz A, Jiang C-L, Lewis R, Parikh A, Parker P D, Rehm K E, Sinha S and Thomas J S 2004 *Phys. Rev. Lett.* **92** 162501
[31] Bertsch G, Hamamoto I 1982 *Phys. Rev. C* **26** 1323
[32] Wakasa T 2010 *Preprint* arXiv:1004.5220 [nucl-ex]
[33] Cao L, Colò G and Sagawa H 2010 *Phys. Rev. C* **81** 044302
[34] Bortignon P F, Colò G and Sagawa H 2010 *J. Phys. G: Nucl. Part. Phys.* **37** 064013
[35] Mahaux C, Bortignon P F, Broglia R A and Dasso C H 1985 *Phys. Rep.* **120** 1
[36] Colò G, Bortignon P F and Sagawa H 2010 *Phys. Rev. C* **82** 064307
[37] Litvinova E and Ring P 2006 *Phys. Rev. C* **73** 044328
[38] Litvinova E V and Afanasjev A V 2011 *Phys. Rev. C* **84** 014305
[39] Ring P and Werner E 1973 *Nucl. Phys. A* **211** 198
[40] Colò G, Bortignon P F, Sagawa H, Moghrabi K, Grasso M and Van Giai N 2011 *J. Phys.: Conf. Series* **321** 012018

Microscopic approach to the shell model: study of nuclei north-east of ^{48}Ca

N Itaco[1,2], L Coraggio[2], A Covello[1,2] and A Gargano[2]

[1]Dipartimento di Scienze Fisiche, Università di Napoli Federico II and [2]Istituto Nazionale di Fisica Nucleare,
Complesso Universitario di Monte S Angelo, Via Cintia - I-80126 Napoli, Italy

E-mail: `nunzio.itaco@na.infn.it`

Abstract. We report on shell-model calculations for nuclei with valence nucleons outside ^{48}Ca wherein a fully microscopic approach is adopted. Namely, the single-particle energies and the matrix elements of the two-body interaction are derived theoretically starting from the high-precision CD-Bonn nucleon-nucleon potential, renormalized by way of the $V_{\rm low-k}$ approach. This has been done within the framework of the time-dependent degenerate linked-diagram perturbation theory. Our results compare satisfactorily with a large amount of experimental data.

1. Introduction

In the last decade interest has grown in so-called *ab initio* calculations for nuclear structure. What is usually meant by an *ab initio* calculation [1] is to start from a realistic nucleon-nucleon (NN) potential and possibly a three-body one and employ them within a many-body framework well suited to handle the short-range correlations induced by the nuclear force. A major requirement is the accuracy of the approximations involved, which should be kept under control by way of convergence checks.

Owing also to the considerable increase in computer power, substantial progress has been achieved in microscopic approaches to the nuclear many-body problem, such as the Green's function Monte Carlo method (GFMC) [2], no-core shell model (NCSM) [3, 4], and coupled-cluster method (CCM) [5, 6].

Despite great efforts to extend the domain of applicability of these approaches, only a limited class of nuclei has been investigated up to now owing to the computational complexity. As a matter of fact, the GFMC and NCSM are presently able to describe nuclei with mass $A \leq 16$, while the CCM has been employed to study closed-shell and one-valence-nucleon systems only.

On the other hand, the advances in experimental techniques to produce exotic beams make now possible to explore the structure of nuclei at and close to the driplines. This is a major challenge of todays nuclear physics, since these studies may reveal new features, such as shell modifications, that go beyond our present knowledge of nuclear structure. There is, consequently, a clear need to study open-shell nuclei in different mass regions, and the realistic shell model is the most advanced tool to perform microscopic calculations for these nuclei. In this approach, the basic ingredients of the shell-model hamiltonian, namely the single-particle energies and

Published under licence by IOP Publishing Ltd

two-body matrix elements, should be derived in the framework of a many-body theory (see for instance [7, 8]) starting from a free NN potential.

These fully realistic shell-model calculations with no adjustable parameters have already proved to be successful in reproducing the spectroscopic properties of the p-shell nuclei [9] and $N = 82$ isotones [10], as well as the neutron dripline for the oxygen and carbon isotopes [11, 12].

In the present work, we focus our attention on nuclei that can be described in terms of valence protons and neutrons outside doubly-magic ^{48}Ca, employing as model space the $0f_{7/2}$ level for protons and the four levels of the $28-50$ shell for neutrons. In this context, it should be mentioned that an issue of current interest is the role played, within the shell-model framework, by the neutron $0g_{9/2}$ orbital in the description of the experimental behavior of the yrast $J^{\pi} = 2^{+}$ states in heavy Cr and Fe isotopes [13–18]. We find worthwhile investigating this problem by performing shell-model calculations without adjustable parameters.

We have derived the shell-model effective hamiltonian starting from a low-momentum realistic potential derived from the high-precision CD-Bonn potential [19] by way of the well-known $V_{\rm low-k}$ approach [20]. As pointed out in [10, 21] a three-body force should be added to our two-body $V_{\rm low-k}$. In this case, our effective hamiltonian would consist of one- and two-body components, including also core-polarization effects arising from the three-body force, and of an effective three-body force that should be explicitly taken into account in the shell-model calculation. The derivation of such a hamiltonian is a very hard task, since we should treat three-body forces up to third-order in perturbation theory, as we usually do for the two-body ones. In this paper, therefore, we neglect the contributions of three-body forces. However, our previous experience with realistic shell-model calculations [8] makes us confident that the spectroscopic properties of these nuclei can be mainly ascribed to the effects of the two-body NN interaction.

In the following section, we outline the perturbative approach to the derivation of our shell-model hamiltonian. In section 3, we present results of calculations for calcium, titanium, chromium, iron, and nickel isotopes. A brief summary is given in the last section.

2. Outline of calculations

Within the framework of the shell model, an auxiliary one-body potential U is introduced in order to break up the hamiltonian for a system of A nucleons as the sum of a one-body term H_0, which describes the independent motion of the nucleons, and a residual interaction H_1:

$$H = \sum_{i=1}^{A} \frac{p_i^2}{2m} + \sum_{i<j=1}^{A} V_{ij}^{NN} = T + V = (T+U) + (V-U) = H_0 + H_1 \quad . \tag{1}$$

Once H_0 has been introduced, it is possible to define a reduced model space in terms of a finite subset of H_0's eigenvectors. In this space, an effective hamiltonian $H_{\rm eff}$ may be constructed and the diagonalization of the many-body hamiltonian (1) in an infinite Hilbert space is then reduced to the solution of an eigenvalue problem in a finite space.

As mentioned in the Introduction, we want to derive an effective shell-model hamiltonian $H_{\rm eff}$ for a model space spanned by the single-proton level $0f_{7/2}$ and the four single-neutron levels $0f_{5/2}$, $1p_{3/2}$, $1p_{1/2}$, $0g_{9/2}$ outside ^{48}Ca core. This has been done in the framework of the time-dependent perturbation theory by way of the Kuo-Lee-Ratcliff (KLR) folded-diagram expansion [22]. More precisely, we first renormalize the high-momentum repulsive components of the CD-Bonn NN potential by way of the $V_{\rm low-k}$ approach [20, 23], which provides a smooth potential preserving exactly the on-shell properties of the original NN potential. Next, we express $H_{\rm eff}$ in terms of the vertex function $\hat{Q}$-box, which is composed of irreducible valence-linked diagrams [24, 25]. In the derivation of the $\hat{Q}$-box we have included one- and two-body Goldstone diagrams through third order in H_1. Since it is computationally prohibitive to perform calculations beyond the third order in perturbation theory, we have calculated the Padé approximant [2|1] [26, 27] of the $\hat{Q}$-box, so as to obtain a value to which the perturbation series should converge, as suggested in [28]. The folded-diagram series is then summed up to all orders using the Lee-Suzuki iteration method [29].

The hamiltonian $H_{\rm eff}$ contains one-body contributions, whose collection is the so-called $\hat{S}$-box [30]. In realistic shell-model calculations a subtraction procedure is commonly used, so as to retain only the two-body terms of $H_{\rm eff}$ ($V_{\rm eff}$) while the SP energies are taken from experiment. Instead, in this work we have employed the SP energies obtained from the $\hat{S}$-box calculation.

A discussion of the convergence properties of $H_{\rm eff}$ is now in order. For the sake of clarity, we first consider the SP energies and then the two-body matrix elements (TBME) of $V_{\rm eff}$.

In figure 1 we report the neutron SP spectrum calculated as a function of the maximum allowed excitation energy of the intermediate states expressed in terms of the oscillator quanta $N_{\rm max}$. The results obtained at third order in perturbation theory are reported with black symbols, while those obtained using the Padé approximants [2|1] are represented by red symbols. We see that an $N_{\rm max} \geq 22$ calculation yields converged values for the SP energies.

Since the comparison between the third-order results and those obtained by means of the Padé approximant is an indicator of the dependence of our results on the higher-order perturbative terms, we conclude that higher-order contributions are not very relevant, and should only slightly affect the relative spectrum.

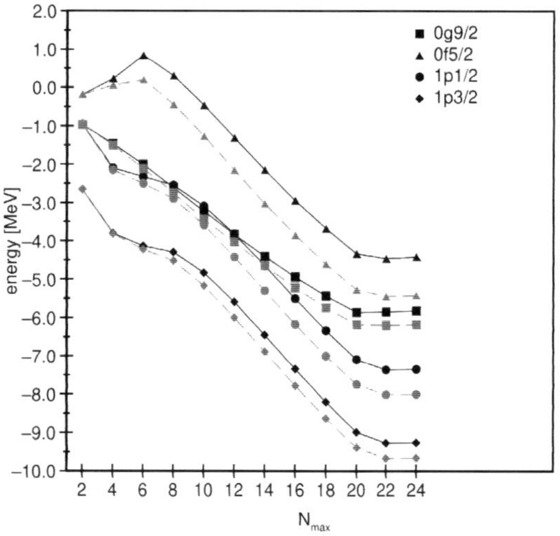

Figure 1. (Color online) Theoretical single-neutron energies as a function of $N_{\rm max}$. The black symbols refer to third-order results, the red ones to those obtained with Padé approximants. See text for details.

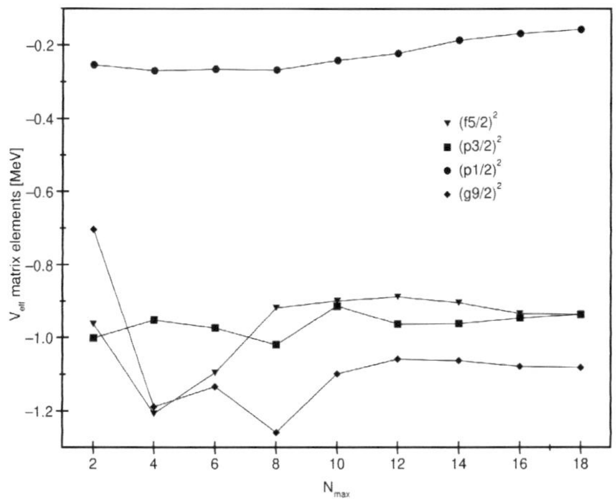

Figure 2. Diagonal neutron-neutron $J^\pi = 0^+$ matrix elements of $V_{\rm eff}$ as a function of $N_{\rm max}$. See text for details.

In figure 2, we report the diagonal neutron-neutron $J^\pi = 0^+$ matrix elements of $V_{\rm eff}$ as a function of $N_{\rm max}$. We have chosen these matrix elements because they are the largest ones in the entire set of $V_{\rm eff}$ TBME and the most sensitive to increases in the number of intermediate states. From figure 2 it is clear that our results have practically achieved convergence at $N_{\rm max} = 18$.

As regards the order-by-order convergence, we compare in table 1 all the $\hat{Q}$-box neutron-neutron $J^\pi = 0^+$ matrix elements calculated at third order with those given by the [2|1] Padé approximant [26]

$$[2|1] = V^0_{Qbox} + V^1_{Qbox} + V^2_{Qbox}(1 - (V^2_{Qbox})^{-1}V^3_{Qbox})^{-1} \; , \qquad (2)$$

where V^n_{Qbox} is the square non-singular matrix representing the nth-order contribution to the $J^\pi = 0^+$ $\hat{Q}$-box in the perturbative expansion. From table 1 we see that our third-order results are in good agreement with those from the [2|1] Padé approximant. This indicates a weak dependence of our results on higher-order $\hat{Q}$-box perturbative terms.

Table 1. $\hat{Q}$-box neutron-neutron $J^\pi = 0^+$ two-body matrix elements at third order in $V_{\rm low-k}$ (in MeV) compared with those obtained by calculating the Padé approximant [2|1].

Configuration	3rd order	Padé [2|1]
$\langle(0f_{5/2})^2\|V_{Qbox}\|(0f_{5/2})^2\rangle$	-1.381	-1.109
$\langle(0f_{5/2})^2\|V_{Qbox}\|(1p_{3/2})^2\rangle$	-1.096	-0.979
$\langle(0f_{5/2})^2\|V_{Qbox}\|(1p_{1/2})^2\rangle$	-0.521	-0.491
$\langle(0f_{5/2})^2\|V_{Qbox}\|(0g_{9/2})^2\rangle$	2.698	2.444
$\langle(1p_{3/2})^2\|V_{Qbox}\|(1p_{3/2})^2\rangle$	-1.083	-1.028
$\langle(1p_{3/2})^2\|V_{Qbox}\|(1p_{1/2})^2\rangle$	-1.742	-1.407
$\langle(1p_{3/2})^2\|V_{Qbox}\|(0g_{9/2})^2\rangle$	1.747	1.567
$\langle(1p_{1/2})^2\|V_{Qbox}\|(1p_{1/2})^2\rangle$	-0.089	-0.236
$\langle(1p_{1/2})^2\|V_{Qbox}\|(0g_{9/2})^2\rangle$	1.312	1.174
$\langle(0g_{9/2})^2\|V_{Qbox}\|(0g_{9/2})^2\rangle$	-1.254	-1.305

3. Results

In this section, we report the results of our shell-model calculations for calcium, titanium, chromium, iron and nickel isotopes with mass number $A > 48$. The calculations have been performed by using the Oslo shell-model code [31].

First, it is worth commenting on our theoretical SP energies as compared to the experimental low energy spectra of ^{49}Sc and ^{49}Ca. The former are reported in table 2. Since we employ the sole SP level $0f_{7/2}$ as proton model space, we can compare the theoretical value, -8.2 MeV, with the experimental ^{49}Sc ground-state (g.s.) energy with respect to ^{48}Ca, -9.63 MeV. The observed energy spacings in ^{49}Ca are $E_{1/2^-} - E_{3/2^-} = 2.023$ NeV , $E_{5/2^-} - E_{3/2^-} = 3.585$ MeV and $E_{9/2^+} - E_{3/2^-} = 4.017$ MeV, while the g.s. energy with respect to ^{48}Ca is -5.147 MeV.

Table 2. Theoretical SP energies (in MeV). See text for details.

nlj	Single-particle energies
$\pi 0f_{7/2}$	-8.2
$\nu 0f_{5/2}$	-5.4
$\nu 1p_{3/2}$	-9.7
$\nu 1p_{1/2}$	-8.0
$\nu 0g_{9/2}$	-6.2

From table 2, we see that our relative single-neutron energies are not in good agreement with experiment. In particular the theoretical $0g_{9/2}$ SP state is about 0.5 MeV lower than the observed yrast $J^\pi = \frac{9}{2}^+$ state in ^{49}Ca, while the theoretical $0f_{5/2}$ SP energy is about 0.7 MeV greater than the experimental excitation energy of the yrast $J^\pi = \frac{5}{2}^-$ state. Moreover, a large discrepancy occurs when comparing the experimental g.s. energies with respect to ^{48}Ca with our calculated ones. Since, as shown in the previous section, the convergence properties of our $H_{\rm eff}$ seem to leave very little room for improvement of our calculation of SP energies and TBME, the above discrepancies can be probably traced to the lack of a three-body force in addition to the two-body $V_{\rm low-k}$ potential.

In the following, besides presenting the results obtained with theoretical SP energies (labelled as I), we also report on calculations performed using, as in Ref. [16], SP energies deduced from the experimental spectra of ^{49}Sc and ^{49}Ca (labelled as II). In this way, we should empirically take into account three-body force effects in the one-body component of our effective hamiltonian.

3.1. Calcium isotopes

In figure 3, we report the calculated excitation energies of the yrast 2^+ states as a function of the neutron number up to $N = 36$. It can be seen that the available observed energies are well reproduced.

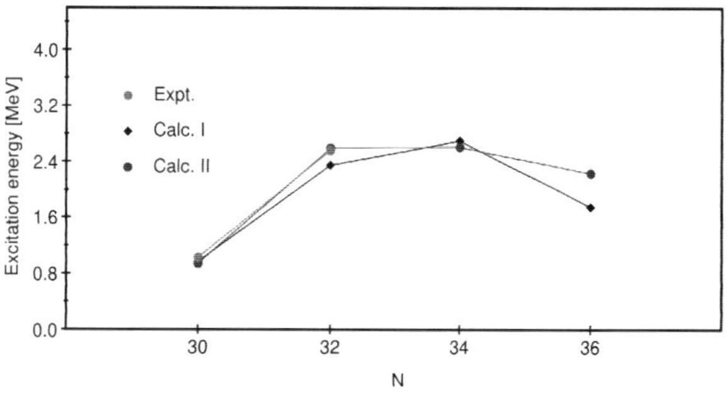

Figure 3. (Color online) Experimental [32] and calculated excitation energies of the yrast $J^\pi = 2^+$ states in calcium isotopes from $N = 30$ to 36.

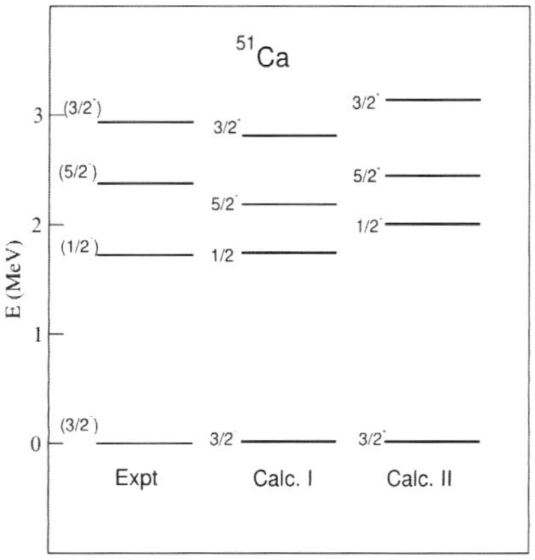

Figure 4. Experimental [33, 34] and calculated low-energy spectra of ^{51}Ca. See text for details.

In figure 4, the experimental and calculated energy spectra of the odd-mass nucleus ^{51}Ca up to 3 MeV excitation energy are shown. In this connection it should be mentioned that calculations I and II provide a seniority-one yrast $J^\pi = \frac{9}{2}^+$ state at excitation energies of 3.603 and 4.131 MeV, respectively. Experimentally, a $J^\pi = \frac{9}{2}^+$ state has been observed at 4.155 MeV, which has been interpreted in Ref. [34] as a $1p - 1h$ $Z = 20$ cross-shell excitation.

3.2. Titanium isotopes

The calculated excitation energies of the yrast 2^+ states in titanium isotopes are reported in figure 5 up to $N = 34$. It can be seen that the observed excitation energies are well reproduced by both calculations I and II up to $N = 34$. The drop of the experimental 2_1^+ excitation energy in ^{56}Ti evidences the lack of a $N = 34$ shell closure in titanium isotopes, which is well reproduced by both calculations.

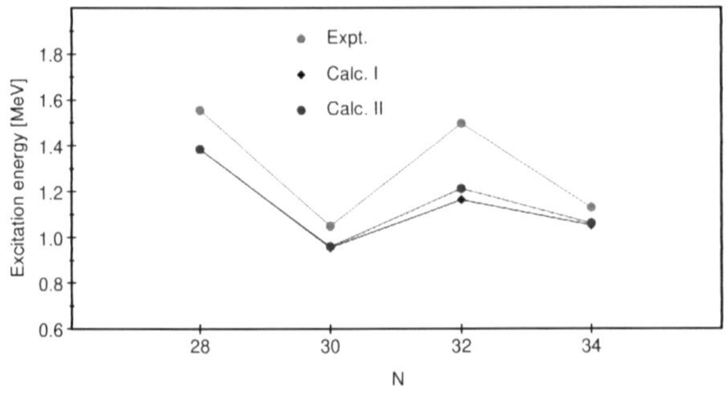

Figure 5. (Color online) Same as in figure 3, but for titanium isotopes from $N = 28$ to 34.

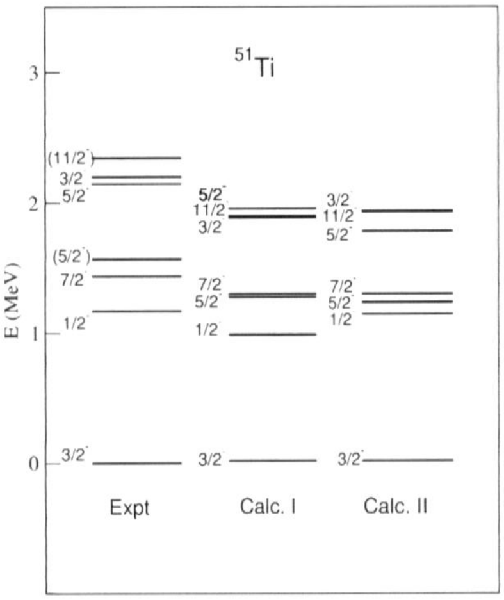

Figure 6. Experimental [32] and calculated low-energy spectra of ^{51}Ti. See text for details.

In figure 6, we report the two calculated energy spectra of the odd-mass nucleus ^{51}Ti up to 2.5 MeV excitation energy. We see that the agreement with the experimental spectrum is quite satisfactory for both calculations. It should be mentioned that I and II predict a yrast $J^{\pi} = \frac{9}{2}^{+}$ state of single-particle nature at excitation energies of 3.096 and 3.620 , respectively. Experimentally, a $J^{\pi} = \frac{9}{2}^{+}$ state has been observed at 3.771 MeV.

3.3. Chromium isotopes

The excitation energies of the yrast 2^{+} states for the chromium isotopes are reported in figure 7 up to ^{64}Cr.

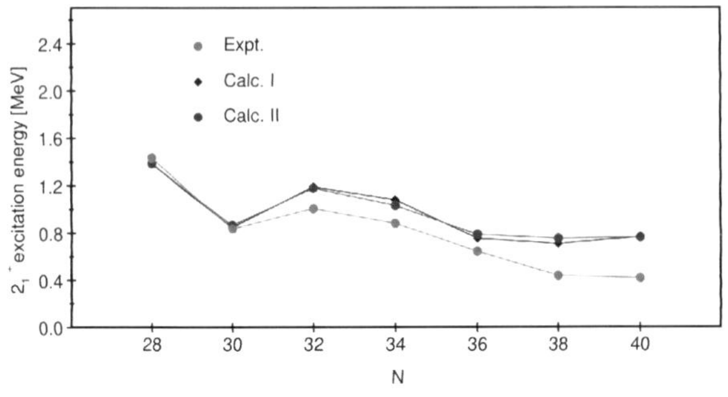

Figure 7. (Color online) Same as in figure 3, but for chromium isotopes from $N = 28$ to 40.

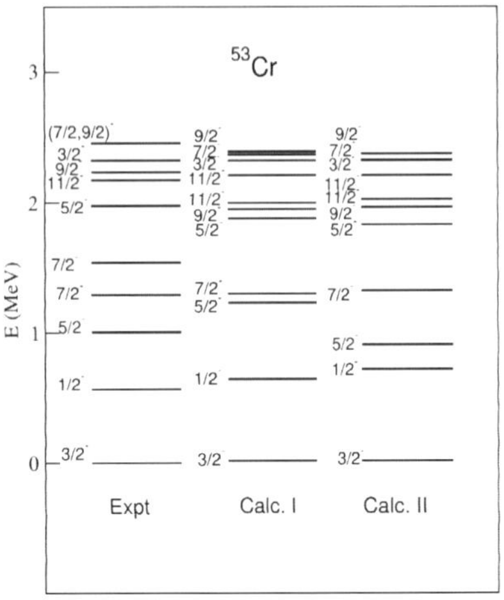

Figure 8. Experimental [32] and calculated low-energy spectra of ^{53}Cr. See text for details.

It is worth noting that the observed lowering of the yrast 2^{+} states is interpreted as a signature of the development of a pronounced collectivity near $N = 40$ [14]. Both calculations

I and II reproduce qualitatively the experimental behavior of the observed $J^\pi = 2_1^+$ energies, even if from ^{62}Cr on there is a sizeable overestimation of the experimental data. This may be a consequence of our limited model space, which is spanned only by one single-proton level and does not take into account explicitly the $1d_{5/2}$ single-neutron state, whose inclusion has been recently suggested to reproduce the large quadrupole collectivity in this mass region [15].

In figure 8, the calculated energy spectra of the odd-mass nucleus ^{53}Cr are shown and compared with the experimental one up to 2.5 MeV excitation energy. It can be seen that the agreement is quite good for both calculations.

3.4. Iron isotopes

In figure 9 the 2_1^+ excitation energies of even iron isotopes are shown as a function of the neutron number N.

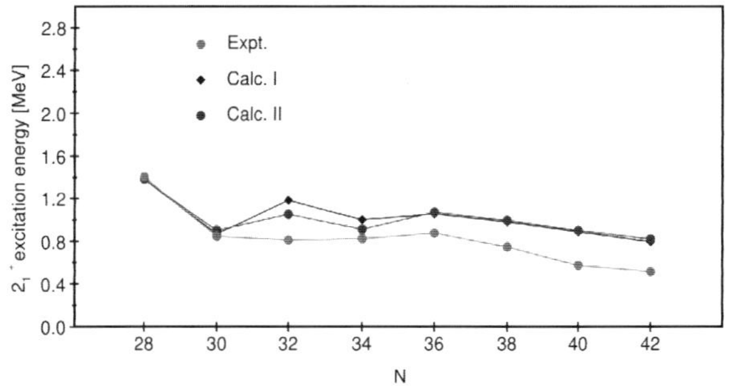

Figure 9. (Color online) Same as in figure 3, but for iron isotopes from $N = 28$ to 42.

As for chromium isotopes, the results with calculations I and II are very similar. We see that the observed decreasing trend, which may be interpreted as the onset of collectivity around $Z = 24$, is reproduced by our calculations, although, owing to the limited size of our model space, from $N = 36$ on our results sistematically overestimate the observed 2_1^+ excitation energies.

3.5. Nickel isotopes

Our model space allows to describe the spectroscopic properties of Ni isotopes with mass $A > 56$. It is worth noting that in this case only the neutron degrees of freedom are explicitly taken into account, the eight valence protons being frozen in the $f_{7/2}$ orbital. This means that the effective proton-neutron interaction comes into play only via its monopole component that makes the ^{49}Ca SP spectrum evolve into the ^{57}Ni SP one. This of course reduces the scope of our description. In figure 10 we report the SP energy levels of ^{57}Ni obtained with calculations I and II, and compare them with the corresponding observed ones. As can be seen, the agreement is rather satisfactory, which indicates the reliability of the proton-neutron monopole components of our effective hamiltonian.

In figure 11 the 2_1^+ excitation energies of even nickel isotopes are shown as a function of the neutron number N. The agreement of the calculated $J^\pi = 2_1^+$ excitation energies with the experimental ones is rather unsatisfactory. In particular, both calculations provide a subshell closure at $N = 32$, which has no experimental counterpart, while the observed $N = 40$ subshell closure is not well reproduced. This is mainly due to the fact that our effective hamiltonian does not take explicitely into account $Z = 28$ cross-shell excitations, which indeed may be not negligible.

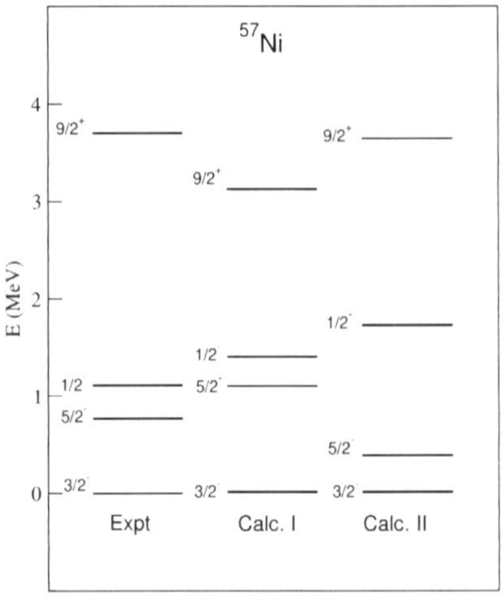

Figure 10. Calculated SP energy levels of ^{57}Ni compared with the corresponding experimental low-lying ones [32, 35].

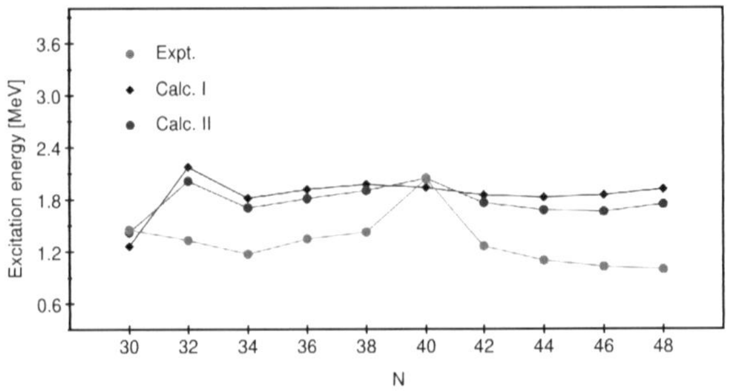

Figure 11. (Color online) Same as in figure 3, but for nickel isotopes from $N = 28$ to 48.

3.6. Ground-state energies

In figure 12 the calculated g.s. energies relative to ^{48}Ca of the even Ca, Ti, Cr and Fe isotopes are reported and compared with the available experimental (continuous red line) and estimated (dashed red line) data [36].

As can be observed, a common feature of all figures is that calculation I fails to reproduce the experimental behavior, providing a collapse of the g.s. energies as function of N. This is not the case of calculation II, whose results are always in a quite good agreement with experiment, apart from a certain overbinding that becomes sizeable only for $N > 36$.

From the comparison of the results of the two calculations, it can be deduced that the monopole component of our effective shell-model interaction is not responsible for the collapse with calculation I and can be therefore considered quite reliable. Furthermore, it is worth pointing out that the main responsibility of this collapse cannot be traced back to a deficiency of the whole theoretical SP spectrum. As a matter of fact, if the theoretical proton and neutron SP spectra are shifted so as to reproduce the experimental ^{49}Sc and ^{49}Ca binding energies respectively, the experimental g.s. energies are well reproduced (black dotted line) as in calculation II.

4. Summary

We have presented here an extensive shell-model study of nuclei north-east of ^{48}Ca within the framework of a fully microscopic approach. This means that, starting from a $V_{\rm low-k}$ derived from the CD-Bonn NN potential, an effective hamiltonian has been derived by way of perturbation theory and then employed in shell-model calculations without any adjustable parameters.

On the whole, when compared with experiment our results exhibit an agreement that may be considered quite good, expecially if one takes into account that our effective hamiltonian consists only of five single-particle energies and less than one hundred two-body matrix elements. In this connection, it should be noted that our choice of a proton model space including only the $0f_{7/2}$ level reduces the scope of a detailed description of spectroscopic properties, especcially when approaching the filling of the $\pi 0f_{7/2}$ level.

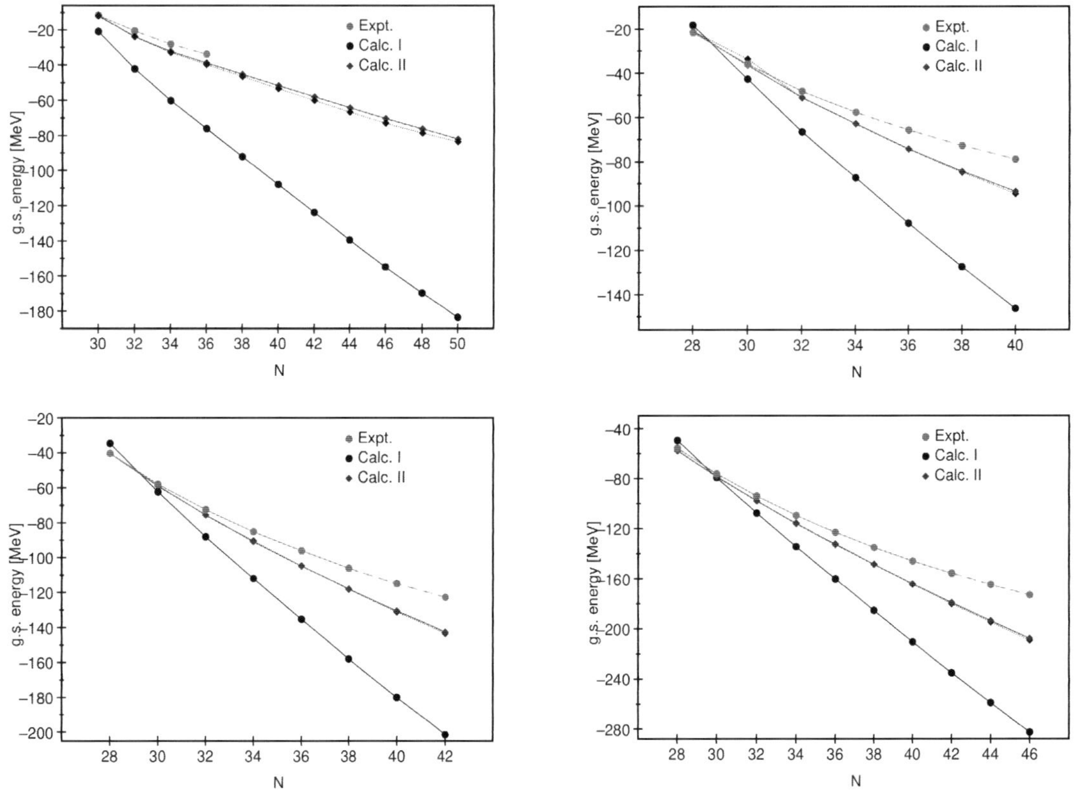

Figure 12. (Color online) Experimental [36] and calculated ground-state energies for calcium, titanium, chromium and iron isotopes as a function of the number of neutrons N. See text for details.

Acknowledgements
This research used the ENEA CRESCO computational resources.

References

[1] Barrett B R, Mihaila B, Pieper S C and Wiringa R B 2003 *Nucl. Phys. News* **13** 17
[2] Pieper S C and Wiringa R B 2001 *Ann. Rev. Nucl. Part. Sci.* **51** 53
[3] Navràtil P, Vary J P and Barrett B R 2000 *Phys. Rev. Lett.* **84** 5728
[4] Navràtil P, Vary J P and Barrett B R 2000 *Phys. Rev. C* **62** 054311
[5] Suzuki K and Okamoto R 1994 *Prog. Theor. Phys.* **92** 1045
[6] Dean D J and Hjorth-Jensen M 2003 *Phys. Rev. C* **69** 054320
[7] Hjorth-Jensen M, Kuo T T S and Osnes E 1995 *Phys. Rep.* **261** 125
[8] Coraggio L, Covello A, Gargano A, Itaco N and Kuo T T S 2009 *Prog. Part. Nucl. Phys.* **62** 135
[9] Coraggio L and Itaco N 2005 *Phys. Lett. B* **616** 43
[10] Coraggio L, Covello A, Gargano A, Itaco N and Kuo T T S 2009 *Phys. Rev. C* **80** 044320
[11] Coraggio L, Covello A, Gargano A, Itaco N, Kuo T T S, Entem D R and Machleidt R 2007 *Phys. Rev. C* **75** 024311
[12] Coraggio L, Covello A, Gargano A and Itaco N 2010 *Phys. Rev. C* **81** 064303
[13] Kaneko K, Sun Y, Hasegawa M and Mizusaki T 2008 *Phys. Rev. C* **78** 064312
[14] Gade A, Janssens R V F, Baugher T, Bazin D, Brown B A, Carpenter M P, Chiara C J, Deacon A N, Freeman S J, Grinyer G F, Hoffman C R, Kay B P, Kondev F G, Lauritsen T, McDaniel S, Meierbachtol K, Ratkiewicz A, Stroberg S R, Walsh K A, Weisshaar D, Winkler R and Zhu S 2010 *Phys. Rev. C* **81** 051304
[15] Lenzi S M, Nowacki F, Poves A and Sieja K 2010 *Phys. Rev. C* **82** 054301
[16] Hoteling N, Chiara C J, Broda R, Walters W B, Janssens R V F, Hjorth-Jensen M, Carpenter M P, Fornal

B, Hecht A A, Królas W, Lauritsen T, Pawłat T, Seweryniak D, Wang X, Wöhr A, Wrzesiński J and Zhu S 2010 *Phys. Rev. C* **82** 044305

[17] Ljungvall J, Görgen A, Obertelli A, Korten W, Clément E, de France G, Bürger A, Delaroche J P, Dewald A, Gadea A, Gaudefroy L, Girod M, Hackstein M, Libert J, Mengoni D, Nowacki F, Pissulla T, Poves A, Recchia F, Rejmund M, Rother W, Sahin E, Schmitt C, Shrivastava A, Sieja K, Valiente-Dobón J J, Zell K O and Zielińska M 2010 *Phys. Rev. C* **81** 061301

[18] Rother W, Dewald A, Iwasaki H, Lenzi S M, Starosta K, Bazin D, Baugher T, Brown B A, Crawford H L, Fransen C, Gade A, Ginter T N, Glasmacher T, Grinyer G F, Hackstein M, Ilie G, Jolie J, McDaniel S, Miller D, Petkov P, Pissulla T, Ratkiewicz A, Ur C A, Voss P, Walsh K A, Weisshaar D and Zell K O 2011 *Phys. Rev. Lett.* **106** 022502

[19] Machleidt R 2001 *Phys. Rev. C* **63** 024001

[20] Bogner S, Kuo T T S, Coraggio L, Covello A and Itaco N 2002 *Phys. Rev. C* **65** 051301(R)

[21] Otsuka T, Suzuki T, Holt J D, Schwenk A and Akaishi Y 2010 *Phys. Rev. Lett.* **105** 032501

[22] Kuo T T S and Osnes E 1990 *Lecture Notes in Physics, vol. 364* (Springer-Verlag, Berlin)

[23] Bogner S, Kuo T T S and Coraggio L 2001 *Nucl. Phys. A* **684** 432c

[24] Kuo T T S, Lee S Y and Ratcliff K F 1971 *Nucl. Phys. A* **176** 65

[25] Kuo T T S, Shurpin J, Tam K C, Osnes E and Ellis P J 1981 *Ann. Phys. (NY)* **132** 237

[26] Baker G A and Gammel J L 1970 *The Padé Approximant in Theoretical Physics* (*Mathematics in Science and Engineering* vol 71) (Academic Press, New York)

[27] Ayoub N and Mavromatis H A 1979 *Nucl. Phys. A* **323** 125

[28] Hoffmann H M, Starkand Y and Kirson M W 1976 *Nucl. Phys. A* **266** 138

[29] Suzuki K and Lee S Y 1980 *Prog. Theor. Phys.* **64** 2091

[30] Shurpin J, Kuo T T S and Strottman D 1983 *Nucl. Phys. A* **408** 310

[31] Engeland T the Oslo shell-model code 1991-2006, unpublished

[32] Data extracted using the NNDC On-line Data Service from the ENSDF database, file revised as of February 16, 2010.

[33] Rejmund M, Bhattacharyya S, Navin A, Mittig W, Gaudefroy L, Gelin M, Mukherjee M, Rejmund F, Roussel-Chomaz P and Theisen C 2007 *Phys. Rev. C* **76** 021304(R)

[34] Fornal B, Janssens R V F, Broda R, Marginean N, Beghini S, Corradi L, Carpenter M P, Angelis G D, Vedova F D, Farnea E, Fioretto E, Gadea A, Guiot B, Honma M, Królas W, Lauritsen T, Lunardi S, Mantica P F, Mason P, Montagnoli G, Napoli D R, Otsuka T, Pawłat T, Pollarolo G, Scarlassara F, Stefanini A, Seweryniak D, Szilner S, Ur C A, Trotta M, Valiente-Dobón J J, Wrseniński J and Zhu S 2008 *Phys. Rev. C* **77** 014304

[35] Rudolph D, Weisshaar D, Cristancho F, Eberth J, Fahlander C, Iordanov O, Skoda S, Teich C, Thelen O and Thomas H 1999 *Eur. Phys. J. A* **6** 377–380

[36] Audi G, Wapstra A H and Thibault C 2003 *Nucl. Phys. A* **729** 337

Nuclear spectra in a large scale shell model approach

D Bianco[1], **F Andreozzi**[1], **N Lo Iudice**[1], **A Porrino**[1] **and F Knapp**[2,3]
[1]Dipartimento di Scienze Fisiche, Università di Napoli Federico II, and INFN, Sezione di Napoli, Napoli, Italy
[2] INFN, Sezione di Napoli, Napoli, Italy

E-mail: loiudice@na.infn.it

Abstract. An iterative matrix diagonalization algorithm is briefly outlined. Its implementation in the spin uncoupled scheme for large scale shell model calculations is discussed with special attention at the convergence properties of the iterative process which establish the domain of applicability of the method. Within this domain, the method is able to generate a large number of eigenstates and, therefore, to provide an exhaustive description of low-lying nuclear spectra. The performance of the method is illustrated by its application to $^{130-134}$Xe isotopes.

1. Introduction

The advent of increasingly powerful computer facilities has stimulated the development of highly sophisticated and efficient matrix diagonalization codes for facing large scale shell model (SM) calculations.

Practically all successful SM codes are implementations of the Lanczos [1] algorithm. Some of them, like Nathan [2, 3] and Nushell [4], use a spin-coupled scheme, others, like Antoine [3], adopt the uncoupled m-scheme.

The coupled scheme yields SM Hamiltonian matrices of relatively small dimensions. These matrices, however, are highly dense and quite involved. Those obtained in the m-scheme, though of much larger dimensions, are sparse and easy to compute.

Sampling techniques for truncating effectively the SM space have also been developed. The quantum Monte Carlo diagonalization (QMCD) method [5], for instance, adopts the Monte Carlo technique, first introduced for studying the nuclear ground state properties [6], to generate stochastically a truncated basis for the diagonalization of the many-body Hamiltonian. Total spin and other symmetries, broken by the stochastic sampling procedure, are restored by specific techniques.

Another method for truncating the basis is based on the density matrix renormalization group (DMRG) [7] and applied recently to medium nuclei [8].

Few years ago, a new iterative algorithm for diagonalizing large matrices was proposed [9, 10]. The method is simple to implement and is endowed with an importance sampling. Its first application was made in the coupled $j-j$ scheme, which came out to be too much time consuming in the construction of the Hamiltonian matrix at each iterative step.

We have now implemented the algorithm in the the m-scheme and proposed a new importance sampling. The method was presented and discussed in [11, 12]. The attention was focused on the

[3] Present address: Faculty of Mathematics and Physics Charles University, Prague, Czech Republic

Published under licence by IOP Publishing Ltd

convergence properties of the iterative process which allow to establish the extent of applicability of the method [12].

In the nuclei where it can be applied, the method yields a large number of levels for each angular momentum and, therefore, offers a complete description of low-lying nuclear spectroscopy. An example was offered by the study of the spectroscopic properties of $^{132-134}$Xe and, partly, of ^{130}Xe [12].

Here, we will review briefly the iterative algorithm and its convergence properties. We will also show how it is applied to $^{130-134}$Xe and discuss some important properties of these nuclei.

2. The algorithm

Let us consider an operator $\hat{A}$, assumed for simplicity to be self-adjoint, and the matrix $A = \{a_{ij}\}$ representing $\hat{A}$ in an orthonormal basis $\{\mid 1\rangle, \mid 2\rangle, \dots, |i >, \dots, \mid N\rangle\}$. The diagonalization procedure goes through several iteration loops. The first loop is composed of the following steps:

1a) Consider the lowest n_0 basis states $|i >$ $(n_0 << N)$, construct and diagonalize the $n_0 \times n_0$ submatrix $A_0 = (a_{ij})$,

1b) select the lowest v eigenvalues $\lambda_1^{(0)}, \dots, \lambda_k^{(0)}, \dots, \lambda_v^{(0)}$ and the corresponding eigenvectors

$$\mid \varphi_k^{(0)}\rangle = \sum_{i=1}^{n_0} c_i^{(0)} \mid i\rangle\rangle, \tag{1}$$

1c) construct and diagonalize the matrix

$$A_k^{(1)} = \begin{pmatrix} \Lambda_{k-1}^{(1)} & B_k^{(1)} \\ B_k^{(1)}(T) & A_k^{(1)}(C) \end{pmatrix}, \tag{2}$$

where $\Lambda_k^{(1)}$ is a v-dimensional diagonal matrix

$$\Lambda_{k-1}^{(1)} = \begin{pmatrix} \lambda_{k-1}^{(1)}(1) & 0 & \dots & 0 \\ 0 & \lambda_{k-1}^{(1)}(2) & \dots & 0 \\ 0 & \dots & \dots & 0 \\ 0 & \dots & 0 & \lambda_{k-1}^{(1)}(v) \end{pmatrix}, \tag{3}$$

and

$$A_k^{(1)}(C) = \{a_{ij}\} \quad (i, j = (k-1)p + 1, \dots, kp) \tag{4}$$

is a p-dimensional submatrix. The two submatrices are coupled by $B_k^{(1)}$ and its transpose, whose matrix elements are

$$b_{ij}^{(k)} = \langle \lambda_{(k-1)}^{(1)}(i) \mid \hat{A} \mid j\rangle, \qquad (i = 1, \dots, v;\ j = (k-1)p + 1, \dots, kp), \tag{5}$$

1d) diagonalize A_1 and extract the new lowest v eigenvalues $\lambda_k^{(1)}$ $(k = 1, v)$ and the corresponding eigenvectors $\mid \varphi_1^{(1)}\rangle, \dots \mid \varphi_v^{(1)}\rangle$,

1e) consider now the new subspace spanned by $\mid \varphi_1^{(1)}\rangle, \dots \mid \varphi_v^{(1)}\rangle$ plus the basis states $|j >= |n_1 + 1 >, \dots |n_2 >$, construct the new submatrix A_2 just as done in point 1c) for A_1 and, after its diagonalization, extract the new lowest v eigenvalues $\lambda_1^{(2)}, \dots \lambda_v^{(2)}$ and the corresponding eigenvectors $\mid \varphi_1^{(2)}\rangle, \dots \mid \varphi_v^{(2)}\rangle$.

The just outlined procedure is iterated until the full basis is exhausted. This zero approximation loop yields the approximate v eigenvalues and eigenvectors

$$E_k^{(1)} \equiv \lambda_k^{(N)}, \quad \mid \psi_k^{(1)}\rangle \equiv\mid \varphi_k^{(N)}\rangle = \sum_{i=1}^{N} c_k^{(N)}(i) \mid i\rangle. \tag{6}$$

These eigensolutions are the new entries for a new iteration. More specifically, we consider the basis composed of the eigenvectors $\psi_k^{(1)}$ plus the original basis states $\{\mid j\rangle\}$. Since the vectors $\psi_k^{(1)}$ are linear combinations of the $\mid j\rangle$ states, this new basis is no longer orthonormal and may be even redundant. We have therefore to solve an eigenvalue problem of the general form and resort to the Choleski decomposition method. With this modifications, the subsequent iteration loops proceed as the first one and generate a sequence of v vectors $\psi_1^{(i)}, \ldots, \psi_v^{(i)}$ that converge to the exact eigensolutions [9].

3. Implementation of the algorithm in the m-scheme

We start with the modified Hamiltonian

$$H_J = H + c[\mathbf{J}^2 - J(J+1)]^2, \tag{7}$$

where H is a two-body Hamiltonian of general form, $\mathbf{J}$ the total spin operator, and c a positive constant. The Hamiltonian acts in a model space spanned by the states $|i> = |\alpha_1, \alpha_2, \ldots, \alpha_p>$, where p denotes the number of valence nucleons, $\alpha_i = \{a_i m_i\}$ the single particle (s.p.) quantum numbers and $a_i = \{n_i l_i j_i\}$ the s.p. shells. The states $|i>$ have a good magnetic quantum number $M = m_1 + \ldots + m_i + \ldots + m_p$.

The full space can be thought to be decomposed into several subspaces

$$\mathcal{H} = \mathcal{H}_0 \oplus \mathcal{H}_1 \ldots \oplus \mathcal{H}_k \ldots \oplus \mathcal{H}_F. \tag{8}$$

Each subspace $\mathcal{H}_k$ is composed of a set of partitions $\{n_i\}_k = \{a_1^{n_1}, \ldots a_i^{n_i}, \ldots\}_k$, where $\sum_i n_i = p$. It is, therefore, invariant with respect to $\mathbf{J}$. The partitions in $\mathcal{H}_k$ differ from those in $\mathcal{H}_{k-1}$ by at most two single particle shells a_i.

We first diagonalize the Hamiltonian H_J in $\mathcal{H}_0$ obtaining v lowest eigenvalues $E_1^{(0)}, \ldots, E_v^{(0)}$ and eigenvectors $\psi_1^{(0)}, \ldots, \psi_v^{(0)}$ spanning a subspace $\mathcal{E}_0$. These eigensolutions are exact in this subspace and have all the same spin J if the constant c is chosen so as to push the states with $J' \neq J$ up in energy.

Because of its two-body nature, the Hamiltonian couples the subspace $\mathcal{E}_0$ to $\mathcal{H}_1$ only. It is, therefore, sufficient to diagonalize H_J in the subspace $\mathcal{E}_0 \oplus \mathcal{H}_1$ to generate new updated eigenvalues $E_1^{(1)}, \ldots, E_v^{(1)}$ and eigenvectors $\psi_1^{(1)}, \ldots, \psi_v^{(1)}$, defining the subspace $\mathcal{E}_1$. We proceed iteratively. Once the updated eigensolutions defining the subspace $\mathcal{E}_k$ are obtained, we diagonalize the Hamiltonian in the upgraded subspace $\mathcal{E}_k \oplus \mathcal{H}_{k+1}$. We cover eventually the full space obtaining the exact v eigensolutions $\{E_i, \psi_i\}$.

Even after having exploited the sparsity of the Hamiltonian matrix, the diagonalization procedure might become prohibitively lengthy as the dimensions of the space increase. In order to cut effectively the basis we adopted the follow sampling procedure.

Let us fix a sequence of positive small numbers of decreasing values $\epsilon_1 > \ldots \epsilon_k > \ldots > \epsilon_F$. Having generated the lowest v eigenvectors $\psi_1^{(0)}, \ldots, \psi_v^{(0)}$ in $\mathcal{H}_0$, we proceed as in the exact case with one constraint: In going from $\mathcal{H}_{k-1}$ to $\mathcal{H}_k$, we pick up only the basis states $|j>$ that fulfill the condition

$$\frac{| < j|H_J|\varphi_i^{(k-1)} > |^2}{a_{jj} - E_i^{(k-1)}} > \epsilon_k. \tag{9}$$

More specifically, in the first step ($k = 1$), the above condition selects a set of states $\{|j>\}$ forming a subspace $\mathcal{H}_1^{(\epsilon_1)} \in \mathcal{H}_1$. The eigenvalue problem is thus solved in $\mathcal{E}_0 \oplus \mathcal{H}_1(\epsilon_1)$ yielding new v eigensolutions $E_k^{(1)}(\epsilon_1)$, $\psi_k^{(1)}(\epsilon_1)$ defining the subspace $\mathcal{E}_1^{(\epsilon_1)}$. We now explore the full subspace complementary to $\mathcal{H}_0 \oplus \mathcal{H}_1(\epsilon_1)$ and select all the states $|j>$ that fulfill the condition (9) with ϵ_2 replacing ϵ_1. The states so selected span a subspace $\mathcal{H}_2^{(\epsilon_2)} \in \mathcal{H}_0 \oplus \mathcal{H}_1 \oplus \mathcal{H}_2$. The above procedure is iterated with updated eigensolutions and decreasing sampling values ϵ_k until the full space is covered.

The sampling procedure is stable and yields orthonormal eigenfunctions. These are obtained, at each step, by the diagonalization of a symmetric submatrix. The subspaces selected by the sampling are not strictly invariant with respect to **J**. The invariance, however, is restored as the sampling value ϵ becomes sufficiently small. We will show, in fact, that all sampled eigenstates reach soon a good J.

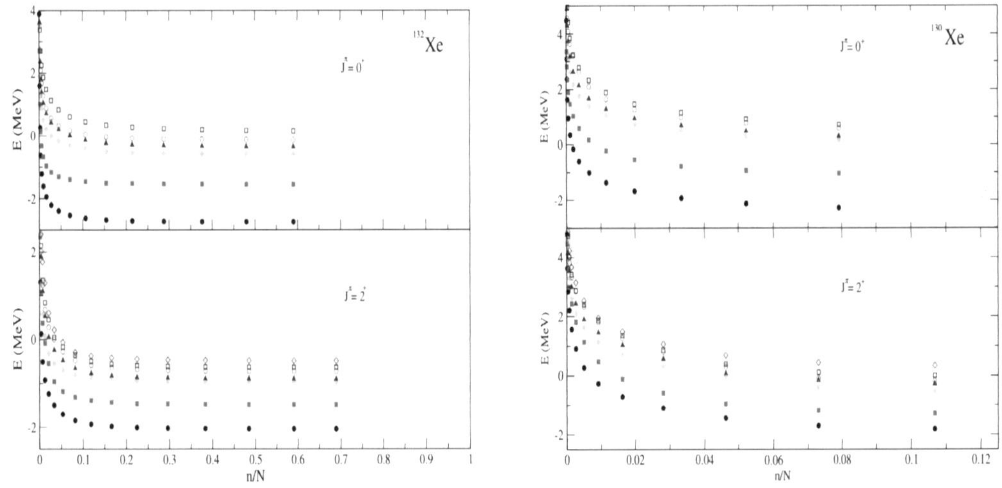

Figure 1. (Color online) Convergence rate of the eigenvalues in ^{130}Xe and ^{132}Xe.

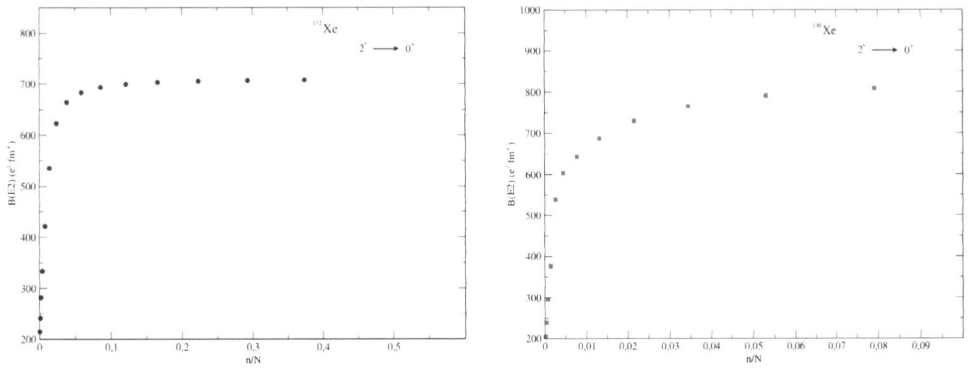

Figure 2. (Color online) Convergence rate of the $B(E2; 2_1^+ \rightarrow 0_1^+)$ strengths in ^{130}Xe and ^{132}Xe.

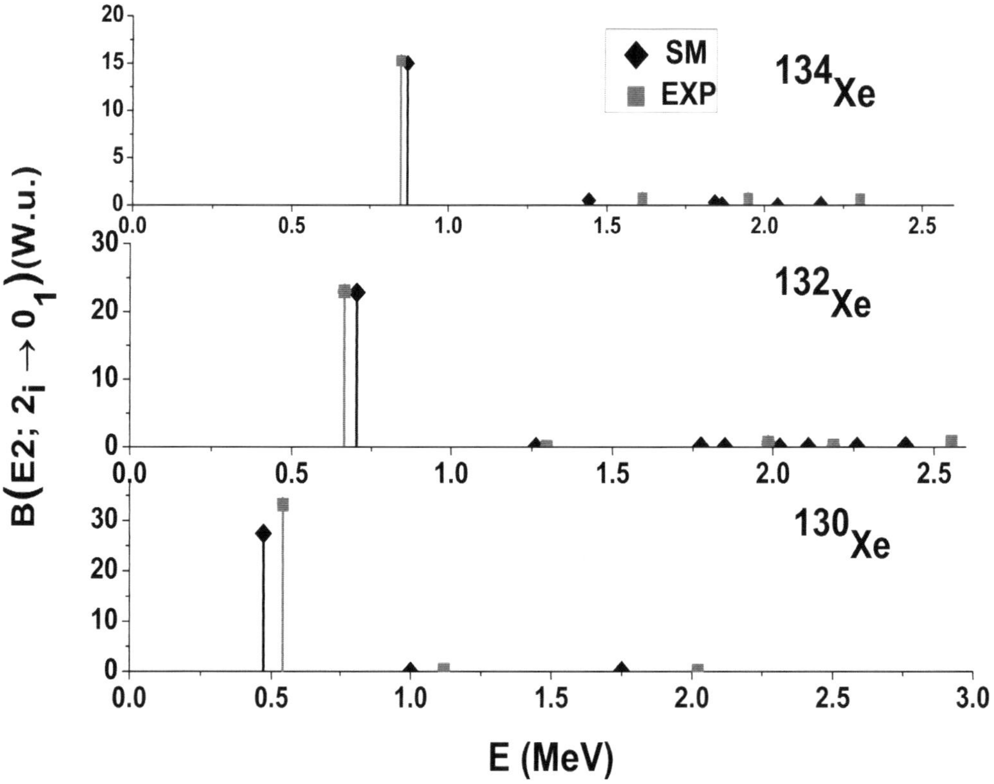

Figure 3. (Color online) $B(E2; 0_1^+ \rightarrow 2_1^+)$ strengths in $^{130-134}$Xe isotopes.

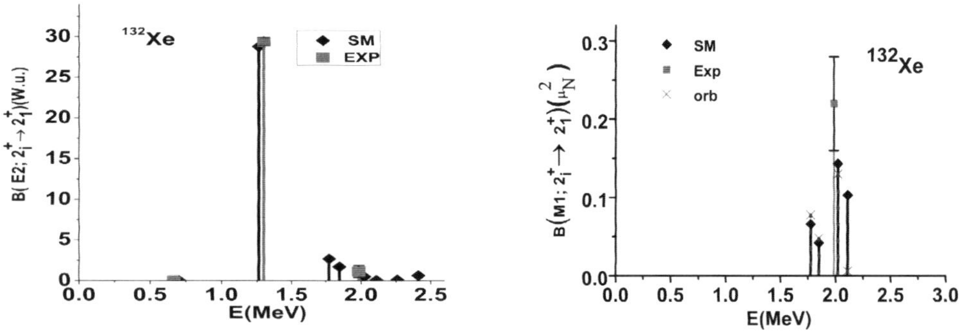

Figure 4. (Color online) $B(E2; 2_i^+ \rightarrow 2_1^+)$ and $B(M1; 2_i^+ \rightarrow 2_1^+)$ strengths in ^{132}Xe.

4. Application to Xe isotopes

The sampling algorithm was applied to the even isotopes $^{134-130}$Xe using a configuration space determined by the shells $\{2d_{5/2}, 1g_{7/2}, 2d_{3/2}, 3s_{1/2}, 1h_{11/2}\}$. The valence protons are particles external to the Z=50 core, while the valence neutrons are described as holes referred to the N=82 core. An effective two-body interaction based on a G matrix [13] derived from the CD-Bonn potential [14] has been employed. This includes the core polarization contributions appropriate for the model space under consideration. We used the proton and neutron effective charges $e_p = 1.6$ and $e_n = 0.7$, respectively, for the $E(2)$ strengths and a spin gyromagnetic quenching factor $g_s = 0.5$ for the $M1$ transition probabilities.

Having chosen an initial subspace $\mathcal{H}_0$ of dimensions $n_0 \simeq 100$ to generate ten lowest eigenstates with good J, we started the iterative sampling procedure using a set of decreasing values of ϵ_k. Each ϵ_k determines uniquely the dimension n_k of the Hamiltonian matrix to be diagonalized.

The dimensions of the Hamiltonian matrices in ^{132}Xe are of the order $N \sim 3.7\times10^7$. Although the full space could be covered easily, it was largely sufficient to carry the iterative procedure up to $\sim 70\%$ of the basis states.

In the case of ^{130}Xe, the dimensions of the Hamiltonian matrices are $N \sim 0.8 \times 10^9$. In this case, we were forced to stop the sampling at $\sim 10\%$ of the basis states. Proceeding further is possible but too time consuming on a desktop. Thus, the sampling will produce reliable results only in case of fast convergence.

This is the case for the energy eigenvalues and the $B(E2; 2_1^+ \to 0_1^+)$ [11]. As shown in figure 1, about 10% of the states are enough to reach the convergence of the two quantities.

On the other hand, the convergence of the strengths of the transitions between excited states was not sufficiently fast to allow to reach the asymptotic values with only $\sim 10\%$ of the sampled states [12].

Thus, the convergence analysis states clearly the domain of applicability of the algorithm. In our specific case, we can carry out an exhaustive study of the spectroscopic properties of ^{132,134}Xe, while, for ^{130}Xe, we can provide a reliable description of the energy levels and the $B(E2; 2_1^+ \to 0_1^+)$ strengths.

Xe isotopes have been recently under intensive experimental investigations [15, 16, 17]. Their attention was focused mainly on the evolution of the MS states [18]. These are states non symmetric with respect to the exchange of valence proton and neutron pairs and describe quadrupole vibrations of protons versus neutrons. They were observed unambiguously and copiously in ^{94}Mo [19] and, since then in several other nuclei in the vicinity of the N=50 and N=82 shell closures.

A review which includes also the theoretical studies may be found in [20]. Mixed Symmetry (MS) states in the vicinity of N=82 were investigated recently within the QPM [21, 22] and in a large scale shell model calculation [23].

The $E2$ and $M1$ transitions between excited states were studied extensively in [12]. Here we report some meaningful results.

The strengths of the $E2$ transitions to the ground state in $^{130-134}$Xe are shown in figures 3. The SM peaks are close to the experimental ones. The SM 2_1^+ collects by far the largest strength of the $E2$ transitions to the ground state. This is shown to be isoscalar indicating that the SM 2_1^+ corresponds to the IBM symmetric one-boson 2^+ or, in microscopic terms, to the QPM isoscalar one-phonon quadrupole vibrational mode. The residual $2_i^+ \to 0_1^+$ $E2$ strength is collected by the 2_3^+ state. This corresponds to the IBM MS state. Indeed, the $2_3^+ \to 0_1^+$ $E2$ transition was shown to be isovector

The strengths of the $E2$ and $M1$ transitions between excited 2^+ states in ^{132}Xe are shown in figure 4. The agreement between SM and experimental strengths of the $2_i^+ \to 2_1^+$ transitions is generally good. Only the 2_2^+ is strongly coupled to the 2_1^+ by the isoscalar $E2$ operator. It is therefore strongly collective and corresponds to a proton-neutron symmetric two-boson state.

The measured $M1$ strength is concentrated mostly, if not solely, on a single 2^+ state, while the SM strength is distributed among few 2^+ states, all clustered around a pronounced peak, lying close to the experimental one.

It is interesting to notice that all transitions are almost purely orbital except for the highest in energy which is due to spin. If we exclude this transition, the total SM strength coincides with the experimental value within the errors.

5. Concluding remarks

When it is implemented in the m-scheme, the importance sampling algorithm becomes certainly more efficient. It requires, in fact, an execution time which grows linearly with the dimensions N of the Hamiltonian matrix.

It becomes, nonetheless, too slow, at least on a desktop, as the dimensions of the Hamiltonian matrix approach $N \sim 10^9$. It follows that, in nuclei like ^{130}Xe, the method is able to sample up to 10% of the basis states. This fraction of states is sufficient to bring to convergence the energy levels and the $B(E2; 2_1^+ \rightarrow 0_1^+)$ but not the strengths of the $E2$ and, especially, the $M1$ transitions among excited states.

In order to improve the performance we may try to find out a sampling criterion which exploits more efficiently the sparsity of the Hamiltonian matrix and/or to develop a parallel version of the code. Such a parallelization should, hopefully, reduce drastically the execution time.

In its present status, the algorithm can be applied to space of dimensions of several hundred millions. Within these limits, the method yields at once an arbitrary number of orthogonal eigenstates with a given J. It provides therefore an exhaustive description of the low-lying spectroscopic properties in the nuclei at reach.

The calculations on $^{132-134}$Xe isotopes offer a meaningful example. We get complete spectra and a full scheme of electromagnetic transitions. The spectra as well as the $E2$ and $M1$ transitions are generally consistent with the experiments. Their analysis allows to determine the collectivity and the proton-neutron symmetry of the low-lying states.

References

[1] Lanczos C 1950 *J. Res. Natl. Bur. Stand.* **45** 252
[2] Caurier E and Nowacki F 1999 *Acta Phys. Pol. B* **30** 705
[3] Caurier E *et al* 2005 *Rev. Mod. Phys.* **77** 427
[4] B. A. Brown and W. D. M. Raen 2007 *NUSHELL@MSU, MSUNSCL Report*(unpublished).
[5] Otsuka T *et al* 2001 *Prog. Part. Nucl. Phys.* **47** 319
[6] Koonin S E, Dean D J and Langanke K 1997 *Phys. Rep.* **278** 1
[7] Dukesky J and Pittel S 2001 *Phys. Rev. C* **63** 061303
[8] Thakur B, Pittel S, Sandulescu N 2008 *Phys. Rev. C* **78** 041303
[9] Andreozzi F, Porrino A and Lo Iudice N 2002 *J. Phys. A: Math. Gen.* **35** L61
[10] Andreozzi F, Lo Iudice N and Porrino A 2003 *J. Phys. : Nucl. Part. Phys.* **29** 2319
[11] Bianco D, Andreozzi F, Lo Iudice N, Porrino A and Knapp F 2011 *J. Phys. G: Nucl. Part. Phys.* **38** 025103
[12] Bianco D, Andreozzi F, Lo Iudice N, Porrino A and Knapp F 2011 *Phys. Rev. C*, accepted for publication
[13] Hjorth-Jensen M *et al* 1995 *Phys. Rep.* **261** 125
[14] Machleidt R 1989 *Adv. Nucl. Phys.* **19** 189
[15] von Garrel H *et al.* 2006 *Phys. Rev. C* **73** 054315
[16] Ahn T *et al.* 2009 *Phys. Lett. B* **679** 19
[17] Coquard L *et al.* 2010 *Phys. Rev. C* **82** 024317
[18] Arima A, Otsuka T, Iachello F and Talmi I 1977 *Phys. Lett. B* **66** 205
[19] Pietralla N *et al.* 1999 *Phys. Rev. Lett.* **83** 1303
[20] Kneissl U, Pietralla N and Zilges A 2006 *J. Phys. G: Nucl. Part. Phys.* **32** R217
[21] Lo Iudice N, Stoyanov Ch D. Tarpanov D 2008 *Phys. Rev. C* **77** 044310
[22] Lo Iudice N, Stoyanov Ch and Pietralla N 2009 *Phys. Rev. C* **80** 024311
[23] Sieja K, Martinez-Pinedo G, Coquard L and Pietralla N 2009 *Phys. Rev. C* **80** 054311

Journal of Physics: Conference Series **336** (2011) 012010 doi:10.1088/1742-6596/336/1/012010

Gamma decay of giant resonances by using Skyrme functionals

Marco Brenna, Gianluca Colò and Pier Francesco Bortignon

Dipartimento di Fisica, Università degli Studi, and INFN, Via Celoria 16, I-20133 Milano, Italy

E-mail: marco.brenna@mi.infn.it, gianluca.colo@mi.infn.it, pierfrancesco.bortignon@mi.infn.it

Abstract. A microscopic model, entirely based on Skyrme energy density functionals, has been developed to study the γ-decay of giant resonances. In particular, it treats the ground-state decay within the fully self-consistent Random Phase Approximation (RPA) and the decay to low-lying states at the lowest order beyond RPA. This model is applied to ^{208}Pb and the results are compared with experimental data and with previous theoretical results obtained, in the past decades, using phenomenological models.

1. Introduction

Nuclear giant multipole resonances had been widely investigated in the past decades both theoretically and experimentally (see, e.g., [1, 2]). These states have finite lifetime and consequently they carry a width of 3-5 MeV. They lie in general above the threshold for emitting a nucleon, so they can decay directly by emission of particles, mainly neutrons in medium-heavy nuclei (being proton emission hindered by the large Coulomb barrier). Then, they can emit photons by coupling to the electromagnetic field. Eventually, the resonances can also relax by coupling to progressively more complicated states lying at the same energy (of 2p-2h, 3p-3h, ..., np-nh character), and this is the major contribution to the width, called $\Gamma^{\downarrow}$. The particle decay and the γ-emission processes are characterized by a width $\Gamma^{\uparrow}$ (escape width) and Γ_γ, respectively. While the escape width is not negligible, although small with respect to the total width ($\Gamma^{\uparrow}/\Gamma \approx 10^{-1}$, in medium-heavy nuclei), the γ decay has an even lower probability ($\Gamma_\gamma/\Gamma \approx 10^{-4}$). In this paper we will focus on this latter process because, although small, it can convey some useful information and it has been the object of recent experimental studies. We will not consider the actual strength function of the giant resonance and the compound gamma decay [3].

A completely microscopic model, based on Skyrme functionals, has been developed to face this problem. In particular, the ground state decay is treated within the fully self-consistent Random Phase Approximation (RPA); for the decay to low-lying states, on the other hand, we include the lowest order correction beyond RPA.

The model has been applied to ^{208}Pb, in particular we focus on the γ-decay of the isoscalar giant quadrupole resonance (ISGQR) to the ground-state and the low-lying collective octupole vibration. In section 2 we briefly summarize the basic features of our formalism and in section 3 we give our results for the two processes at hand.

Published under licence by IOP Publishing Ltd

2. Formalism

In this section we briefly sketch our theoretical framework, referring the reader to a forthcoming paper [4]. The transition amplitude for the emission of a photon of given multipolarity from the nucleus, is proportional to the matrix element of the electric multipole operator $Q_{\lambda\mu}$. The long wavelength approximation is appropriate in our case, then this latter operator takes the form

$$Q_{\lambda\mu} \equiv \frac{1}{2}\sum_{i=1}^{A} e_i^{eff} i^\lambda r_i^\lambda Y_{\lambda\mu}(\hat{\boldsymbol{r}}_i) \tag{1}$$

where the effective charges due to the recoil of the center of mass of the nucleus are defined e.g. in Ref. [5].

The gamma decay width, summed over the magnetic substates of the photon and of the final nuclear state, is then given by

$$\Gamma_\gamma(E\lambda; i \to f) = \frac{8\pi(\lambda+1)}{\lambda[(2\lambda+1)!!]^2}\left(\frac{E}{\hbar c}\right)^{2\lambda+1} B(E\lambda; i \to f), \tag{2}$$

where E is the energy of the transition and the reduced transition probability B associated with the above operator $Q_{\lambda\mu}$ is

$$B(E\lambda; i \to f) = \frac{1}{2J_i+1}|\langle J_f \| Q_\lambda \| J_i \rangle|^2. \tag{3}$$

2.1. The γ-decay to the ground state

We consider in this subsection the decay of an excited RPA state to the ground-state. We limit ourselves to spherical systems, so that the RPA states have quantum numbers JM (only natural parity, or non spin-flip, are considered); in addition, they are labelled by an index n. At the RPA level, in the case of the decay of the state $|nJ\rangle$ to the ground-state, we obtain for the reduced matrix of Eq. (3),

$$\langle 0 \| Q_J \| nJ \rangle = \sum_{ph} \left(X_{ph}^{nJ} + Y_{ph}^{nJ}\right) e_{ph}^{eff} \langle j_p \| i^J r^J Y_J \| j_h \rangle, \tag{4}$$

where X (Y) are the forward (backward) RPA amplitudes (see, e.g., [6]). It is possible to give a diagrammatic representation of the ground state decay (see Fig. 1).

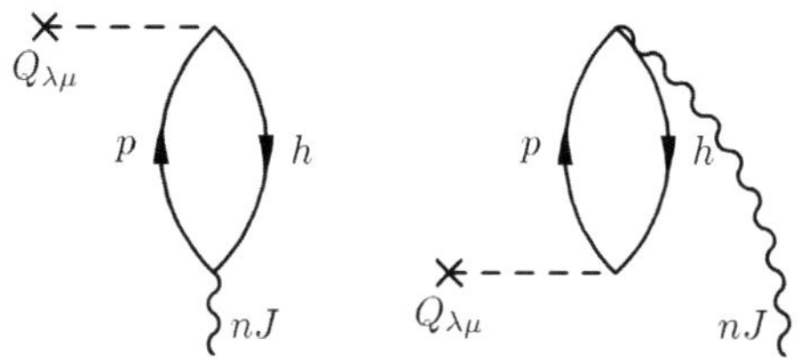

Figure 1. Diagrams representing the decay of the vibrational state $|nJ\rangle$ state to the ground state.

2.2. The γ-decay to low-lying states

By construction RPA is an appropriate theory to describe transition amplitudes between states that differ only by one vibrational (phonon) state. For this reason, the description has to be improved to describe the case of a decay between two vibrational states, extending to a treatment beyond RPA. A consistent framework which is available is the one provided by the Nuclear Field Theory (NFT) [7, 8], since this framework takes into account the interweaving, by means of particle-vibration vertices, between phonons and single-particle degrees of freedom, considered as the relevant independent building blocks of the low-lying spectrum of finite nuclei. The particle-vibration coupling idea [9] has been already applied to the study of γ-decay in Ref. [10]. However, in the present work a consistent use of a microscopic effective interaction of Skyrme type is undertaken for the first time. We consider all the lowest-order contributions to the γ-decay of a phonon state to a different phonon, i.e., all the diagrams involving single-particle states and phonon states, that can lead from the initial to the final state by the action of the external electromagnetic field.

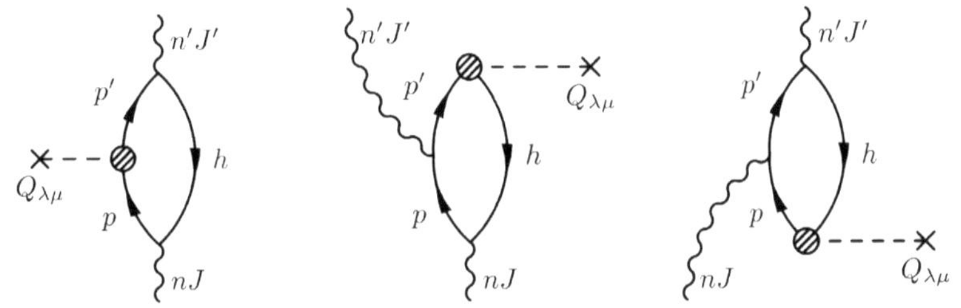

Figure 2. Processes associated with the decay of the $|nJ\rangle$ state to the $|n'J'\rangle$ state. A similar group of diagram are obtained by interchanging the role of particles and holes.

The perturbative diagrams associated with the λ-pole decay of the initial RPA state $|nJ\rangle$ (at energy E_J) to the final state $|n'J'\rangle$ (at energy $E_{J'}$) can be grouped into three families. A representative of each family is shown in Fig. 2 and corresponding analytical expressions read

$$\langle n'J' \| Q_\lambda \| nJ\rangle_{\text{A}} = \sum_{pp'h} (-)^{J+\lambda+J'+1} \begin{Bmatrix} J & \lambda & J' \\ j_{p'} & j_h & j_p \end{Bmatrix} \frac{\langle p\|V\|h, nJ\rangle\langle h, n'J'\|V\|p'\rangle Q^{\lambda pol}_{p'p}}{(E_J - \epsilon_{ph} + i\eta)\,(E_{J'} - \epsilon_{p'h})}, \qquad (5.\text{A})$$

$$\langle n'J' \| Q_\lambda \| nJ\rangle_{\text{B}} = \sum_{pp'h} (-) \begin{Bmatrix} J & \lambda & J' \\ j_{p'} & j_p & j_h \end{Bmatrix} \frac{\langle p\|V\|h, nJ\rangle\langle p', n'J'\|V\|p\rangle Q^{\lambda pol}_{hp'}}{(E_J - \epsilon_{ph} + i\eta)\,(E_J - E_{J'} - \epsilon_{p'h} + i\eta')}, \qquad (5.\text{B})$$

$$\langle n'J' \| Q_\lambda \| nJ\rangle_{\text{C}} = \sum_{pp'h} \begin{Bmatrix} J & \lambda & J' \\ j_h & j_{p'} & j_p \end{Bmatrix} \frac{\langle p'\|V\|p, nJ\rangle\langle h, n'J'\|V\|p'\rangle Q^{\lambda pol}_{ph}}{(E_{J'} - \epsilon_{p'h})\,(E_J + \epsilon_{ph} - E_{J'} + i\eta)}. \qquad (5.\text{C})$$

In these equations ϵ_{ph} is equal to the difference of the Hartree-Fock (HF) single-particle energies $\epsilon_p - \epsilon_h$, and V is the residual particle-hole interaction. In all the energy denominators we include finite imaginary parts η to avoid accidental divergences and to take into account the coupling to more complicated configurations, not included in the model space. The other contributions are obtained interchanging the role of particles and holes, applying the usual rules of diagrammatic expansion.

In all the above equations, the matrix elements of the operator Q_λ include the contribution from the nuclear polarization (consequently they carry the label *pol*).

$$Q_{ij}^{\lambda pol} = \langle i\|Q_\lambda\|j\rangle + \\ + \sum_{n'} \frac{1}{\sqrt{2\lambda+1}} \left[\frac{\langle 0\|Q_\lambda\|n'\lambda\rangle\langle i, n'\lambda\|V\|j\rangle}{(E_J - E_{J'}) - E_{n'} + i\eta} - \frac{\langle i\|V\|j, n'\lambda\rangle\langle n'\lambda\|Q_\lambda\|0\rangle}{(E_J - E_{J'}) + E_{n'} + i\eta} \right], \tag{6}$$

where $|n'\lambda\rangle$ are the RPA states having multipolarity λ (and lying at energy $E_{n'}$), while the bare operator Q_λ has been defined in Eq. (1). The polarization contribution, that is, the second and third term in the latter equation, has the effect of screening partially the external field. In a diagrammatic way, the bare and the polarization contributions to Eq. (6) are drawn in Fig. 3.

Figure 3. Polarization contribution to the operator $Q_{\lambda\mu}$

As mentioned above, in the present implementation of the formalism we use consistently different zero-range Skyrme interactions. The single-particle energies ϵ_i, and the corresponding wavefunctions, come from the solution of the HF equations. The energies (and X and Y amplitudes) of the vibrational states are obtained through fully self-consistent RPA [11]. These quantities enter the reduced matrix elements associated with the PVC vertices. The basic one, that couples the single-particle state i to the particle-vibration pair j plus nJ, is

$$\langle i\|V\|j, nJ\rangle = \sqrt{2J+1}\sum_{ph} X_{ph}^{nJ} V_J(ihjp) + (-)^{j_h - j_p + J} Y_{ph}^{nJ} V_J(ipjh). \tag{7}$$

V_J is the particle-hole coupled matrix element

$$V_J(ihjp) = \sum_{\{m\}} (-)^{j_j - m_j + j_h - m_h} \langle j_i m_i j_j - m_j | JM\rangle \langle j_p m_p j_h - m_h | JM\rangle v_{ihjp}, \tag{8}$$

while v_{ihjp} is a shorthand notation for $\langle j_i m_i j_h m_h | V | j_j m_j j_p m_p\rangle$. For the detailed derivation of the the reduced matrix element of Eq. (7) we refer to the Appendix of Ref. [12].

3. Results

In this section, the results obtained from our numerical calculations in ^{208}Pb are discussed. In particular, we focus on the γ-decay width Γ_γ associated with the decay of the isoscalar giant quadrupole resonance (ISGQR) either to the ground-state or to the 3^- low-lying state. We have employed four different Skyrme forces: SLy5 [13], SGII [14], SkP [15] and LNS [16].

In all cases, we start by solving the HF equations in a radial mesh that extends up to 20 fm with a radial step of 0.1 fm. Once the HF solution is found, the RPA equations are solved in the usual matrix formulation. Vibrations (or phonons) with multipolarity L ranging from 1 to 3, and with natural parity, are calculated. The RPA model space consists of all the occupied states, and all the unoccupied states lying below a cutoff energy E_C equal to 50 MeV. The states at positive energy are obtained by setting the system in a box of 20 fm, that is, the continuum is discretized. These states have increasing values of the radial quantum number n, and are calculated for those values of l and j that are allowed by selection rules. With this choice of the model space the energy-weighted sum rules (EWSRs) satisfy the double commutator values at the level of about 99%; moreover, the energy and the fraction of EWSR of the states which are relevant for the following discussion are well converged.

Table 1. Energy of the ISGQR and γ-decay width associated with its transition to the ground state. The first four rows correspond to the present RPA calculations performed with different Skyrme parameter sets. In this case we show both the bare Γ_γ from Eq. (2) as well as the renormalized value which is discussed in the main text. The next two rows report the results of previous theoretical calculations [10, 17]. In the last row the experimental value from [18] is displayed.

	$E_{\rm GQR}$ [MeV]	Γ_γ [eV]	Γ_γ^{ren} [eV]
SLy5	12.28	231.54	127.58
SGII	11.72	163.22	113.57
SkP	10.28	119.18	159.72
LNS	12.10	176.57	104.74
Ref. [17]	10.60	112	
Ref. [10]	11.20	142	
Ref. [18]	10.60	130±40	

3.1. Ground-state decay

We group in table 1 the results obtained for the decay of the ISGQR to the ground state. In general, our calculations reproduce the experiment quite well, without any parameter adjustment. They tend to overestimate the decay width, and this is true in particular for SLy5; however, even in this worst case, our result are still compatible with the experimental value.

More importantly, this discrepancy is entirely due to the fact that the properties of the giant resonance (energy and fraction of EWSR) do not fit accurately the experimental findings. In particular, the critical quantity turns out to be the resonance energy, since in Eq. (2) the energy of the transition is raised to the fifth power: consequently, an increase of the energy by 1 MeV produces an increase of the gamma decay width by about 50% (at 10 MeV). To substantiate this point, in the last column of table 1 we report the values obtained for the decay width after having rescaled the ISGQR energy to the experimental value (shown in table 2).

Since for all interactions the experimental value of the ground-state decay width can be obtained simply by scaling to the experimental value the energy and the fraction of EWSR, we conclude that this kind of measurement is not able to discriminate between models more than other integral observables (e.g. energy and strength).

For completeness, in table 1 the previous theoretical values found in the literature [10, 17] are listed as well. In Ref. [17], the theory of finite Fermi systems [19] is implemented with a separable interaction in order to obtain the decay width. In Ref. [10], in which the nuclear field theory with a separable interaction at the particle-vibration vertex is used, the result is obtained from the values of energy and EWSR in table 1.

3.2. ISGQR decay to the low-lying 3^- state

In table 3, the results obtained for the decay of the ISGQR to the low-lying octupole state are shown, together with theoretical values from Refs. [10, 17]. These latter are obtained using different models: in [17], the theory of finite Fermi systems (cf. Ref. [19]) with a separable interaction is used, while in [10], the nuclear field theory with a separable interaction at the particle-vibration coupling vertex is implemented.

It can be noticed that only two interactions, namely SLy5 and SkP, can reasonably reproduce the experimental value for the decay width. Nevertheless, all these forces are able to produce a

Table 2. Detailed properties of the ISGQR and of the low-lying 3^- state in ^{208}Pb.

	ISGQR		3^-	
	E [MeV]	EWSR [%]	E [MeV]	B(E3↑) [10^5e^{2}fm^6]
Experiment	10.9 ± 0.3	100	2.6145 ± 0.0003	6.11 ± 0.09
SLy5	12.28	69.27	3.62	6.54
SGII	11.72	72.31	3.14	6.58
SkP	10.28	81.79	3.29	5.11
LNS	12.10	66.98	3.19	5.67

Table 3. γ-decay width associated with the transition from the ISGQR to the low-lying 3^- state. The first four rows correspond to the present calculation. The energy of the transition is also reported in the second column. The next two rows report the results of previous theoretical calculations [10, 17] are also listed. In the last row, the experimental value from [18] is provided.

Interaction	E_{trans} $[MeV]$	Γ_γ $[eV]$
SLy5	8.66	3.39
SGII	8.58	29.18
SkP	6.99	8.34
LNS	8.90	39.87
Ref. [17]	7.99	4.00
Ref. [10]	8.59	3.50
Ref. [18]	7.99	5.00±5.00

Table 4. Quenching factors that combine to produce the decay width Γ_γ from the typical particle-hole dipole transition at the energy of interest of ≈ 8.5 MeV: they are discussed in the main text. The same quantities from [10] are displayed.

	SLy5	SGII	SkP	LNS	Ref. [10]
p − h transition [eV]	$1.2{\cdot}10^3$	$3{\cdot}10^3$	$1.5{\cdot}10^3$	$3.0{\cdot}10^3$	$1.5{\cdot}10^3$
Recoupling coefficient	3	3	3	3	3
$\pi - \nu$ cancellation	5	4	3–4	4	4
p − h cancellation	3–4	2–3	2–3	3–4	2–3
Polarization	6	3	7–8	4	15
Γ_γ[eV]	3.39	29.18	8.34	39.87	3.50

total $\Gamma_\gamma(\text{GQR} \to 3^-)$ which is only a few percent of $\Gamma_\gamma(\text{GQR} \to gs)$, as the experiment indicates.

The factors in the amplitude, obtained summing all the contributions reported in section 2.2, combine to produce a rather small width in the end: they are displayed in table 4, for the four forces used, together with similar factors from [10]. A typical particle-hole dipole transition would produce at the energy of interest of ≈ 8.5 MeV the γ-decay width reported in the first row. However, several effects intervene to quench this strong coupling. The Wigner $6-j$ symbol times the other geometrical and statistical factors in Eq. (5) reduces the width by a factor 3 (second row). The third row displays the quenching deriving from the fact that proton and neutron amplitudes tend to cancel, being the initial and final state predominantly isoscalar and the electromagnetic operator (1) isovector. As it is well known [20], another source of interference is the cancellation between particles and holes: the associated quenching factor is displayed in the fourth row. Eventually, the polarization contribution (6) reduces the width by the factors list in the fifth entry, to get a width of the order of electronvolts (or tens of electronvolts), as reported in the last row. In particular, the most important effect is the polarization of the nuclear medium.

4. Conclusions

In our work we deal with the de-excitation of nuclear excited RPA states via electromagnetic emission, by using a fully microscopic approach: in fact, the γ-decay has been studied in the past decades using only phenomenological models. In particular, in our scheme, within the Skyrme framework, the single particle states are obtained within HF, the vibrations are calculated using fully self-consistent RPA and the whole Skyrme force is employed at the particle-vibration vertices. We treat the ground-state decay within the fully self-consistent RPA and the decay to low-lying collective vibrations at the lowest contributing order of perturbation theory beyond RPA.

We have applied our model to the γ-decay of the isoscalar giant quadrupole resonance in ^{208}Pb into the ground-state and the first low-lying octupole vibration. Concerning the ground-state decay, we found that our outcomes are consistent with previous theoretical calculations, based on phenomenological models, and with the experimental data. In particular, all the Skyrme parametrizations give a decay width of the order of hundreds of electronvolts, though, at the same time, they tend to overestimate it: these discrepancies are due to the fact that the energy of the resonance does not totally agree with the experimental data. For this reason, we conclude that the γ-decay to the ground-state is not so able to discriminate between different models, at least not more than other inclusive observable (as energy and strength).

On the other hand, the γ-decay to low-lying collective states is more sensitive to the interaction used. As a matter of fact, only two interactions (namely SLy5 and SkP) manage to achieve a decay width of few electronvolts, consistently with the experimental finding. Nevertheless, the other interactions give a width that is some percent of the one corresponding to the ground-state decay. In particular, the description of the dipole spectrum is a crucial point, because small differences in the strength of the dipole states introduced as intermediate states change significantly the polarization of the nuclear medium, as it is discussed in [4].

References

[1] Bortignon P F, Bracco A and Broglia R A 1998 *Giant Resonances: Nuclear structure at finite temperature* Contemporary Concepts in Physics (Harwood Academic publisher)
[2] Harakeh M N and van der Woude A 2001 *Giant Resonances: fundamental high-frequency modes of nuclear excitation* Oxford Studies in Nuclear Physics (Oxford University Press Inc.)
[3] Beene J R, Bertsch G F, Bortignon P F and Broglia R A 1985 *Phys. Lett. B* **164** 19
[4] Brenna M, Colò G and Bortignon P F 2011 *Preprint* arXiv:1111.0619v1 [nucl-th]
[5] de Shalit A and Feshbach H 1974 *Theoretical Nuclear Physics: Nuclear Structure* vol 1 (John Wiley and Sons Inc.)
[6] Rowe D J 1970 *Nuclear Collective Motion* (Methuen and Co. Ltd.)
[7] Bes D R, Dussel G G, Broglia R A, Liotta R J and Mottelson B R 1974 *Phys. Lett. B* **52** 253
[8] Bortignon P F, Broglia R A, Bes D R and Liotta R 1977 *Phys. Rep.* **30** 305
[9] Bohr A and Mottelson B R 1975 *Nuclear Stucture* vol II (W. A.Benjamin Inc.)
[10] Bortignon P F, Broglia R A and Bertsch G F 1984 *Phys. Lett. B* **148** 20
[11] Colò G, Bortignon P F, Fracasso S and Van Giai N 2007 *Nucl. Phys. A* **788** 173c
[12] Colò G, Sagawa H and Bortignon P F 2010 *Phys. Rev. C* **82** 064307
[13] Chabanat E, Bonche P, Haensel P, Meyer J and Schaeffer R 1998 *Nucl. Phys. A* **635** 231

[14] Van Giai N and Sagawa H 1981 *Phys. Lett. B* **106** 379
[15] Dobaczewski J, Flocard H and Treiner J 1984 *Nucl. Phys. A* **422** 103
[16] Cao L G, Lombardo U, Shen C W and Van Giai N 2006 *Phys. Rev. C* **73** 014313
[17] Speth J, Cha D, Klemt V and Wambach J 1985 *Phys. Rev. C* **31** 2310
[18] Beene J R, Bertrand F E, Halbert M L, Auble R L, Hensley D C, Horen D J, Robinson R L, Sayer R O and Sjoreen T P 1989 *Phys. Rev. C* **39** 1307
[19] Speth J, Werner E and Wild W 1977 *Phys. Rep.* **33** 127
[20] Bertsch G F, Bortignon P F and Broglia R A 1983 *Rev. Mod. Phys.* **55** 287

Journal of Physics: Conference Series **336** (2011) 012011 doi:10.1088/1742-6596/336/1/012011

Skyrme second random-phase approximation description of low-lying dipole states in ^{48}Ca isotope

D Gambacurta[1,2], M Grasso[3] and F Catara[1,2]

[1] Dipartimento di Fisica e Astronomia dell'Universitá di Catania
[2] Istituto Nazionale di Fisica Nucleare, Sezione di Catania, Via S.Sofia 64, I-95123 Catania, Italy
[3]Institut de Physique Nucléaire, Université Paris-Sud, IN2P3-CNRS, F-91406 Orsay Cedex, France

E-mail: danilo.gambacurta@ct.infn.it

Abstract. We apply the second random phase approximation (RPA) with Skyrme interaction to the study of low-lying dipole strength distributions of ^{48}Ca, in the energy region of the excitation spectrum between 5 and 10 MeV. RPA models do not usually predict any presence of strength in this energy region, while experimentally a significant amount of strength is found. The inclusion of the 2 particle -2 hole configurations within the second RPA is found to be very important. Several states are obtained and many of them, despite being essentially 2 particle - 2 hole states, show quite large B(E1) values. These results are in reasonable agreement with other beyond mean-field predictions and with the corresponding experimental measurements. A detailed analysis of the properties of the most collective state is done in terms of its 1 particle -1 hole nature and its transition densities.

1. Introduction

In the last years the properties of the low-lying dipole response in neutron rich nuclei have been studied with particular attention to the appeareance of the so-called Pygmy Dipole Resonance. A recent review about the main theoretical results and experimental measurements can be found in Ref. [1]. In very light nuclei, the development of a strong dipole response at low energies is usually related to the extremely small value of the separation energies. These low-energy states are not collective excitations and have mainly a single-particle character [2]. Low-energy dipole excitations have been observed also for the isotopes $^{16-22}$O [3]. From the experimental point of view, the character of these excitations (collective or single-particle) has not yet been clearly elucidated.

For heavier stable and unstable nuclei, the development of a pygmy dipole response in neutron-rich systems is currently related to the formation of a thick neutron skin at the surface of the nucleus: the low-lying dipole modes are interpreted in terms of oscillations of the skin against the core composed by both neutrons and protons. From the theoretical point of view, pygmy resonances have been analyzed with several models. Some examples are the relativistic and non-relativistic (Q)RPA approach (see, for instance, Ref. [1] and references therein, and Refs. [4]-[8]), the particle-phonon-coupling models (Ref. [1] and references therein), a semiclassical coupled-channels approach for Sn isotopes [9, 10], hydrodynamical models [11], the phonon-damping model [12], and the so-called Extended Theory of Finite Fermi Systems (ETFFS) [13].

Published under licence by IOP Publishing Ltd

Discrepancies among the different theoretical predictions are found concerning in particular the collective character and the fragmentation of the low-lying modes.

It is well known that each excitation mode is characterized by an escape ($\Gamma^{\uparrow}$) and a spreading ($\Gamma^{\downarrow}$) width as well as by a Landau damping. The latter can already be described by a standard discrete RPA calculation. The spreading width cannot be described by the standard RPA since more complex configurations (collective coordinates or multiparticle-multihole configurations) have to be included within a beyond mean-field model. Among the different beyond mean-field models that allow one to describe, at least partially, the width and the fragmentation of the excitation modes, the second random-phase approximation (SRPA) is a powerful theoretical tool where the coupling with 2 particle- 2 hole ($2p2h$) configurations is included within an RPA-like formalism. In this way, the so-called spreading widths can be described together with the Landau damping. Escape widths are missing if the coupling to the continuum is not included. To include higher multiparticle-multihole configurations different directions may be followed [14, 15].

Due to the heavy numerical effort required, the SRPA equations have been usually solved resorting to some approximations. Recently, full SRPA calculations have been performed for some O and Ca isotopes [16, 17]. In particular, in Ref. [17] calculations with the density-dependent Skyrme interaction have been performed adopting two currently used approximations for treating the rearrangement terms of the residual interaction appearing in beyond-RPA matrix elements. The two approximations consist in either neglecting these rearrangement terms or treating them with the standard RPA procedure. Important differences have been found between the corresponding two sets of results. The same authors of Ref. [17] have addressed this point in a more recent work [18] where a procedure to derive the expressions of all the rearrangement terms within the SRPA framework has been presented and applied to calculations for the nucleus ^{16}O. In this first application, the importance of the proper treatment of the rearrangement terms in SRPA for the description of the fragmentation of the excited modes has been shown.

In this work, the low-lying excitation spectrum of the stable isotopes ^{48}Ca is analyzed within the Skyrme-SRPA model .

The paper is organized as follows. In section 2 the main formal aspects of the SRPA model are briefly recalled. In section 3 the low-lying strength distributions are analyzed for the ^{48}Ca and the transition densities associated to some states are displayed. We draw our conclusions in section 4.

2. Formalism

The excited states in SRPA are superpositions of 1 particle-1 hole ($1p1h$) and $2p2h$ configurations. The SRPA equations can be written in the compact form

$$\begin{pmatrix} \mathcal{A} & \mathcal{B} \\ -\mathcal{B}^* & -\mathcal{A}^* \end{pmatrix} \begin{pmatrix} \mathcal{X}^\nu \\ \mathcal{Y}^\nu \end{pmatrix} = \omega_\nu \begin{pmatrix} \mathcal{X}^\nu \\ \mathcal{Y}^\nu \end{pmatrix}, \tag{1}$$

where:

$$\mathcal{A} = \begin{pmatrix} A_{11} & A_{12} \\ A_{21} & A_{22} \end{pmatrix}, \mathcal{B} = \begin{pmatrix} B_{11} & B_{12} \\ B_{21} & B_{22} \end{pmatrix},$$

$$\mathcal{X}^\nu = \begin{pmatrix} X_1^\nu \\ X_2^\nu \end{pmatrix}, \quad \mathcal{Y}^\nu = \begin{pmatrix} Y_1^\nu \\ Y_2^\nu \end{pmatrix}.$$

In the above equations, '1' and '2' stand for $1p1h$ and $2p2h$, respectively. Thus, A_{11} and B_{11} represent the usual RPA matrices, whereas the matrices A_{12} and B_{12} couple $1p1h$ with $2p2h$ configurations and the matrices A_{22} and B_{22} couple among themselves $2p2h$ configurations. The detailed expressions of these matrices can be found for example in Ref. [17]. If a density-dependent interaction like the Skyrme force is employed, rearrangement terms appear in the residual interaction. The usual RPA rearrangement terms appear in the matrices A_{11} and B_{11}.

Table 1. Total $B(E1)$ ($e^2 fm^2$) and EWSRs ($e^2 fm^2 MeV$) integrated up to 10 MeV as a function of the energy cutoff $ECUT$ (MeV) on the $2p2h$ configurations for ^{48}Ca.

^{48}Ca		
$ECUT$	$\sum B(E1)$	EWSRs
40	0.184	1.623
45	0.218	1.895
50	0.226	1.944
55	0.240	2.049
60	0.230	1.964

New types of rearrangement terms have been obtained for the other matrix elements in Ref. [18] within a variational derivation of the SRPA equations. The expressions of these rearrangement terms are reported in Ref. [18] and are used in this work.

3. Results

The experimental low-lying dipole response of the ^{48}Ca isotope has been recently analyzed [19] and the development of a low-energy strength, between 5 and 10 MeV, has been observed. From the theoretical point of view, it has been found that relativistic and non-relativistic (Q)RPA models are not able to well describe the low-lying response in ^{48}Ca because they either do not provide the good excitation energies (too high energies) or do not predict the experimental fragmentation of the peaks. For example, in recent calculations performed with the relativistic RPA model no strength has been found in the response below the excitation energy of 10 MeV [20]. The same kind of results is obtained in our Skyrme SGII-RPA calculations. On the other hand, a reasonable agreement (energies and fragmentation) with the experimental results has been found within the ETFFS model [13, 21] where a quasiparticle-phonon coupling is included.

We perform SRPA calculations with the Skyrme interaction SGII. More details about the calculations can be found in Ref. [22]. Because of the zero-range of the Skyrme interaction, a natural energy cutoff is not provided. Different procedures to treat this problem may be envisaged for future studies (see, for instance, the exploratory work presented in Ref. [23]). In this work, we have introduced an energy cutoff ($ECUT$) on the $2p2h$ configurations. By varying it from 40 to 60 MeV we have verified that a reasonable stability of the results is achieved around a cutoff of 50-55 MeV. The total $B(E1)$ and EWSRs values, integrated up to an energy of 10 MeV, are shown in table 1 as a function of the energy cutoff $ECUT$ on the $2p2h$ configurations.

The $B(E1)$ distributions for different choices of the energy cutoff are plotted in figure 1 up to an excitation energy of 10 MeV. The employed transition operator is

$$F_{10} = e_p \sum_{i=1}^{Z} r_i Y_{10}(\Omega_i) - e_n \sum_{i=1}^{N} r_i Y_{10}(\Omega_i) \tag{2}$$

where e_p and e_n are the kinematic charges, $e_p = Ne/A$ and $e_n = Ze/A$, respectively. One can observe that the results do not change strongly starting from a cutoff of 45 MeV. In what follows, we will analyze the results obtained with a cutoff of 60 MeV (bottom panel of figure 1).

These results are qualitatively of the same type as those found in Ref. [21] with the ETFFS approach which is also a beyond-mean-field model where the coupling is done with collective phonons instead of $2p2h$ configurations (as is done in SRPA). We can compare the location of the theoretical peaks with the experimental distribution (see, for instance, figure 2 of Ref. [21]). We can distinguish two regions: from 6 to 8 MeV and from 8 to 10 MeV. Experimentally, the highest peak in the first region is found at $\sim$ 7 MeV whereas in our case we have several small peaks

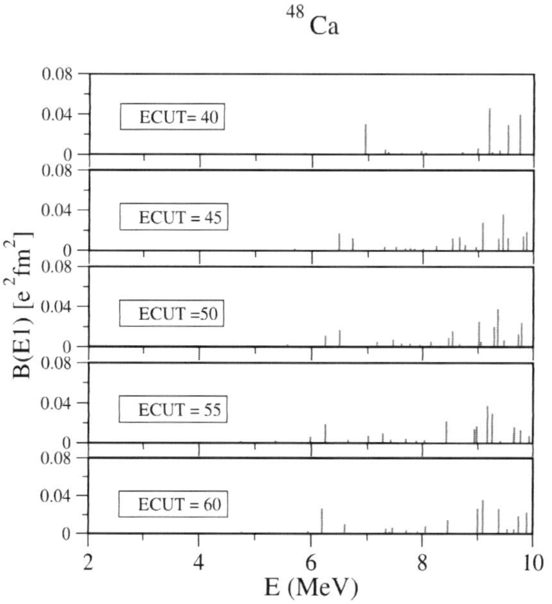

Figure 1. (Color online) SRPA dipole strength distribution for the ^{48}Ca isotope for increasing values of the energy cutoff $ECUT$ (MeV) on the $2p2h$ configurations included in the calculations.

between 6 and 8 MeV and the highest peak is located around 6.2 MeV. In the interval between 8 and 10 MeV, our response is more fragmented than the experimental one. Experimentally, peaks are found around 8.5, 9 and 9.5 MeV. Our highest peak is located at $\sim$ 9.1 MeV. We have also evaluated the response in the nucleus ^{40}Ca. Experimentally, a negligible strength has been found for this nucleus between 5 and 10 MeV (figure 2 of Ref. [21]). Some peaks are actually found at low energy but the corresponding strength is much lower than in ^{48}Ca (almost a factor 10).

An interesting information that can be analyzed in connection with the strength distribution is the composition of the excitation modes in terms of $1p1h$ and $2p2h$ configurations. By extracting the expression of N_1 from the SRPA normalization condition,

$$\sum_{ph}(\mid X^{\nu}_{ph}\mid^2 - \mid Y^{\nu}_{ph}\mid^2) + \sum_{p<p',h<h'}(\mid X^{\nu}_{php'h'}\mid^2 - \mid Y^{\nu}_{php'h'}\mid^2)$$

$$= N_1 + N_2 = 1, \tag{3}$$

we plot in figure 2 the $B(E1)$ values corresponding to a cutoff of 60 MeV (upper panel, same as in bottom panel of figure 1) and the quantity N_1 (lower panel) for each excitation. One observes that all the excitations present a mixing of $1p1h$ and $2p2h$ configurations. Those which have the highest $1p1h$ content (around 50%) may be interpreted as excitations that already exist in the RPA spectrum at higher energies (the first excitations in RPA are located around 11 MeV) and that are shifted down to lower energies due to the coupling with $2p2h$ configurations. In the SRPA spectrum we also see several states which present a dominant $2p2h$ nature and show a relatively large $B(E1)$ despite their very low content of $1p1h$ configurations (see for instance the energy region around 9 MeV).

In the following we focus our attention to the most collective state located at about 9.09 MeV. In figure 3 we show the transition densities corresponding to this peak. We show separately the

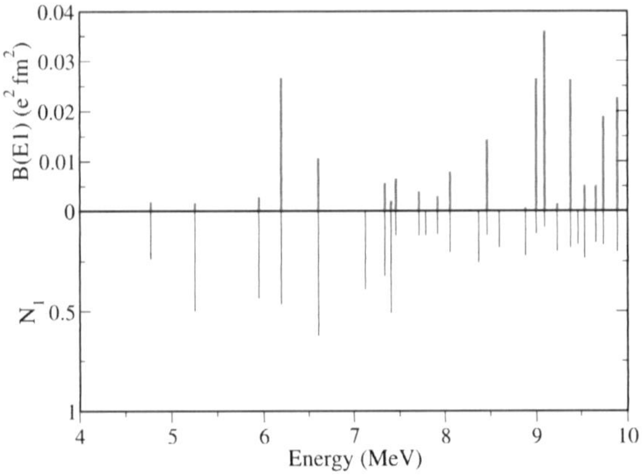

Figure 2. (Color online) For each state the $B(E1)$ value corresponding to a cutoff of 60 MeV (upper panel) and the total $1p1h$ contribution N_1 to the norm of the state defined in Eq. (3) (lower panel), are shown for ^{48}Ca.

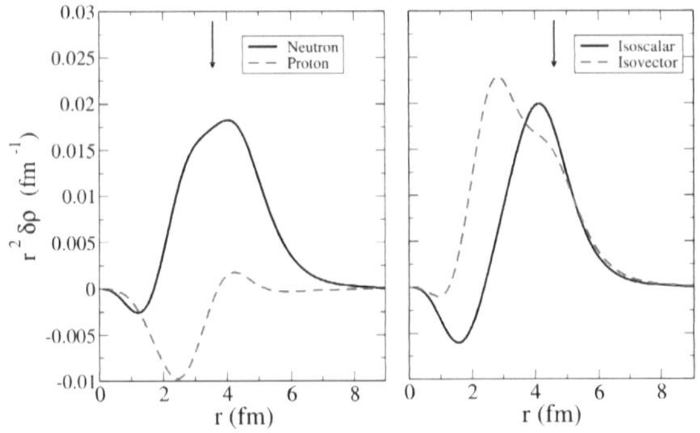

Figure 3. (Color online) Neutron and proton transition densities (left) and the corresponding isoscalar and isovector ones (right) associated to the most collective peak located at 9.06 MeV. The arrows roughly indicate the position of the neutron r.m.s radius.

neutron $\delta\rho_n(r)$ and the proton $\delta\rho_p(r)$ (left), the isoscalar $\delta\rho_n(r) + \delta\rho_p(r)$ and the isovector $\delta\rho_n(r) - \delta\rho_p(r)$ (right) transitions densities.

We see that the neutron transition density dominates over the proton one that is almost vanishing in the external part of the nucleus while in the interior the two densities oscillate out of phase. This leads to a strong mixing of isoscalar and isovector components. At this regard, we notice that isoscalar $B(E1, T = 0)$ and isovector $B(E1, T = 1)$ transition probabilities obtained by integrating the corresponding transition densities of figure 3 are 0.099 e^2 fm^2 and 0.186 e^2 fm^2, respectively.

In SRPA as well as in RPA the reduced transition probability for a one-body operator describing the excitation from the ground state to a state ν can be written as

$$B(E\lambda) = |\sum_{ph} b_{ph}(E\lambda)|^2 = |\sum_{ph}(X^{\nu}_{ph} - Y^{\nu}_{ph})F^{\lambda}_{ph}|^2 \tag{4}$$

where F^{λ}_{ph} are the multipole transition amplitudes associated to a $1p1h$ configuration. We remark that also in the case of SRPA only $1p1h$ amplitudes appear in the expression of the transition

Table 2. Particle-hole configurations which give the major contributions to the dipole low-lying state located at 9.09 MeV. For each ph configuration, the energy, the contribution to the norm of the state A_{ph}, the partial contribution to the reduced transition amplitude b_{ph} (e fm) and the matrix element of the transition operator F^{λ}_{ph} are reported. The superscripts π, ν refer to proton and neutron states, respectively.

		$E = 9.09$ MeV		
ph conf.	E (MeV)	A_{ph}	$b_{ph}(E1)$	F^{λ}_{ph}
$(2p_{3/2}, 1d_{5/2})^{\pi}$	17.499	0.002	-0.039	1.343
$(1f_{7/2}, 1d_{5/2})^{\pi}$	11.732	0.001	-0.098	3.304
$(1f_{5/2}, 1d_{5/2})^{\pi}$	19.246	0.006	-0.032	-0.734
$(2p_{1/2}, 2s_{1/2})^{\pi}$	14.133	0.005	0.045	-1.233
$(2p_{3/2}, 1d_{3/2})^{\pi}$	12.120	0.007	-0.023	0.451
$(2p_{1/2}, 1d_{3/2})^{\pi}$	13.809	0.000	-0.014	1.053
$(1f_{5/2}, 1d_{3/2})^{\pi}$	13.867	0.017	0.197	2.756
$(2p_{3/2}, 1d_{5/2})^{\nu}$	15.683	0.006	-0.044	1.410
$(2p_{3/2}, 2s_{1/2})^{\nu}$	11.773	0.006	-0.055	1.737
$(2p_{1/2}, 2s_{1/2})^{\nu}$	13.556	0.001	-0.015	-1.248
$(2p_{3/2}, 1d_{3/2})^{\nu}$	10.329	0.011	-0.020	0.456
$(2p_{1/2}, 1d_{3/2})^{\nu}$	12.112	0.001	0.018	1.084
$(1f_{5/2}, 1d_{3/2})^{\nu}$	14.072	0.009	0.105	2.675
$(2d_{5/2}, 1f_{7/2})^{\nu}$	12.698	0.001	0.014	1.212
$(1g_{9/2}, 1f_{7/2})^{\nu}$	11.364	0.008	0.149	4.171
Partial Sum		0.082	0.189	
Total Sum		0.083	0.190	

probability.

In table 2 we report the particle-hole configurations which provide the major contributions to the dipole peak. For each configuration we report the unperturbed energy, the contribution A_{ph} to the norm of the state,

$$A_{ph} = |X^{\nu}_{ph}|^2 - |Y^{\nu}_{ph}|^2, \quad \text{with} \quad \sum_{ph} A_{ph} = N_1, \tag{5}$$

the partial contribution b_{ph} to the reduced transition amplitude (see Eq. (4)) and the matrix element of the transition operator. By looking at the the partial contribution b_{ph} to the reduced transitions we can see that strong cancellations of the proton contributions are present while a quite coherent behavior is exhibited by the neutron $1p1h$ configurations. In particular, we observe strong contributions coming from the outermost neutrons. This result, together with the profile of the corresponding transition density indicates that this state shows some features usually associated to pygmy resonances.

In table 3 we compare the total $B(E1)$ and EWSRs integrated up to 10 MeV and the corresponding centroid energies with the experimental values [21] for the ^{48}Ca. We see that the SRPA total $B(E1)$ is much larger, almost by a factor 4, than the experimental value. This is due to the larger number of states obtained in SRPA with respect to the experimental spectrum and, at the same time, to their higher strength (figure 4). The same kind of discrepancies is found for the EWSRs while the SRPA centroid energy is very close to the experimental value. Regarding these large differences some comments are in order. First we recall that, because of the non-local terms of the Skyrme interaction, the double commutator sum rule is

Table 3. Total $B(E1)$ and EWSRs integrated up to 10 MeV and corresponding centroid energies obtained in SRPA compared with the experimental values [21] for the ^{48}Ca isotope.

		^{48}Ca
$\sum B(E1)$ (10^{-3} e^2 fm^2)	SRPA	230
	Exp	68.7 ± 7.5
$\sum_i E_i B_i(E1)$ 10^{-3} e^2 fm^2 MeV	SRPA	1964
	Exp	570 ± 62
$E_{centroid}$ MeV	SRPA	8.54
	Exp	8.40

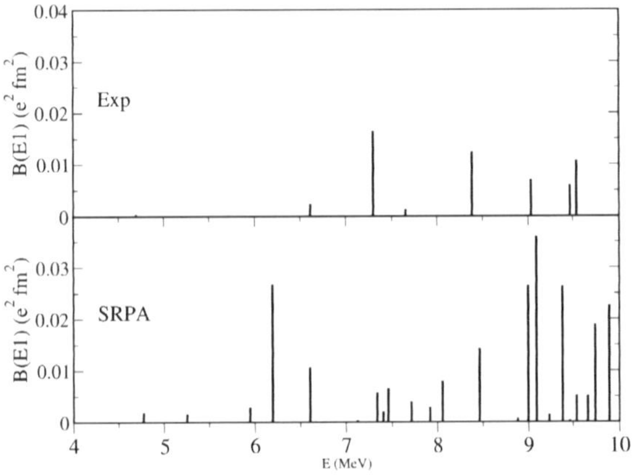

Figure 4. Comparison of the experimental B(E1) strength distribution [21] (upper panel) with the SRPA calculations (lower panel) for ^{48}Ca.

enhanced with respect to the classical sum rule by a factor 1.35 for SGII. However, this is not enough to explain the strong deviation with respect to the experimental values. It could be also interesting to analyze whether or not this kind of discrepancy may depend on the choice of the Skyrme interaction. As a check, we have performed SRPA calculations by using the parametrization SLy4 and the same kind of deviations has been found. In a recent work [24] it has been suggested that the low-lying dipole strength distribution could be be eventually related to the slope of the symmetry energy. This kind of analysis is however beyond our present scopes and it will be performed in future investigations. Finally, we recall that theoretical $B(E1)$ values much larger than the corresponding experimental results have also been found within the ETFFS model for the nucleus ^{44}Ca; it has been shown that this discrepancy is related to the use of some approximations in the treatment of pairing and continuum coupling [13]. We also mention that in previous experimental measurements a much larger strength had been found in the energy region from 5 to 10 MeV [25]. Finally, as mentioned above, since the low-lying dipole states obtained in SRPA have a strong $2p2h$ nature, it could more appropriate the use of a transition operator containing both one-body and two-body terms. Of course, the use of such a more general operator would affect the total strength associated to this energy region. This investigation is however left as a subject for a future work.

4. Conclusions

In this work, we have analyzed the low-energy dipole spectrum (from 5 to 10 MeV) of the ^{48}Ca isotope in the framework of the Skyrme-SRPA model. The Skyrme interaction SGII is used. The distribution and the fragmentation of the peaks is in reasonable agreement with the corresponding experimental measurements. This kind of results cannot be provided by the standard RPA model: SGII-RPA calculations do not lead to any strength in the energy region from 5 to 10 MeV. However, we have found a $B(E1)$ value integrated up to 10 MeV which is quite larger than the corresponding experimental result.

The inclusion of $2p2h$ configurations and their coupling with the $1p1h$ ones and among themselves within SRPA has a two-fold effect on the low-lying dipole strength. From one side the states that already exist in RPA are shifted to lower energies. These states maintain a quite strong $1p1h$ nature. On the other side several other states of almost pure $2p2h$ character appear. They are excited by the one-body dipole operator through their $1p1h$ components.

A detailed analysis of the most collective peak is done, by looking in particular to the corresponding transition densities. A strong mixing of its isoscalar and isovector components has been found showing thus some features generally associated with a dipole pygmy resonance .

References

[1] Paar N, Vretenar D, Khan E and Colò G 2007 *Rep. Prog. Phys.* **70** 691
[2] Catara F, Dasso C H and Vitturi A 1991 *Nucl. Phys. A* **602** 181
[3] Aumann T, *et al.* 1996 *Nucl. Phys. A* **649** 297c ; Leistenschneider A, *et al.* 2001 *Phys. Rev. Lett.* **86** 5442; Tryggestad E, *et al.* 2001 *Nucl. Phys. A* **687** 231c; Tryggestad E, *et al.* 2002 *Phys. Lett. B* **541** 52; Tryggestad E, *et al.* 2003 *Phys. Rev. C* **67** 064309
[4] Péru S, Goutte H and Berger J F 2007 *Nucl. Phys. A* **788** 44
[5] Paar N, Niu Y F, Vretenar D and Meng J 2009 *Phys. Rev. Lett.* 103 032502
[6] Cò G, De Donno V, Maieron C, Anguiano M and Lallena M A 2009 *Phys. Rev. C* **80** 014308
[7] Daoutidis I and Ring P 2011 *Phys. Rev. C* **83** 044303
[8] Martini M, Péru S and Dupuis M 2011 *Phys. Rev. C* **83** 034309
[9] Lanza EG, Catara F , Gambacurta D, Andrés MV and Chomaz P 2009 *Phys. Rev. C* **79** 054615
[10] Vitturi A, Lanza EG, Andres MV, Catara F and Gambacurta D 2010 *Pranama Journal of Physics* **75** 73
[11] Mohan R, Danos M and Biedenharn L C 1971 *Phys. Rev. C* **3** 1740
[12] Dang N D and Arima A 1998, *Phys. Rev. Lett.* **80** 4145
[13] Tertychny G *et al.* 2007 *Nucl. Phys. A* **788** 159c
[14] Pillet N , Berger J F and Caurier E 2008 *Phys. Rev.C* **78** 024305
[15] Andreozzi F, Knapp F, Lo Iudice N , Porrino A and Kvasil J 2007 *Phys. Rev. C* **75** 044312
[16] Papakonstantinou P and Roth R 2010 *Phys. Rev. C* **81** 024317; 2009 *Phys. Lett. B* **671** 356
[17] Gambacurta D , Grasso M and Catara F 2010 *Phys. Rev. C* **81** 054312
[18] Gambacurta D , Grasso M and Catara F 2011 *J. Phys. G: Nucl. and Part. Phys.* **38** 035103
[19] Hartmann T , Enders J , Mohr P , Vogt K , Volz S and Zilges A 2002 *Phys. Rev. C* **65** 034301
[20] Vretenar D , Paar N , Ring P and Lalazissis G A 2001 *Nucl. Phys. A* **692** 496
[21] Hartmann T , Babilon M , Kamerdzhiev S , Litvinova E, Savran D , Volz S and Zilges A 2004, *Phys. Rev. Lett.* **93** 192501
[22] Gambacurta D, Grasso M and Catara F 2011 *Phys. Rev. C* **84**, 034301. .
[23] Moghrabi K , Grasso M , Colò G and Van Giai N 2010 *Phys. Rev. Lett.* **105** 262501
[24] Carbone A, Coló G , Bracco A , Cao L G , Bortignon P F, Camera F and Wieland O 2010 *Phys. Rev. C* **81** 041301(R)
[25] Ottini-Hustache S *et al.* 1999 *Phys. Rev. C* **59** 3429

Contribution of Nuclear Field Theory medium effects to the pairing gap

A Idini[1,2], F Barranco[3], R A Broglia [1,2,4], E Vigezzi [2]

[1] Dipartimento di Fisica, Università degli Studi di Milano, Italy
[2] INFN sezione di Milano, Italy
[3] Departamento de Fisica Aplicada III, Universidad de Sevilla, Spain
[4] The Niels Bohr Institute, University of Copenhagen, Denmark

E-mail: andrea.idini@mi.infn.it

Abstract. We calculate the different contributions to the pairing gap for the case of the superfluid nucleus ^{120}Sn, making use of the Dyson equation (also known as the Nambu-Gor'kov equation in the case of superfluid systems) to propagate leading medium polarization Nuclear Field Theory (NFT) processes to all orders. Starting from a mean field obtained with the SLy4 effective force ($m_k \approx 0.7$), we renormalize the single-particle states and the pairing gap obtained from $^1S_0$ component of the Argonne v_{14} $N-N$ interaction, through the interweaving of single-particle motion and collective vibrations of the system. The results provide a detailed description of the fragmentation of the quasiparticle strength and of the state-dependent pairing gap.

1. Dyson equation as a framework for Self Energy renormalization

Nuclear Field Theory (NFT) provides an exact theoretical framework to describe nuclear structure (see Refs. [1,2] and references therein and within the context of the present contribution, see also Ref.[3]) and reactions [4] in terms of elementary modes of excitation which, in the case of spherical nuclei, correspond to single-particle (quasiparticle) motion and collective vibrations. In keeping with the overcompleteness of the basis (vibrations are built out of the same quasiparticle degrees of freedom as those involved in independent single-particle motion), there is a (coupling) term H_c in the NFT Hamiltonian which depends both on the single-particle and on the collective coordinates. Following the NFT theory rule stating that the collective vibrations are to be calculated within the framework of the RPA, the radial part of the matrix element associated with scattering process which brings a nucleon from a single-particle state of quantum numbers $a \equiv \{n_a, l_a, j_a\}$ to a state b (mean field solutions) and a phonon vibration, can be written as [5]

$$h(a, b\lambda_\nu) = -\frac{i^{l_a - l_b + \lambda}}{\sqrt{4\pi}} \langle j_a \frac{1}{2} \lambda 0 | j_b \frac{1}{2} \rangle \beta_{\lambda_\nu} \langle n_a l_a j_a | R_0 \frac{\partial U}{\partial r} | n_b l_b j_b \rangle, \quad (1)$$

In the above expression $U(r)$ is the mean field, ν labels the various phonons of angular momentum λ and β_{λ_ν} is the (dynamical) deformation parameter associated with the vibration. One can now systematically calculate, making use of (1), the variety of NFT contributions to the matrix elements of H_c, to any selected order of perturbation. Diagonalizing the resulting

Published under licence by IOP Publishing Ltd

matrix gives rise to dressed particles and vibrations (self-energy and vertex corrections), as well as to renormalized interactions.

We first perform a HF+BCS calculation, employing the Argonne v_{14} bare $N-N$ interaction as the pairing force in the $^1S_0$ channel. In this way we obtain a set of quasiparticles with energies E_a and amplitudes u_a, v_a. We then renormalize the quasiparticle properties carrying out the diagonalization of H_c making use of the Dyson equation. Because in superfluid nuclei one has to take into account both the normal and anomalous self-energies namely Σ^{11} and Σ^{12} respectively [6], the Dyson equation becomes a set of two coupled equations known as the Nambu-Gor'kov equations:

$$\left[\begin{pmatrix} E_a & 0 \\ 0 & -E_a \end{pmatrix} + \begin{pmatrix} \Sigma^{11}(a,\tilde{E}_{a(n)}) & \Sigma^{12}(a,\tilde{E}_{a(n)}) \\ \Sigma^{21}(a,\tilde{E}_{a(n)}) & \Sigma^{22}(a,\tilde{E}_{a(n)}) \end{pmatrix}\right] \begin{pmatrix} x_{a(n)} \\ y_{a(n)} \end{pmatrix} = \tilde{E}_{a(n)} \begin{pmatrix} x_{a(n)} \\ y_{a(n)} \end{pmatrix}, \tag{2}$$

where

$$\begin{aligned} \Sigma^{11}(a,\tilde{E}_{a(n)}) &= -\Sigma^{22}(a,-\tilde{E}_{a(n)}) = \\ &= \sum_{b,m,\lambda,\nu}\left[\frac{|h(a,b\lambda_\nu)(u_a\widetilde{u}_{b(m)} - v_a\widetilde{v}_{b(m)})|^2}{\tilde{E}_{a(n)} - \tilde{E}_{b(m)} - \hbar\omega_{\lambda_\nu}} + \frac{|h(a,b\lambda_\nu)(u_a\widetilde{v}_{b(m)} + v_a\widetilde{u}_{b(m)})|^2}{\tilde{E}_{a(n)} + \tilde{E}_{b(m)} + \hbar\omega_{\lambda_\nu}}\right], \end{aligned} \tag{3}$$

and

$$\begin{aligned} \Sigma^{12}(a,\tilde{E}_{a(n)}) &= \Sigma^{21}(a,\tilde{E}_{a(n)}) = \Sigma^{12}(a,-\tilde{E}_{a(n)}) = \\ &= -\sum_{b,m}(u_a\widetilde{u}_{b(m)} - v_a\widetilde{v}_{b(m)})(u_a\widetilde{v}_{b(m)} + v_a\widetilde{u}_{b(m)}) \cdot \\ &\cdot\left[\sum_{\lambda,\nu}\left(\frac{|h(a,b\lambda_\nu)|^2}{\tilde{E}_{a(n)} - (\tilde{E}_{b(m)} + \hbar\omega_{\lambda_\nu})} - \frac{|h(a,b\lambda_\nu)|^2}{\tilde{E}_{a(n)} + (\tilde{E}_{b(m)} + \hbar\omega_{\lambda_\nu})}\right)\right], \end{aligned} \tag{4}$$

from which the amplitudes $x_{a(n)}, y_{a(n)}$ and the renormalized quasiparticle energies $\widetilde{E}_{a(n)}$ can be obtained selfconsistently considering the following relation for the renormalized quasiparticle amplitudes $\widetilde{u}_{a(n)}, \widetilde{v}_{a(n)}$:

$$\widetilde{u}_{a(n)} = [x_{a(n)}u_a - y_{a(n)}v_a], \quad \widetilde{v}_{a(n)} = [x_{a(n)}v_a + y_{a(n)}u_a]. \tag{5}$$

These quasiparticle amplitudes obey the normalization condition

$$1 = x^2_{a(n)} + y^2_{a(n)} - \frac{\partial\Sigma^{11}}{\partial E}x^2_{a(n)} - \frac{\partial\Sigma^{22}}{\partial E}y^2_{a(n)} - 2\frac{\partial\Sigma^{12}}{\partial E}x_{a(n)}y_{a(n)}. \tag{6}$$

The above relation arises from the energy-dependence of Eq. (2), embodied in the energy denominators of Eqs. (3) and (4). Consequently, the quasiparticle strength associated with a given fragment $a(n)$, namely $N_{a(n)} = x^2_{a(n)} + y^2_{a(n)} = \widetilde{u}^2_{a(n)} + \widetilde{v}^2_{a(n)}$, can be smaller than one. In other words, the original HF single-particle strength, after the interweaving with collective vibrations in terms of NFT processes, can become fragmented over a number quasiparticle states.

Making use of these elements one can define, for each quasiparticle fragment $a(n)$, a state-dependent pairing gap (see Ref.[6],[7]) according to

$$\tilde{\Delta}_{a(n)} = Z_{a(n)}\tilde{\Sigma}^{12}(a,\tilde{E}_{a(n)}) = Z_{a(n)}\left(\Delta^{BCS}_a + \tilde{\Sigma}^{12,pho}(a,\tilde{E}_{a(n)})\right) \equiv \tilde{\Delta}^{bare}_{a(n)} + \tilde{\Delta}^{pho}_{a(n)}, \tag{7}$$

where

$$Z_{a(n)} = \left(1 + \frac{\Sigma^{11}(a, \tilde{E}_{a(n)}) + \Sigma^{22}(a, \tilde{E}_{a(n)})}{2\tilde{E}_{a(n)}}\right)^{-1}, \tag{8}$$

provides a measure of the discontinuity of the single-particle strength at the Fermi energy. The term $\tilde{\Sigma}^{12,pho}(a, \tilde{E}_{a(n)})$ is given by

$$\tilde{\Sigma}^{12,pho}(a, \tilde{E}_{a(n)}) = \Sigma^{12}(a, \tilde{E}_{a(n)})(u_a^2 - v_a^2) + u_a v_a(\Sigma^{11}(a, \tilde{E}_{a(n)}) - \Sigma^{22}(a, \tilde{E}_{a(n)}). \tag{9}$$

1.1. Mean Field and QRPA

The mean field and the single-particle levels used in the present contribution were obtained from a HF calculation based on the effective SLy4 interaction [8]. The collective modes have been calculated in QRPA using the gap extracted from the odd-even mass difference ($\Delta \approx 1.4$MeV), together with a separable interaction associated with a surface peaked formfactor given by $R_0 dU/dr$ (see Eq.(1)). The coupling constants have been determined so as to reproduce the experimental energy and transition strength of the low-lying $2^+, 3^-, 4^+$ and 5^- collective surface vibrations. Of course this is not consistent with NFT in keeping with the fact that phonons should also be renormalized to the same order to which particles and interactions are. In other words this should be done through self-energy and vertex corrections (see e.g. [3] and refs. therein). In principle, this would have consequences not only on the properties of the calculated phonons, but also on their ability to renormalize the single-particle strength and the bare $N - N$ interaction.

In keeping with the fact that the work presented here can be considered as a pilot calculation to quantitatively assess the degree of quasiparticle fragmentation resulting from the interweaving of quasiparticle and vibrational states, we have preferred in this first step to use phenomenological phonons, constraining the effective interaction to reproduce the experimental properties of the low-lying collective vibrational states [13] (more precisely, we impose that the experimental value of the polarizability $\beta_{\lambda_1}^2/\hbar\omega_{\lambda_1}$ is reproduced).

We are of course aware of the dangers of treating at different levels of accuracy the renormalization processes of the different elementary modes of excitation entering in NFT, and will remedy these shortcomings in a future publication. Once this will be done, we will have the elements to calculate the absolute single-particle transfer cross sections associated with each quasiparticle fragment, and thus be able to directly compare experimental observations with theoretical predictions, avoiding the question of whether or not one can define spectroscopic factors (see e.g. [9] and refs. therein).

2. Results for the ^{120}Sn nucleus

As mentioned above, the coupling of nucleons to collective vibrations leads to a fragmentation of the quasiparticle states. In most cases and near the Fermi energy, the quasiparticle strength remains concentrated in a single peak carrying typically 70% of the single particle strength. Thus, in such cases, the one-pole (quasiparticle) approximation is reasonably good, as for the $h_{11/2}$ orbital (see figure 1 (a)). It may however happen, that a level becomes strongly fragmented, as in the case of the quasiparticle strength associated with the $d_{5/2}$ orbital (see figure 1 (b)). In such a case the calculation becomes rather dependent on the mean field input. Arguably, this result may reflect the lack of consistency at the NFT level, implied by the use of phenomenological phonons but, most likely, to the fact of not having considered the coupling of single-particle states to vibrational modes associated with $\vec{\sigma} \cdot \vec{\sigma}$, $\vec{\tau} \cdot \vec{\tau}$ and $(\vec{\sigma} \cdot \vec{\sigma})(\vec{\tau} \cdot \vec{\tau})$ (spin, isospin, and spin-isospin) degrees of freedom (see below). If this was the case, one would have a unique opportunity to assess, from a non-conventional point of view, the role of these degrees of freedom in the nuclear spectrum.

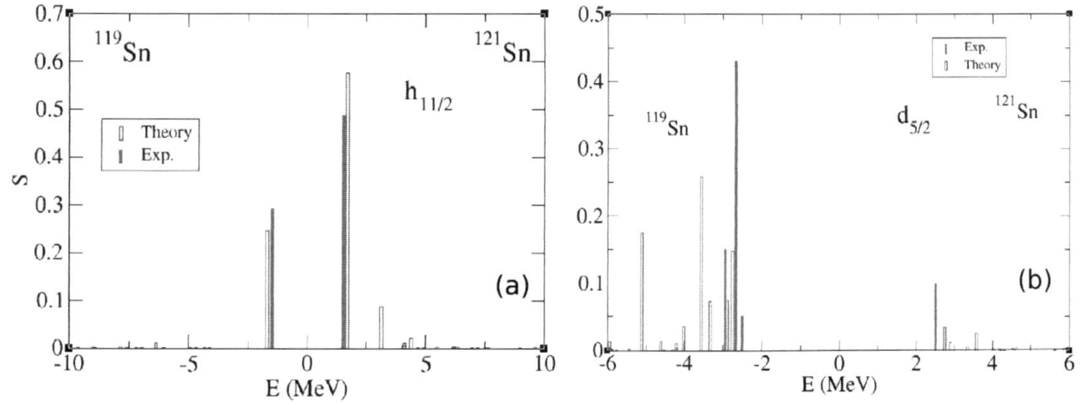

Figure 1. (Color online) Strength distribution of $\tilde{u}_a^2$ and $\tilde{v}_a^2$ for the $h_{11/2}$ (a) and $d_{5/2}$ (b) orbitals resulting from the coupling to collective vibrations compared with available experimental spectroscopic factors [14, 15, 16].

As seen from figure 2, the quasiparticle energy spectrum resulting from the fragments carrying the largest single-particle strength is considerably denser than that obtained from Hartree-Fock plus BCS theory, in overall agreement with the experimental findings.

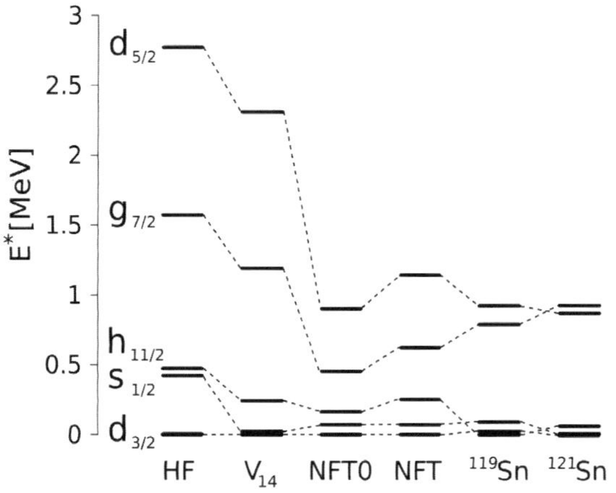

Figure 2. Single-particle spectrum obtained from HF theory with the SLy4 interaction (first column) and used as starting point for solving the BCS equations (second column) with the Argonne v_{14} pairing interaction. In the third and fourth column we give respectively the solution of the Dyson equation which takes into account the particle-vibration coupling at the first iteration and at convergence. The experimental results [14, 15, 16] for the energy of the main quasiparticle peaks of ^{119}Sn and ^{121}Sn are displayed in the last two columns.

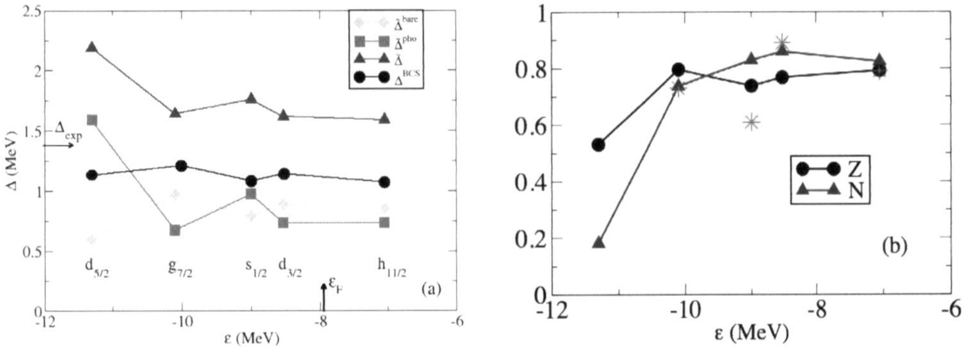

Figure 3. (Color online) (a) State-dependent pairing gap obtained from the BCS calculation performed with the v_{14} residual pairing interaction (circles); this is compared with the induced (squares) and bare (diamonds) contributions to the total renormalized paring gap (triangles) calculated by the Dyson equation. The labels indicate the quantum number of the five valence orbitals of ^{120}Sn. (b) $Z-$ (circles) and $N-$ factors (blue triangles). The red stars report, with due caution, the experimental spectroscopic factors [14, 15, 16].

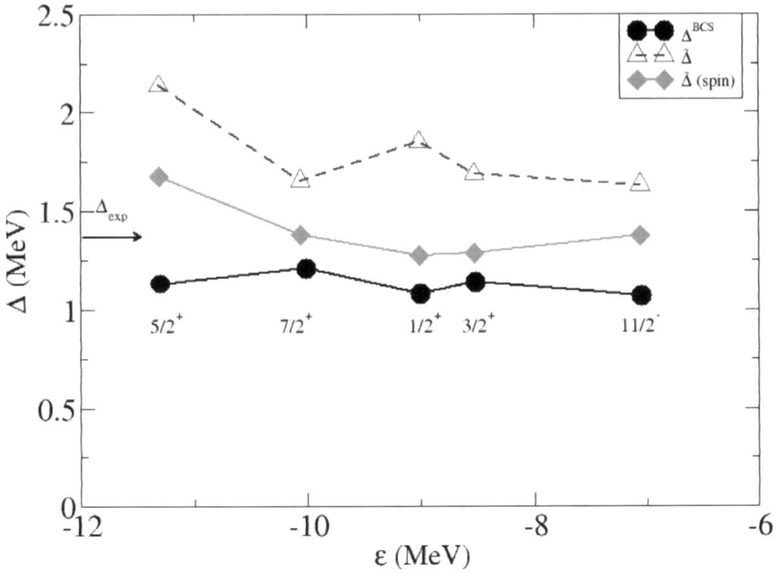

Figure 4. (Color online) Comparison between different contributions to the pairing gap. The circles show the bare BCS-v_{14} pairing gap, while the empty triangles show the gap renormalized by the coupling to S=0 (density) modes, already shown in figure 3(a); the brown diamonds show the gap obtained including also the coupling to S=1 (spin) modes.

Of notice that the pairing gap receives contributions from both the bare interaction and the induced one (phonon-mediated). In the case under discussion the induced and the bare contributions each account for about one half of the total gap, respectively (see figure 3(a)). In the determination of the pairing gap the Z-factor plays a crucial role, since its value is about 0.75 (see figure 3(b)), thus renormalizing Σ^{12} in an important way (see Eq.(1), see also [10, 11]).

Another effect to be considered is the coupling with the spin modes. In fact, it is rather well established in the literature that this coupling represents the leading renormalization effect in uniform nuclear matter, providing a repulsive contribution to the pairing gap. While the effect in finite nuclei seems to be not as crucial as it is in nuclear matter, it has been estimated in [12] that the contribution of spin modes to the induced pairing interaction leads to a reduction by about 30% of the pairing gap obtained renormalizing the bare pairing interaction by only including density fluctuations (see figure 4).

3. Conclusions

The Dyson equation constitutes a powerful framework to propagate to all orders the basic NFT processes which renormalize the motion of nucleons and their interaction. This is particularly important in connection with self-energy processes, which give the distribution of the full single-particle strength and thus of the quasiparticle lifetime, the abnormal component of the self-energy providing information concerning the phonon-induced interaction. Quasiparticle fragmentation and the induced pairing interaction can affect the structure of nuclear Cooper pairs and thus nuclear superfluidity in an important way. Such effects can be specifically probed through two-particle transfer reactions and made quantitative, and thus eventually subject to experimental test, in terms of the associated absolute differential cross sections.

References

[1] Bes D R Broglia R A Dussel G G Liotta R J and Sofia H M 1976 *Nucl. Phys. A* **260** 1
[2] Bortignon P F Broglia R A Bes D R and Liotta R 1977 *Phys. Rep.* **120** 305
[3] Barranco F Broglia R A Colò G Vigezzi E and Bortignon P F 2004 *Eur. Phys. J. A* **21** 57
[4] Broglia R A and Winther A 2005 *Heavy Ion Reactions*, Westview Press, Perseus Books, Boulder, 2nd ed.
[5] Bohr A and Mottelson B R 1975 *Nuclear Structure Vol.II* (New York: Benjamin)
[6] Schrieffer J R 1964 *Theory of Superconductivity* (New York: Benjamin)
[7] Terasaki J Barranco F Vigezzi E and Broglia R A 2002 *Prog. Theor. Phys.* **108** 495
[8] Chabanat E Bonche P Haensel P Meyer J and Schaeffer R 1997 A *Nucl. Phys. A* **627** 710
[9] Jennings BK *Preprint* arXiv:1102.3721
[10] Idini A Barranco F Broglia R A Vigezzi E 2010 *Journ. Phys. Conf.* INPC2010 Proceeding
[11] Idini A Barranco F Vigezzi E 2011 *Preprint* arXiv:1107.0251v1
[12] Gori G Ramponi F Barranco F Bortignon P F Broglia R A Colò G and Vigezzi E 2005 *Phys. Rev. C* **72** 011302
[13] Beer O El Behay A Lopato P Terrien Y Vallois G and Seth K K 1970 *Nucl. Phys. A* **147** 326
[14] Gerlic E Berrier-Ronsin G Duhamel G Galès S Hourani E Langevin-Joliot H Vergnes M and Van de Wiele J 1980 *Phys. Rev. C* **21** 124
[15] Dickey S A Kraushaar JJ Ristinen R and Rumore M A 1982 *Nucl. Phys. A* **377** 137
[16] Bechara M J and Dietzch O 1975 *Phys. Rev. C* **12** 90

Linear response in light deformed nuclei

C Losa[a], **A Pastore**[b], **T Døssing**[c], **E Vigezzi**[e] **and R A Broglia**[c,d,e]

[a] International School for Advanced Studies, SISSA Via Bonomea 265 34136 Trieste, Italy.
[b] Université de Lyon, F-69003 Lyon, France; Université Lyon 1, 43 Bd. du 11 Novembre 1918, F-69622 Villeurbanne cedex, France CNRS-IN2P3, UMR 5822, Institut de Physique Nucleaire de Lyon.
[c] The Niels Bohr Institute, University of Copenhagen, Blegdamsvej 17, 2100 Copenhagen Ø, Denmark.
[d] Dipartimento di Fisica, Università degli Studi di Milano,via Celoria 16, 20133 Milano, Italy.
[e] INFN, Sezione di Milano, via Celoria 16, 20133 Milano, Italy.

Abstract. We develop of a fully consistent Quasparticle Random Phase Approximation (QRPA) method that employs the canonical harmonic oscillator (HO) Hartree–Fock–Bogoliubov (HFB) basis. Skyrme energy density functionals and density–dependent pairing functionals are used for both the HFB mean field and the QRPA approaches. We accurately describe multipole strength functions in axially–symmetric deformed even–even nuclei. Isoscalar and isovector responses in the deformed $^{24-26}$Mg and ^{34}Mg isotopes are presented investigating the consequences of neglecting the spin–orbit and Coulomb residual interactions in QRPA.

1. Introduction

Correlations among particles composing many–body systems may be analyzed in terms of small amplitude oscillations such as RPA and Quasiparticle–RPA (QRPA) in superfluid systems where quasiparticle excitations play a role.

Several self-consistent RPA calculations based on the Skyrme–Hartree–Fock (SHF) method [1, 2] and QRPA including pairing correlations [3, 4, 5, 6, 7, 8] have been performed in spherical nuclei. The QRPA approach has been recently applied to light deformed nuclei (see e.g. [9, 10] and references therein) and heavy deformed nuclei [11, 12, 13]. At the same time new iterative methods have been developed to calculate QRPA strength functions for both spherical [14, 15] and deformed [16, 17, 18] nuclear systems.

Within this context we present a self–consistent HFB+QRPA approach for light deformed nuclei already presented in Ref. [9], discussing the role played by the residual interaction in the response of these nuclear systems to an external electric field.

2. Method

We start by solving HFB equations in a finite canonical harmonic oscillator (HO) basis with $N_{sh} = 15$ shells. See Ref. [9] for details on the adopted method. We take the Skyrme SkM* force together with a density–dependent pairing contact interaction [19]

$$V_{pair}\left(\boldsymbol{r},\boldsymbol{r}'\right) = \frac{1-P_\sigma}{2}\left[V_0 + \frac{V_1}{6}\rho_{00}^{\gamma}\left(\boldsymbol{r}\right)\right]\delta\left(\boldsymbol{r}-\boldsymbol{r}'\right), \tag{1}$$

Published under licence by IOP Publishing Ltd

($\rho_{00}(\boldsymbol{r})$ is the associated isoscalar density and P_σ the spin exchange operator) with quasiparticle energy cutoff $E_{cut} = 50$ MeV and parameters $V_0 = -280$ MeV fm^{-3}, $V_1 = 18.75V_0$, $\gamma = 1$ such that a mixed surfaced–volume type of pairing potential is reproduced. The cutoff parameters adopted in the QRPA calculations on the single–particle energies and on the occupation probabilities are $\varepsilon_{crit} = 200$ MeV and $v_{crit} = 10^{-2}$ respectively.

The strength functions

$$S_J^\tau(E) = \sum_\lambda \sum_\Omega \frac{\Gamma/2}{\pi} \frac{\left|\langle\lambda|\,\hat{F}_{J\Omega}^\tau\,|0\rangle\right|^2}{(E - E_\lambda)^2 + \Gamma^2/4}, \tag{2}$$

are calculated for $\Gamma = 1$ MeV with transition operator

$$\hat{F}_{2\Omega}^{IS} = \frac{eZ}{A} \sum_{i=1}^{A} r_i^2 Y_{2\Omega}(\hat{r}_i), \tag{3}$$

$$\hat{F}_{1\Omega}^{IV} = \frac{eN}{A} \sum_{i=1}^{Z} r_i Y_{1\Omega}(\hat{r}_i) - \frac{eZ}{A} \sum_{i=1}^{N} r_i Y_{1\Omega}(\hat{r}_i), \tag{4}$$

in the IS quadrupole and IV dipole modes, respectively for ^{24,26}Mg and ^{34}Mg.

3. Results

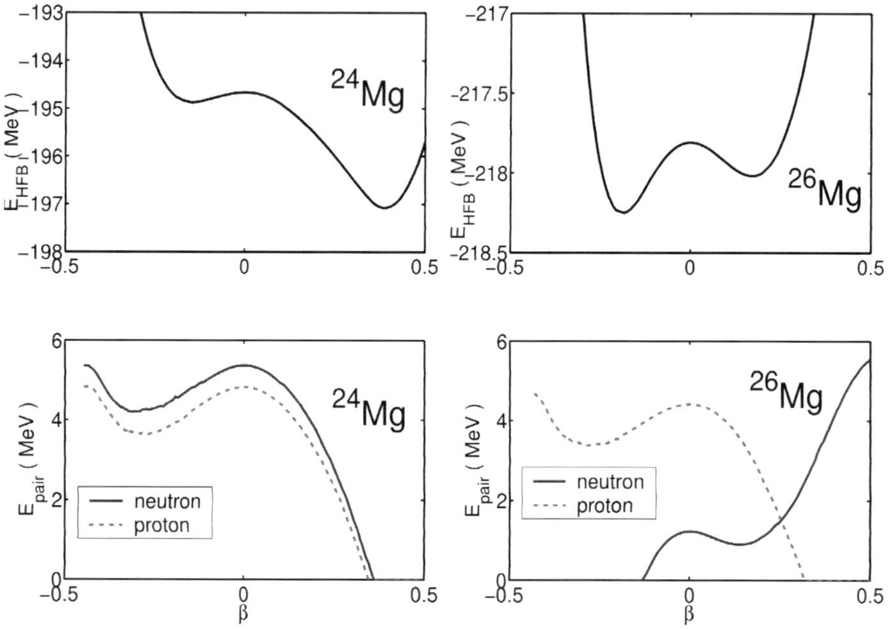

Figure 1. (Color online) HFB potential energy curves E_{HFB} (upper panels), neutron and proton pairing energies E_{pair} (lower panels) as functions of axial deformation parameter β in ^{24}Mg (left–hand panels) and ^{26}Mg (right–hand panels).

3.1. $^{24,26}Mg$

In figure 1 the potential energy curves for ^{24}Mg and ^{26}Mg are plotted as function of the deformation parameter

$$\beta = \sqrt{\frac{\pi}{5}} \frac{< \hat{Q} >_n + < \hat{Q} >_p}{\langle r^2 \rangle_n + \langle r^2 \rangle_p}, \tag{5}$$

$<\hat{Q}>_q$ being the average value of the quadrupole–moment operator $\hat{Q} = 2z^2 - r_\perp^2$ for protons ($q = p$) and neutrons ($q = n$). The QRPA calculations are obtained using the HFB solutions with minimum energy as mean field basis.

In the upper panels of figure 1, ^{24}Mg is found to be prolately deformed with deformation $\beta = 0.39$, while ^{26}Mg presents an oblate deformation at $\beta = -0.18$. The lower panels of figure 1 show that for ^{24}Mg the total pairing energy is zero at $\beta = 0.39$, so QRPA calculations reduce to RPA ones. A full QRPA treatment is needed for ^{26}Mg with HFB ground state having proton correlations.

The upper panels of figure 2 display the fraction EWSR for the IS 2^+ mode in ^{24}Mg and ^{26}Mg. They present a giant resonance in the energy region 15–25 MeV. In ^{24}Mg, the three–peaks structure corresponds to the Ω–splitting, with $\Omega^+ = 0^+$ component along the longest axis at the lowest energy. The Ω–splitting is not pronounced in ^{26}Mg due to the smaller value of the deformation parameter [20]. Because of the oblate shape of ^{26}Mg the peak at the lowest energy belongs to the $\Omega^+ = 2^+$ component. The quadrupole response in ^{24}Mg is compared with the experimental results of Youngblood *et al.* [21]. The shape of the experimental curve is quite well reproduced, however in our calculations the central peak at around 20 MeV is too pronounced and too little strength is found at lower energies. The sharp experimental peak at 15 MeV is of notice and it may be connected to the $\Omega^+ = 0^+$ vibration which is calculated at 17.5 MeV. These differences are apparent from the values of the mean energies

$$E(GR) = \frac{m_1 \left[E_{\min}, E_{\max}\right]}{m_0 \left[E_{\min}, E_{\max}\right]}, \tag{6}$$

with moments

$$m_\lambda \left[E_{\min}, E_{\max}\right] = \sum_k \sum_\Omega E_k^\lambda \left| \langle k | \hat{F}_{J\Omega} | 0 \rangle \right|^2, \tag{7}$$

shown in table 1 calculated in the energy interval $[E_{\min}, E_{\max}] = [9, 41]$ MeV. The theoretical centroid of 19.4 MeV overshoots the experimental mean energy by about 2 MeV. The calculated mean value of the giant resonance is also compared with that obtained by S. Péru *et al.* [22] with the use of a Gogny force. The difference of about 1 MeV may be addressed to the lower effective mass $m^*/m = 0.7$ for the Gogny force, having the values of $m^*/m = 0.8$ for the SkM* force. A calculation performed with a SLy4 force ($m^*/m = 0.7$) shows an increasing of the centroid energy of about 0.6 MeV, indicating that half of the difference between the two theoretical results is due to the effective mass.

The IV dipole fraction of EWSR of ^{24}Mg and ^{26}Mg are shown in the lower panels of figure 2 where they are compared to the experimental results of Irgashev *et al.* [23] and Fultz *et al.* [24] respectively. For ^{24}Mg, both curves display the same two–peaked structure, but the calculated peaks at 16 MeV and 21–22 MeV appear too low in energy of about 3 MeV. This difference is found also in the mean energy (see table 1) extracted from the experimental data in the energy range $[E_{\min}, E_{\max}] = [10, 29]$ MeV and equal to 22.1 MeV, to be compared to the theoretical values of 19.6 MeV. The centroid energy of 23.0 MeV obtained by the Gogny–force calculation [22] approaches more to the observed one. In ^{26}Mg, there is agreement between the calculated curve and the experimental one [24], both of them presenting two maxima at around 18 MeV and 21–22 MeV separated by a minimum at about 20 MeV. Table 1 shows that our calculations differ from experimental centroid energies by 300 keV. Our calculated mean energies differ from those obtained by S. Péru *et al.* [22] by about 2.5 MeV.

In both ^{24}Mg and ^{26}Mg, the effects of the spin–orbit and Coulomb parts of the residual interaction are stronger for the IS 2^+ mode than for the IV dipole response. The quadrupole calculations performed without these terms shift the strength function down in energy by about 0.9 MeV. The IS 2^+ peaks calculated in ^{24}Mg by K. Yoshida *et al.* [25], who neglected these terms, are further lowered in energy of about 0.9 MeV with respect to those obtained by our calculations without spin–orbit and Coulomb terms. This may be due to the different basis used in the two calculations and to the renormalization of the interaction performed in Ref. [25].

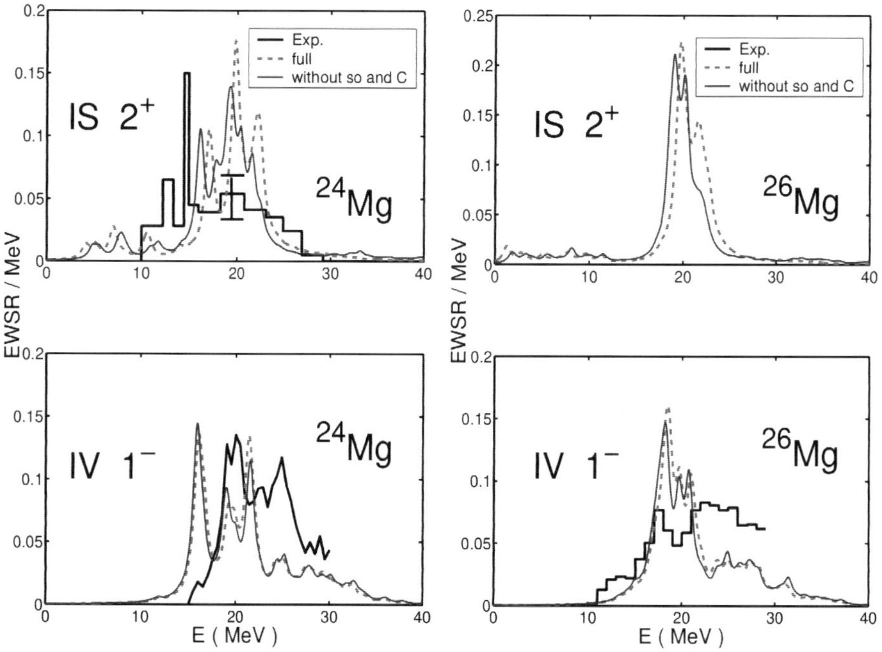

Figure 2. (Color online) Fractions of IS 2^+ (upper panels) and IV 1^- (lower panels) EWSR in ^{24}Mg (left–hand panels) and ^{26}Mg (right–hand panels) without the residual spin–orbit (so) and Coulomb (C) interaction (thin solid curve) and with all the terms included (dashed curve). These are compared with experimental data by Youngblood *et al.*, Irgashev *et al.* and Fultz *et al.* for IS 2^+, IV 1^- in ^{24}Mg and IV 1^- in ^{26}Mg respectively (thick solid curve).

3.2. ^{34}*Mg*

In the upper panel of figure 3 ^{34}Mg is found to be prolately deformed with $\beta = 0.36$. Full QRPA calculations are performed for this nuclear system, with HFB ground state having neutron pairing correlations as shown in the lower panel of figure 3.

The upper panel of figure 4 shows the IS quadrupole response for the $\Omega^\pi = 0^+, 1^+, 2^+$ components. A low–lying peak and a giant resonance are given at 2–3 MeV and 15–22 MeV, respectively. The main contribution to the first low–lying peak belonging to the $\Omega^\pi = 0^+$ excitation, is given by the neutron particle–particle (pp) transitions $[N, n_z, \Lambda, \pm\Omega^\pi]^2 \equiv \left[2, 0, 2, \pm\frac{3}{2}^+\right]^2$, $\left[3, 2, 1, \pm\frac{3}{2}^+\right]^2$, $\left[3, 3, 0, \pm\frac{1}{2}^-\right]^2$, the first of them coming from $\left(1d_{3/2}\right)^2$ the latter

Journal of Physics: Conference Series **336** (2011) 012013 doi:10.1088/1742-6596/336/1/012013

Table 1. Theoretical mean energy values (in MeV) in ^{24,26}Mg obtained with the Gogny force by S. Péru *et al.*, with the SkM* Skyrme force (present work) and experimental energy values by Youngblood *et al.* and Irgashev *et al.* of the IS giant quadrupole resonances (calculated in the energy interval $[E_{\rm min}, E_{\rm max}] = [9, 41]$ MeV), and of the IV giant dipole resonance (calculated in the energy interval $[E_{\rm min}, E_{\rm max}] = [10, 29]$ MeV) in ^{24}Mg. For IV 1^- resonance in ^{26}Mg the experimental mean energy value by Fultz *et al.* is also given.

$E(GR)$ Interval	^{24}Mg $J^\pi = 2^+$ [9,41]	^{24}Mg $J^\pi = 1^-$ [10,29]	^{26}Mg $J^\pi = 2^+$ [9,41]	^{26}Mg $J^\pi = 1^-$ [10,29]
Theor. SkM*	19.4	19.6	20.3	20.3
Theor. SLy4	20.0	19.8		
Theor. D1S (Péru *et al.*)	20.5	23.0	21.0	22.9
Exp. (Youngblood *et al.*)	16.9 ± 0.6			
Exp. (Irgashev *et al.*)		22.1		
Exp. (Fultz *et al.*)				20.6

two from $\left(1f_{7/2}\right)^2$. The data are treated in table 2 in detail and agree with those calculated by K. Yoshida *et al.* (see table I of Ref. [26]). The value of the two–quasiparticle (2qp) transition energies $E_{qpK} + E_{qpK'}$, obtained diagonalizing the HFB Hamiltonian in the canonical basis (cf. Eqs. (4.14b), (4.20) of Ref. [27]) is higher than the energy of the low–lying peak at 2–3 MeV. Thus, the residual interaction shifts down in energy the IS 2^+ strength distribution in the pp channel.

The IV dipole response shown in the lower panel of figure 4 displays a fragmented behavior both in the lower energy peaks of the IV giant dipole resonance at (12–17) MeV belonging to the Ω^π=0^- projection and in the upper ones at (18–25) MeV, belonging to Ω^π=1^-. The general behavior of the resonance displays anyhow the typical two–peak structure due to the Ω^π=$0^-, 1^-$–splitting. The last two columns of table 2 show that the major components contributing to the peak at 15.57 MeV belonging to the Ω^π=0^- component, are the proton particle-hole (ph) and neutron hole-hole (hh) $\left[2, 2, 0, \pm\frac{1}{2}^+\right] \rightarrow \left[3, 3, 0, \pm\frac{1}{2}^-\right]$ transitions, coming from the $1d_{5/2} \rightarrow 1f_{7/2}$ transitions. It is interesting to see that the energy $E_{qpK} + E_{qpK'}$ of the unperturbed 2qp transitions is lower than the energy of the QRPA peak for the proton transitions, viceversa for the neutron transitions. One may conclude that the main contribution of the residual interaction in the ph and hh channels is to shift up and down in energy respectively the IV dipole response.

As seen in Section 3.1 for ^{24,26}Mg, a significant contribution of the spin–orbit and Coulomb terms to the residual interaction is given in the 2^+ modes. In figure 5 the total IS quadrupole strength distribution is displayed. The omission of the spin–orbit and Coulomb residual interaction shifts up in energy of about 0.2 MeV the low–lying peak belonging to the Ω^π=$0^+, 2^+$ components and shifts down in energy the giant resonance by about 0.6 MeV. The general behavior of the response without spin–orbit and Coulomb terms is quite close to the corresponding calculations by K. Yoshida *et al.* [25]. Our giant resonance is shifted up in energy by about 1.3 MeV respect to that in Ref. [25], similarly to the difference discussed in Section 3.1 between the present calculations and those in [25] for ^{24}Mg.

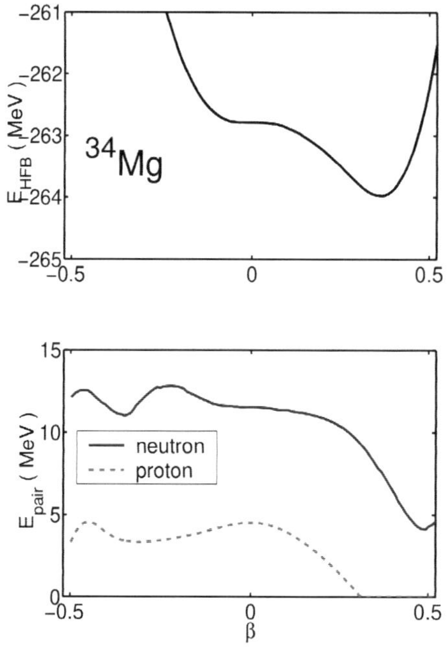

Figure 3. (Color online) HFB potential energy curves E_{HFB} (upper panels), neutron and proton pairing energies E_{pair} (lower panels) as functions of axial deformation parameter β in ^{34}Mg.

4. Conclusions

A fully consistent QRPA method is adopted to calculate the linear response of axially–symmetric–deformed nuclei in the canonical HO HFB basis. We analyzed the IS 2^+ and IV 1^- modes of $^{24-26}$Mg and ^{34}Mg.

For $^{24-26}$Mg we focused on the role of deformation in the energy region of giant resonances. It manifests itself in the splitting between the different projections which is more pronounced for the strongly deformed ^{24}Mg. An other feature of the giant modes is the fragmented behavior, especially of their highest energy projections. In this respect, our calculations confirm earlier theoretical results [22, 25] and are in quite good agreement with experimental data [21, 23, 24]. The inclusion of the spin-orbit and Coulomb parts of the residual interaction only introduces minor changes in the giant resonances in the IS quadrupole modes, shifting the centroid by about 1 MeV and leaving the widths rather unaffected.

In ^{34}Mg we concentrated on the microscopic features of the low-lying states of the IS quadrupole $\Omega^\pi = 0^+$ excitation and of the $\Omega^\pi = 0^-$ component of the IV dipole giant resonance and we found an overall agreement with the theoretical results of K. Yoshida *et al.* [25, 26].

In a general perspective, the present work represents the first, unavoidable step for a consistent and more systematic study of collective modes in nuclei, in particular light exotic nuclei, taking properly into account also medium polarization effects.

References

[1] Hamamoto I, Sagawa H and Zhang X Z 1996 *Phys. Rev. C* **53** 765
[2] Shlomo S and Agrawal B K 2003 *Nucl. Phys. A* **722** 98c
[3] Matsuo M 2001 *Nucl. Phys. A* **696** 371
[4] Hagino K and Sagawa H 2001 *Nucl. Phys. A* **695** 82
[5] Bender M, Dobaczewski J, Engel J and Nazarewicz W 2002 *Phys. Rev. C* **65** 054322

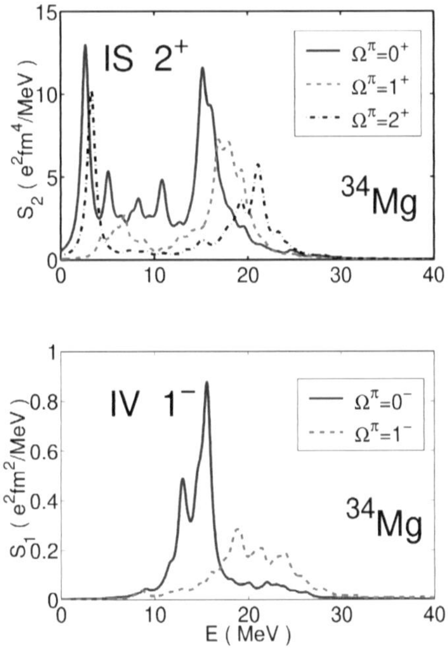

Figure 4. (Color online) IS 2^+ strength functions (upper panel) for the $\Omega^\pi = 0^+$ (solid curve), 1^+ (dashed curve) and 2^+ (dashed–dotted curve) excitations in ^{34}Mg. IV 1^- strength functions (lower panel) for the $\Omega^\pi = 0^-$ (solid curve), 1^- (dashed curve) excitations.

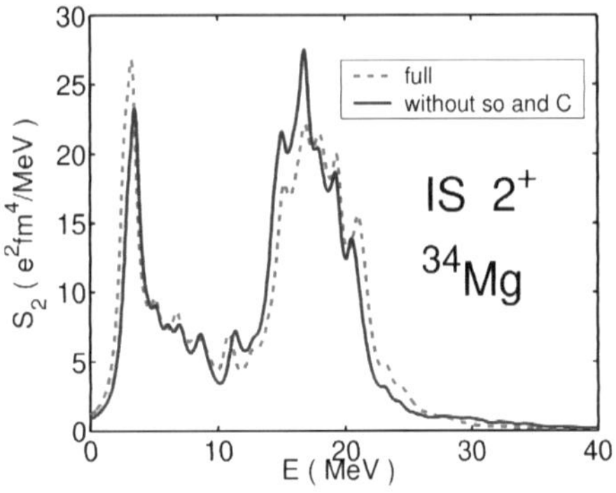

Figure 5. (Color online) IS 2^+ strength functions in ^{34}Mg without the residual spin–orbit (so) and Coulomb (C) interaction (thin solid curve) and with all the terms included (dashed curve).

[6] Khan E, Sandulescu N, Grasso M and Van Giai N 2002 *Phys. Rev. C* **66** 024309
[7] Yamagami M and Van Giai N 2004 *Phys. Rev. C* **69** 034301
[8] Terasaki J, Engel J, Bender M, Dobaczewski J, Nazarewicz W and Stoitsov M 2005 *Phys. Rev. C* **71** 034310
[9] Losa C, Pastore A, Døssing T, Vigezzi E, Broglia R A 2010 *Phys. Rev. C* **81** 064307
[10] Martini M, Péru S and Dupuis M 2011 *arXiv*:1103.1553v1
[11] Péru S, Gosselin G, Martini M, Dupuis M, Hilarie S and Devaux J -C 2011 *Phys. Rev. C* **83** 014314
[12] Terasaki J and Engel J 2011 *arXiv*:1105.3817v1
[13] Yoshida K. 2010 *Phys. Rev. C* **82** 034324
[14] Toivanen J, Carlsson B G, Dobaczewski J, Mizuyama K, Rodríguez-Guzmán R R, Toivanen P, Veselý P 2010 *Phys. Rev. C* **81** 034312

Table 2. QRPA amplitudes $[N, n_z, \Lambda, \pm\Omega^\pi]_K \longrightarrow [N, n_z, \Lambda, \pm\Omega^\pi]_{K'}$ with isospin q ($q = n$ for neutrons, $q = p$ for protons) of the IS quadrupole $\Omega^\pi = 0^+$ mode at 2.64 MeV (2nd–4th columns) and IV dipole $\Omega^\pi = 0^-$ mode at 15.57 MeV (5nd and 6th columns) in ^{34}Mg. The quasi particle occupations v_K^2, the single particle energies ε_K and the 2qp excitation energies $E_{qpK} + E_{qpK'}$, are given. Only components with $X_{KK'}^2 - Y_{KK'}^2 > 0.01$ are listed.

K	$\left[2,0,2,\pm\frac{3}{2}^+\right]$	$\left[3,2,1,\pm\frac{3}{2}^-\right]$	$\left[3,3,0,\pm\frac{1}{2}^-\right]$	$\left[2,2,0,\pm\frac{1}{2}^+\right]$	$\left[2,2,0,\pm\frac{1}{2}^+\right]$
K'	$\left[2,0,2,\pm\frac{3}{2}^+\right]$	$\left[3,2,1,\pm\frac{3}{2}^-\right]$	$\left[3,3,0,\pm\frac{1}{2}^-\right]$	$\left[3,3,0,\pm\frac{1}{2}^-\right]$	$\left[3,3,0,\pm\frac{1}{2}^-\right]$
J^π, Ω^π	2^+, 0^+	2^+, 0^+	2^+, 0^+	1^-, 0^-	1^-, 0^-
E_λ	2.64 MeV	2.64 MeV	2.64 MeV	15.57 MeV	15.57 MeV
q_K	n	n	n	p	n
$q_{K'}$	n	n	n	p	n
v_K^2	0.24	0.70	0.92	0.99	0.99
$v_{K'}^2$	0.24	0.70	0.92	10^{-10}	0.92
ε_K	-3.20	-4.91	-6.84	-23.13	-17.13
$\varepsilon_{K'}$	-3.20	-4.91	-6.84	-11.51	-6.84
$X_{KK'}$	-0.64	0.67	0.27	0.27	0.22
$Y_{KK'}$	-0.09	0.03	0.04	-0.03	-0.01
$E_{qpK} + E_{qpK'}$	3.58	3.49	5.72	10.95	16.05

[15] Avogadro P and Nakatsukasa T 2011 *Phys. Rev. C* **84** 014314
[16] Nakatsukasa T, Inakura T and Yabana K 2007 *Phys. Rev. C* **76** 024318
[17] Inakura T, Nakatsukasa T and Yabana K 2009 *Phys. Rev. C* **80** 044301
[18] Stoitsov M, Kortelainen M, Nakatsukasa T, Losa C and Nazarewicz W 2011 *arXiv:* 1107.3530v1
[19] Chasman R R 1976 *Phys. Rev. C* **14** 1935
[20] Bortignon P F, Bracco A and Broglia R A 1998 *Giant Resonances* (Amsterdam: Harwood Academic Press)
[21] Youngblood D H, Lui Y -W and Clark H L 1999 *Phys. Rev. C* **60** 014304
[22] Péru S and Goutte H 2008 *Phys. Rev. C* **77** 044313
[23] Irgashev K M, Ishkhanov B S and Kapitonov I M 1966 *Phys. Rev. C* **145** 771
[24] Fultz S C, Alvarez R A, Berman B L, Kelly M A, Lasher D R, Phillips T W and McElhinney J 1971 *Phys. Rev. C* **4** 149
[25] Yoshida K and Van Giai N 2008 Phys. Rev. *Phys. Rev. C* **78** 064316
[26] Yoshida K and Yamagami M 2008 *Phys. Rev. C* **77** 044312
[27] Dobaczewski J, Nazarewicz W, Werner T R, Berger J F, Chinn C R and Dechargé J 1996 *Phys. Rev. C* **53** 2809

Journal of Physics: Conference Series **336** (2011) 012014 doi:10.1088/1742-6596/336/1/012014

Recent progress on the accurate determination of the equation of state of neutron and nuclear matter

Paolo Armani, Alexey Yu Illarionov, Diego Lonardoni and Francesco Pederiva

Physics Department, University of Trento, and INFN, Gruppo collegato di Trento, via Sommarive, 14, I-38123 Trento, Italy

Stefano Gandolfi

Theoretical Division, Los Alamos National Laboratory, Los Alamos, NM 87545, USA

Kevin E Schmidt

Department of Physics and Astronomy, Arizona State University, Tempe, AZ 85287, USA

Stefano Fantoni

ANVUR, National Agency for the Evaluation of Universities and Research Institutes, Piazzale Kennedy 20, I-00144 Roma, Italy

E-mail: pederiva@science.unitn.it

Abstract. The problem of accurately determining the equation of state of nuclear and neutron matter at density near and beyond saturation is still an open challenge. In this paper we will review the most recent progress made by means of Quantum Monte Carlo calculations, which are at present the only ab-initio method capable to treat a sufficiently large number of particles to give meaningful estimates depending only on the choice of the nucleon-nucleon interaction. In particular, we will discuss the introduction of density-dependent interactions, the study of the temperature dependence of the equation of state, and the possibility of accurately studying the effect of the onset of hyperons by developing an accurate hyperon-nucleon and hyperon-nucleon-nucleon interaction.

1. Introduction

In the last decade the connection between astrophysical observations and the accurate determination of the inter-nucleon forces has become stricter and stricter. This is partly due to the availability of qualitatively much improved data on radii and masses of neutron stars, which now permit, in principle, to exclude some of the models that have been introduced over time. For instance, the recent discovery of very massive neutron star (with estimated mass $> 2M_\odot$ [1]), and the general agreement on the fact that maximum masses exceed the value $1.4M_\odot$, seem to exclude the hypothesis that at density of order $2\rho_0$, with $\rho_0 = 0.16$ fm^{-3}, a significant fraction of hyperons appears, that, according to existing calculations, would give a maximum mass largely below the currently accepted limit [2, 3]. However, our present knowledge of both many-nucleon

Published under licence by IOP Publishing Ltd

interaction and of the hyperon-nucleon interaction is too scarce to give to these results any prediction meaning. One of the most important shortcomings of our present knowledge lies in the fact that calculations are performed using approximate interaction in conjunction with approximate methods, and this fact essentially prevents to draw definite conclusions.

In this paper we will show some of the recent progress that has been made towards an ab-initio treatment of the many-nucleon problem at high densities, by means of Auxiliary Field Diffusion Monte Carlo (AFDMC) and Fermi Hyper-Netted Chain (FHNC) calculations. In particular, we will focus on the recent rediscovery of the idea of modeling many-body interactions by means of density-dependent potential [4, 5], the inclusion of finite temperature effects, and the preliminary work on the inclusion of three-body hyperon-nucleon-nucleon forces in the treatment of a mixed hyperon/nucleon matter.

2. Auxiliary Field Diffusion Monte Carlo

We briefly recall some notions about the AFDMC method. It is part of a wider class of algorithms implementing a stochastic procedure for projecting an initial trial state of an arbitrary Hamiltonian onto the state of lowest energy belonging to the same subspace in the Hilbert space $\{\phi_n\}$ of the eigenstates of the Hamiltonian itself:

$$\lim_{\tau\to\infty} \Psi^\alpha(R,\tau) = \lim_{\tau\to\infty} \sum_n c_n e^{-\tau(\hat{H}-E_0^\alpha)}\phi_n = c_0^\alpha \phi_0^\alpha. \quad (1)$$

The index α represents a specific symmetry for the state described by a set of quantum numbers, and/or the symmetry imposed by the Pauli principle. The constant E_0^α is a reference energy needed to control the normalization of the propagated states. The propagation is obtained by expanding the initial state on a finite number eigenstates of the position and spin/isospin operators $\{|x\rangle\}$. Each initial point is then propagated by sampling a kernel, which, in general, is approximated by a Trotter-Suzuki break up of the original propagator for small imaginary time-steps:

$$\langle x'|e^{-\Delta\tau\hat{H}}|x\rangle\langle x|\Psi\rangle = \langle x'|e^{-\Delta\tau[\hat{V}-E_0^\alpha]}|x'\rangle\langle x'|e^{-\Delta\tau\hat{T}}|x\rangle\langle x|\Psi\rangle + o(\Delta\tau). \quad (2)$$

The resulting expression for the propagation of a set of configurations of A particles can be summarized in integral form:

$$\Psi(R',\Delta\tau) = \left(\sqrt{\frac{m}{2\pi\hbar^2\Delta\tau}}\right)^{3A} \int dR \exp[-\Delta\tau(V(R')-E_0^\alpha)] \exp\left[-\frac{(R-R')^2}{2\hbar^2/m\Delta\tau}\right]\Psi(R,0). \quad (3)$$

Therefore, the state is propagated by sampling the propagator and applying it to a set of points representative of the initial wave function. The points are iteratively displaced and weighted according to the various terms in the propagator, until a sufficiently long imaginary time is reached. In general the distribution of points is importance sampled, in the sense that it is multiplied by some approximate expression of the wave function of the state searched in order to avoid the large fluctuations in the normalization coming from the possibly divergent behavior of the potential energy. The scheme works flawlessly only if the projected state is the absolute ground state of $\hat{H}$, i.e., if it is a function symmetric under particle exchange. For a many Fermion system, as in the case of nucleons, the calculation is proved to be unstable, in that the variance/average ratio diverges exponentially. In this case it is customary to apply some constraint on the sampled configurations in order to maintain the overlap between the sampled distribution of points and the state onto which we want to project. These procedure are approximate, because they introduce a (usually small) bias on the estimates by effectively changing the Hamiltonian used in the propagation.

An additional source of difficulty is introduced by the fact that, for many-nucleon systems, the potential is usually non-local. Local expressions of the two-body interaction, like the Argonne AV18 potential, are written as a sum over operatorial components:

$$V(r_{ij}) = \sum_{p=1}^{18} v_p(r_{ij}) O_p. \tag{4}$$

For the simple AV6' case, discussed in this paper, the operators considered are:

$$O_p = (1, \tau_i \cdot \tau_j) \otimes (1, \sigma_i \cdot \sigma_j, S_{ij}), \tag{5}$$

where S_{ij} is the tensor operator. From the point of view of the Diffusion Monte Carlo (DMC) algorithm, the use of such potentials imply that the factor $\exp[-\Delta\tau(V(R') - E_0^\alpha)]$ cannot be treated as a local "weight" for the propagated points, but rather behaves as an imaginary time propagator for the spin/isospin degrees of freedom. In the AFDMC algorithm the problem is circumvented by recasting the interaction in a spin/isospin independent part V_{si} and a spin/isospin dependent one V_{sd}. The second is in turn written as a bilinear form in the spin/isospin operators. For simplicity, let us drop the isospin components. The interaction becomes:

$$V = V_{si} + \frac{1}{2} \sum_{i,\gamma,j,\delta} \sigma_{i,\gamma} A_{i,\gamma,j,\delta} \sigma_{j,\delta}. \tag{6}$$

By diagonalizing the matrix A, and indicating as φ_n and λ_n the corresponding eigenvectors and eigenvalues, we have:

$$V = V_{si} + \frac{1}{2} \sum_{n=1}^{3A} \lambda_n O_n^2, \tag{7}$$

where:

$$O_n = \sum_{j,\delta} \varphi_n^{j,\delta} \sigma_{j,\delta}. \tag{8}$$

At this point, in the small $\Delta\tau$ limit, the spin dependent propagator can be recast by using the Hubbard-Stratonovich transformation in order to reduce the operatorial dependence from quadratic to linear:

$$e^{-\Delta\tau \hat{V}_{sd}} \simeq \prod_{n=1}^{3A} e^{-\frac{1}{2}\lambda_n O_n^2 \Delta\tau} = \prod_{n=1}^{3A} \frac{1}{\sqrt{2\pi}} \int dx_n e^{-\frac{x_n^2}{2}} e^{x_n \sqrt{-\lambda_n \Delta\tau} O_n}. \tag{9}$$

The main advantage of this procedure is in avoiding the necessity of a sum over all the components of the wave function corresponding to different two-nucleon states. This number grows exponentially with the number of nucleons, and prevents the straightforward application of Green's Function Monte Carlo (GFMC) techniques for systems larger than $A = 12$ [6]. However, the price to pay is the introduction of a set of auxiliary degrees of freedom by means of which rotation of the spinorial components of the single particle functions are sampled. Treating a sufficiently large number of nucleons in a periodic box is a necessary condition for studying the properties of infinite matter without suffering too heavy finite-size effects. A study of the convergence of the results to the thermodynamic limit can be found in Refs. [7, 8].

The AFDMC algorithm actually in use exploits a number of technicalities that are not reported here, including an extension of the importance sampling concept, and a particular form of constraint to avoid the problem of the exponential growth of the variance. All such details have been largely discussed in the existing literature. For neutron matter and neutron drops the method can be extended to Hamiltonians including spin-orbit (AV8') and three-body interactions [9, 10, 11]. In the case of nuclear matter the method is still limited to the use of the 6 operators mentioned above [12].

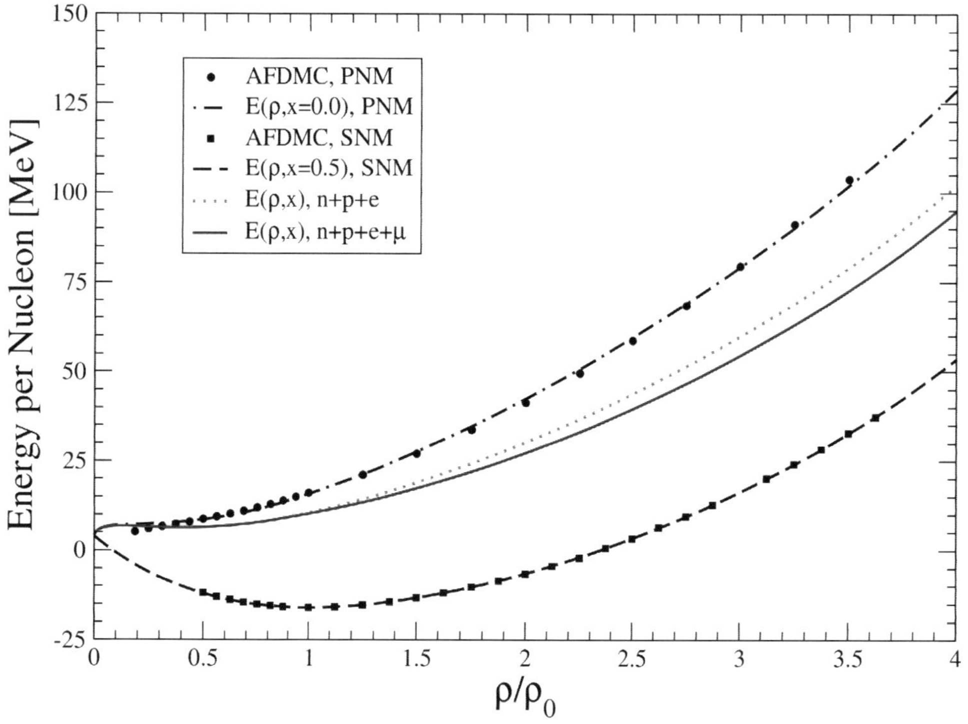

Figure 1. (Color online) Equation of state of symmetric and neutron matter from AFDMC calculations with a density dependent interaction (see legend). The curve computed at β-equilibrium with and without inclusion of muons is also reported.

3. Density dependent interactions

An accurate description of the many-nucleon interaction in nuclear matter should in principle rely on the forces that describe the binding energies of nuclei. However, at present, the simple transposition of the Hamiltonian as is from nuclei to the infinite systems seems not to give satisfactory results. This is obviously due to the difficulty of developing a true ab-initio theory of the many-nucleon interaction. In order to be operative, and provide to astrophysicists some significant estimates of the equation of state, it is possible to implement a two-body potential that effectively includes many-body effects through a density dependence of the coupling constants. Following Lagaris and Pandharipande, the AV6 potential, can be recast by introducing a density dependent intermediate part:

$$v_{ij} = v_\pi + e^{-\gamma_1 \rho} v_I + v_r. \tag{10}$$

In order to reproduce the saturation density and energy, it is then necessary to introduce a purely phenomenological attractive contribution which includes a symmetry term:

$$TNA = \gamma_2 \rho^2 e^{-\gamma_3 \rho} \left[3 - 2 \left(\frac{\rho_n - \rho_p}{\rho} \right) \right] \tag{11}$$

The three parameters are fitted, by means of AFDMC calculations, towards the saturation energy, saturation density and compressibility in nuclear matter [4]. Their values are $\gamma_1 = 0.10$ fm^3, $\gamma_2 = -750$ MeV$\cdot$fm^6, and $\gamma_3 = 13.9$ fm^3 respectively. With these values is then possible to evaluate the energy per nucleon at densities above and below saturation. The resulting equation of state of the Symmetric Nuclear Matter (SNM) can be fitted by the expression:

$$\frac{E_{SNM}(\rho)}{N} = E_0 + b(\rho - \rho_0)^2 + c(\rho - \rho_0)^3 e^{\gamma(\rho - \rho_0)}, \tag{12}$$

with $E_0 = -16.0$ MeV, $\rho_0 = 0.16$ fm^{-3}, $b = 520.0$ MeV$\cdot$fm^6, $c = -1297.4$MeV$\cdot$fm^9, and $\gamma = -2.213$fm^3 [4]. In figure 1 we report the corresponding curve, together with the results of the equation of state of pure neutron matter (PNM) obtained computing the energy per neutron with the same density dependent interaction (DDI). Assuming a quadratic behavior in the proton/neutron number difference for the symmetry energy, it is possible to write an equation of state for a generic proton fraction x_p as follows:

$$E(\rho, x_p) = E_{SNM}(\rho) + C_s \left(\frac{\rho}{\rho_0}\right)^{\gamma_s} (1 - 2x_p)^2, \tag{13}$$

with $C_s = 31.3$MeV, and $\gamma_s = 0.64$. Typical values for these parameters have been quoted as $C_s \approx 31 - 33$ MeV and $\gamma_s \approx 0.55 - 0.69$ by [13] and as $C_s = 31.6$ MeV and $\gamma_s \approx 0.69 - 1.05$ by [14]. It should be noticed that, usually, the symmetry energy is constrained over a range of densities typical of nuclei, whereas here it was fitted the parameters over a very wide density range. This means that the parametrization of Eq. (13) should be accurate up to very high densities.

In this way it is possible, density by density, to compute the equation of state at beta-equilibrium with electrons and muons, by assuming that the chemical potential of the leptons is described by that of an ultra-relativistic Fermi gas. The resulting curves are also reported in figure 1.

4. Finite temperature corrections

In the final stages of a supernova explosion, the dynamics of the resulting proto neutron star is driven by the equation of state at temperatures that initially reach values of about 20 MeV. In astrophysical models the temperature dependence of the equation of state is usually assumed to be very simple. So far, realistic calculations have been performed in the Brueckner-Hartree-Fock framework[15, 16]. In order to have a completely ab-initio description it would be necessary to turn to Path Integral based Monte Carlo methods that allow to compute expectations over the quantum thermal matrix. However, this step implies a number of technical difficulties that have not been faced yet.

An interesting intermediate approach consists of estimating the thermal corrections to the zero temperature EoS by means of of a temperature dependent FHNC calculation[17]. Such corrections can then be applied to the equation of state computed by AFDMC to obtain a result as accurate as possible. Several tests showed that by including the temperature effects using FHNC the contributions of the most important elementary diagrams cancel, and the difference is only weakly affected by the lack of such diagrams.

By extending the variational chain summation to finite temperatures one has to face the so-called orthogonality corrections [17, 18], since wave functions used are not mutually orthogonal. The orthogonality corrections are not unique and, in general, their calculation requires the evaluation of the off-diagonal matrix elements of the Hamiltonian. At present no accurate method exists to efficiently compute such off-diagonal matrix elements. Recently it was proved that the orthogonality corrections to the free energy vanish in the thermodynamic limit [19]. In the spirit of Landau's theory of Fermi liquids, where it is assumed that the low-lying eigenstates of an interacting system have one-to-one correspondence with those of the noninteracting gas, we assume that a good approximation for the eigenstates of the interacting fermion system is

Journal of Physics: Conference Series **336** (2011) 012014 doi:10.1088/1742-6596/336/1/012014

given by the correlated basis states [17]:

$$\Psi_i\left[n_i(\mathbf{k})\right] = \frac{\mathcal{S}\left(\prod_{i<j}\mathcal{F}_{ij}\right)\Phi_i\left[n_i(\mathbf{k})\right]}{\sqrt{\Phi_i^\dagger\left[n_i(\mathbf{k})\right]\left(\mathcal{S}\left(\prod_{i<j}\mathcal{F}_{ij}\right)\right)^2\Phi_i\left[n_i(\mathbf{k})\right]}}, \tag{14}$$

where Φ_i are the single particle states of the non-interacting system . The $n_i(\mathbf{k}) = 0, 1$ are the occupation numbers for single particle states labeled by $\mathbf{k}$. The pair correlation operator defining the correlated wave function $\mathcal{F}_{ij}$ is taken to be

$$\mathcal{F}_{ij} = \sum_{p=1}^{6} f_p(r_{ij})O^p_{ij}\,, \tag{15}$$

where $p = 1-6$ operators are defined as in Eq. (5). Since the operators O^p_{ij} do not commute, the product of correlation operators is symmetrized with the symmetrization operator $\mathcal{S}$ to make the wave function antisymmetric.

The upper bound for the free energy $F(\rho, T)$ can be obtained by using the Gibbs-Bogoliubov variational principle

$$F(\rho, T) \leq F_V(\rho, T) = \mathrm{Tr}\left(\rho_V H\right) - TS_V(\rho, T)\,. \tag{16}$$

ρ_V is any arbitrary density matrix satisfying:

$$\mathrm{Tr}\rho_V = 1, \tag{17}$$

and $S_V(\rho, T) = \mathrm{Tr}\left(\rho_V \ln \rho_V\right)$ is the entropy derived from the density matrix ρ_V at temperature T. The equality in (16) holds when ρ_V is the exact density matrix of the system. In order to obtain ρ_V it is necessary to start from an ansatz for a correlated effective Hamiltonian H_V, such that:

$$\rho_V = \frac{\exp\left(-\beta H_V\right)}{\mathrm{Tr}\exp\left(-\beta H_V\right)}, \tag{18}$$

where $\beta = 1/T$ is the inverse temperature ($k_B = 1$).

In practice H_V is chosen to be a one-body operator such that correlated basis states (14) are eigenstates of it:

$$H_V|\Psi_i\left[n_i(\mathbf{k})\right]\rangle = \left[\sum_{\mathbf{k}} n_i(\mathbf{k})\epsilon_V(\mathbf{k}, \rho, T)\right]|\Psi_i\left[n_i(\mathbf{k})\right]\rangle\,. \tag{19}$$

The eigenvalues of this H_V can be varied by changing the single-particle spectrum $\epsilon_V(\mathbf{k})$, which can be interpreted as the quasi-particle spectrum, and the eigenfunctions by varying the correlation operator $\mathcal{F}_{ij}$.

With this choice of H_V, the calculation of the entropy S_V is trivial. At temperature T the average occupation number of a single-particle state is given by:

$$\bar{n}(\mathbf{k}, \rho, T) = \frac{1}{\exp\left[\beta\left(\epsilon(\mathbf{k}, \rho, T) - \mu(\rho, T)\right)\right] + 1}, \tag{20}$$

where the chemical potential $\mu(\rho, T)$ is required to satisfy the normalization condition:

$$A = \sum_{\mathbf{k}} \bar{n}(\mathbf{k}, \rho, T)\,, \tag{21}$$

where A is the total number of particles in the system, and the entropy is given by:

$$S_V(\rho,T) = -\sum_{\mathbf{k}}\Big[\bar{n}(\mathbf{k},\rho,T)\ln\left(\bar{n}(\mathbf{k},\rho,T)\right) + (1-\bar{n}(\mathbf{k},\rho,T))\ln\left(1-\bar{n}(\mathbf{k},\rho,T)\right)\Big]\,. \tag{22}$$

Since the correlated basis states (14) are not mutually orthogonal, the last equation is only an approximation if the variational Hamiltonian H_V is defined by Eq. (19). It is exact if only orthonormalized correlated basis states are be used.

The calculation of $E_V(\rho,T)/A = \mathrm{Tr}\,(\rho_V H)$ is very similar to the variational calculation of the ground state expectation value E_0 by expanding it in power of $\mathcal{F}^2_{ij}-1$. Schematically, we have:

$$\frac{E_V(\rho,T)}{A} = \frac{\hbar^2 k^2_{\mathrm{av}}}{2m} + \sum \mathrm{diagrams}\big(v,\mathcal{F},\ell(r,\rho,T)\big)\,, \tag{23}$$

where m is the average bare mass of a nucleon and k^2_{av} is a mean square momentum per particle,

$$k^2_{\mathrm{av}} = \frac{1}{A}\sum_{\mathbf{k}}\mathbf{k}^2\bar{n}(\mathbf{k},\rho,T)\,. \tag{24}$$

The diagrams are many-body integrals involving the potential v, the correlation operator $\mathcal{F}$ and the finite temperature Slater function ℓ,

$$\ell(r,\rho,T) = \frac{1}{A}\sum_{\mathbf{k}}\exp(i\mathbf{kr})\bar{n}(\mathbf{k},\rho,T)\,. \tag{25}$$

The single particle spectrum is parametrized as:

$$\epsilon(\mathbf{k},\rho,T) = \frac{\hbar^2 k^2}{2m\Big[1 + A(\rho,T)\exp\big(-B(\rho,T)k^2\big)\Big]}\,. \tag{26}$$

It should be noted that in general it is possible for $\epsilon(\mathbf{k},\rho,T)$ to have higher order terms in k. However, the free energy was found to be insensitive to any such dependence at moderate density range. Under this assumption, if $B(\rho,T)=0$, the spectrum is fully determined by an effective mass:

$$\frac{m^\star(\rho,T)}{m} = \frac{\hbar^2}{m}\left(\frac{1}{k}\frac{d\epsilon}{dk}\right)^{-1} = 1 + A(\rho,T)\,, \tag{27}$$

that can vary with temperature and density.

This scheme has been applied to compute the thermal corrections to SNM and PNM within the FHNC/SOC scheme. The EoS resulting by adding such corrections to the AFDMC result obtained with a DDI is plotted in figure 2. The following functional form provides a good parametrization of the numerical results in the required ranges of density and temperature:

$$\begin{aligned} F(\rho,T,x_p)/A = {} & E(\rho,x_p)/A + \Delta F_0(\rho,T)/A + (1-2x_p)^2\Delta F_S(\rho,T)/A \\ & -\alpha\left(\frac{\rho_0}{\rho}\right)^\beta\left[x_p^{1/3} + (1-x_p)^{1/3}\right]T^2\,, \end{aligned} \tag{28}$$

where α and β are almost independent on the proton fraction x_p. The fit is inspired by the Sommerfeld expansion, and resembles the excitation energy of a hot non-interacting Fermi gas [22]:

$$\begin{aligned} (F-F_0)/A & = -\tfrac{3\pi^2}{8\mu_F}(k_B T)^2\left[x_p^{1/3} + (1-x_p)^{1/3}\right] + O(T^4) \\ & \approx -a_e\left(\tfrac{\rho_0}{\rho}\right)^{2/3}\left[x_p^{1/3} + (1-x_p)^{1/3}\right]T^2\,. \end{aligned} \tag{29}$$

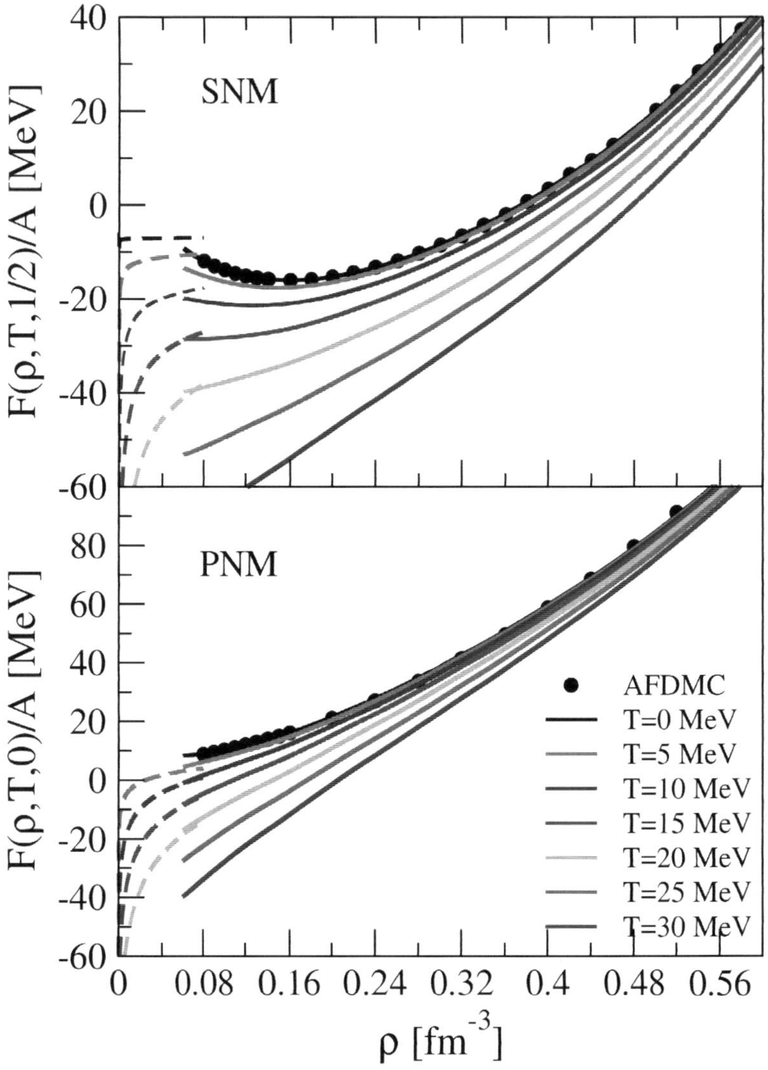

Figure 2. (Color online) Equation of state of Pure Neutron Matter (bottom panel) and Symmetric Nuclear Matter (top panel), computed by adding the estimate of the FHNC free energy estimation to the zero temperature results obtained by means of AFDMC calculations. Results are reported for temperatures ranging from 0 to 30 MeV. The low density PNM EoS is fitted to the virial equation of state of Ref. [20, 21].

For pure neutron matter at the normal density the parameter a_e has the value $3\pi^2/(8\mu_F) = 0.03315\ \mathrm{MeV}^{-1}$.

Other functions, entering the definition of (29), are the following:

$$\Delta F_0(\rho,T)/A = \left[a_T \log\rho + b_T\left(\frac{\rho_0}{\rho}\right)\right]T + \left[c_T \log^2\rho + d_T\left(\frac{\rho_0}{\rho}\right)\right]T^2\,, \quad (30)$$

$$\Delta F_S(\rho,T)/A = e_T\left(\frac{\rho_0}{\rho}\right)T^2\,. \quad (31)$$

The parameters have been fitted against the AFDMC results. Their values are the following:

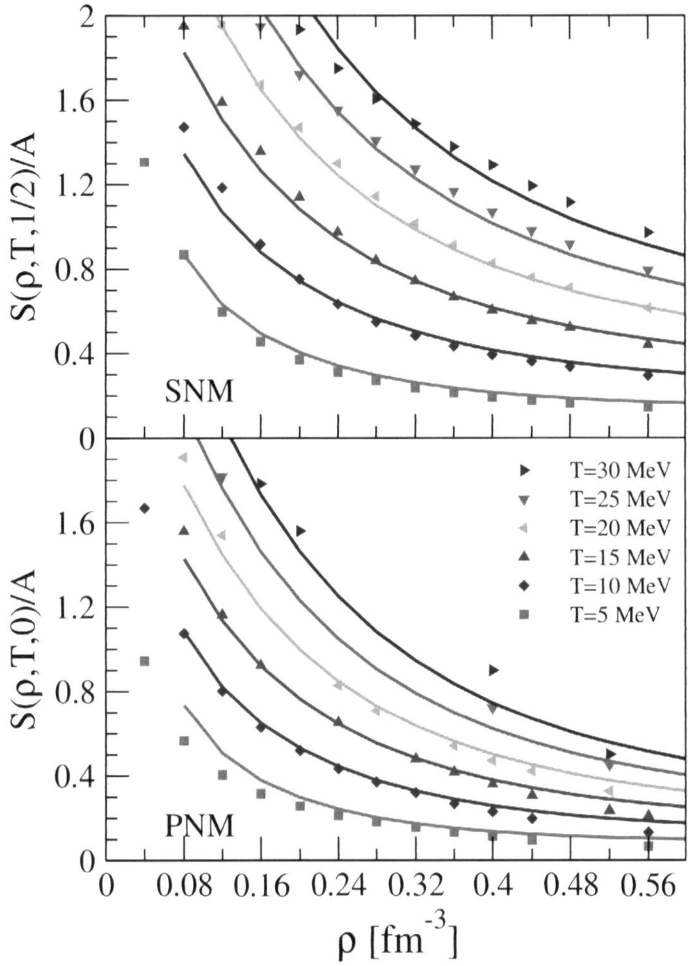

Figure 3. (Color online) Entropy per nucleon in SNM (upper panel) and PNM (lower panel) as a function of the density of the system.

$a_T = -0.15(2)$, $b_T = -0.38(4)$, $c_T = -0.008(1)$, $d_T = 0.06(3)$, $e_T = -0.016(13)$, $\alpha = 0.047(23)$, $\beta = 0.72(14)$ with $\chi^2/n.d.f = 0.54$.

The entropy per nucleon $S(\rho, T, x_p)$, which is a measure of thermal disorder, is calculated from the quasi-particle occupation probabilities $\bar{n}(\mathbf{k}, \rho, T)$ using Eq. (22). It is also possible to compute $S(\rho, T, x_p)$ by means of the following expression:

$$S(\rho, T, x_p) = -\left(\frac{\partial F/A}{\partial T}\right)_V . \qquad (32)$$

The values of the entropy computed by the two different procedures are in excellent agreement. This is a strong test of the quality of the variational calculation of $F(\rho, T, x_p)$ as discussed in [23]. Notice that entropy production in multi-fragmentation events in heavy-ion collisions is a crucial quantity in the determining the mass fragment distribution. The entropy per particle is shown in figure 3 as a function of density and for various temperatures. The entropy increases with temperature, as physically reasonable, and decreases substantially with density. At low T, it is expected to approach a linear dependence due to the fact that, for a Fermi liquid, the relation between S and T should be approximately

$$S \approx \frac{\pi^2}{3\rho} N(T=0) T = \frac{\pi^2 m^\star}{\hbar^2 k_F^2} T \,, \tag{33}$$

in terms of the density of states at the Fermi surface. The specific heat is defined as

$$C_V(\rho, T, x_p) = T \left(\frac{\partial S}{\partial T} \right)_V \,. \tag{34}$$

The energy density $\epsilon(\rho, T, x_p)$, sound velocity $c_s(\rho, T, x_p)$ (in units of c) and adiabatic index $\Gamma(\rho, T, x_p)$ are given by:

$$\epsilon(\rho, T, x_p) = \rho \left(F(\rho, T, x_p)/A + (1 - x_p) m_n c^2 + x_p m_p c^2 \right) \,, \tag{35}$$

$$c_s(\rho, T, x_p) = \sqrt{\frac{\partial P(\epsilon)}{\partial \epsilon}} \,, \tag{36}$$

$$\Gamma(\rho, T, x_p) = \frac{\epsilon}{P} c_s^2 \,. \tag{37}$$

If Γ is constant, then the EOS becomes of the usual polytrope form, $P \sim \epsilon^\Gamma$.

5. Hyperons and hypernuclei

The onset of degrees of freedom with strangeness $S \neq 0$ in nuclei and nuclear matter is a problem of great interest in the study of the properties of dense stars. At densities of about $2\rho_0$, where $\rho_0 = 0.16$ fm^{-3} is the saturation density of the nuclear medium, the chemical potential μ_e of the ultra-relativistic electron gas, determined by the β-equilibrium condition, becomes comparable with the difference between μ_{Σ^-} and μ_n, assuming neutrino-free matter. Under this condition hyperons become stable due to its larger mass and the consequent decrease in kinetic energy. This fact has strong consequences on the EoS, which becomes softer than that predicted by models in which strange degrees of freedom are absent. Present calculations of the equation of state (EOS) of dense matter including hyperons show that this softening leads to an unphysical limitation of the maximum observable mass of a star [2, 3]. This seems to be an indication against the indirect evidence of the occurrence of this mechanism. On the other hand, most calculations neglect important pieces of the interaction, and in particular the hyperon-nucleon-nucleon (YNN) contribution, which, if repulsive on average, might in principle completely change this picture. Due to the fact that the YN interaction must be mediated by at least two pions, the YNN force appears at the same order, and cannot be assumed to be small.

Recently we started an ambitious project, which should lead to a more accurate determination of the YN and YNN interactions combining AFDMC calculations and possibly new available data on separation energies in hyper nuclei. Due to the very limited availability of data on Σ-hypernuclei, it is necessary to focus on the ΛN and ΛNN interactions only. The starting model is that introduced by Usmani [24]. The system under study is a hypernucleus composed by A nucleons interacting through a two body AV6' potential, and one hyperon. We write the Hamiltonian of the system as:

$$H_{N+\Lambda} = H_N + H_\Lambda = -\sum_{i=1}^{A} \frac{\hbar^2}{2m_i} \nabla_i^2 + \sum_{j>i=1}^{A} V^{NN}(i,j) - \frac{\hbar^2}{2m_\Lambda} \nabla_\Lambda^2 + \sum_{i=1}^{A} V^{\Lambda N}(\Lambda, i) + \ldots \tag{38}$$

The explicit form of the Λ-nucleon potential is given in Ref. [24]. It essentially accounts for a two-pion exchange interaction. In principle, it should also include a contribution from Kaon

Table 1. Hyperon separation energy $B(\Lambda)$ as computed by AFDMC simulations for a set of Λ-hypernuclei. Results are reported for Hamiltonians with a two body YN interaction, as described in the text, and for a Hamiltonian including a three-body YNN interaction. Experimental estimates are from Ref. [27]

	YN only	YN+YNN	E(no Λ)	$B(\Lambda)_{YN}$	$B(\Lambda)_{YN+YNN}$	$B(\Lambda)_{exp}$
$^{5}_{\Lambda}$He	-33. 4(1)	-30.4(2)	-26.81(8)	6.6(1)	3.6(2)	3.12
$^{7}_{\Lambda}$He	-	-29.5(3)	-25.05(6)	-	4.46	5.23
$^{9}_{\Lambda}$He	-	-28.3(4)	-24.03(8)	-	4.29	?
$^{17}_{\Lambda}$O	-138.0(4)	-117.3(8)	-105.4(1)	32.6(9)	12.0(9)	(13.5)
$^{18}_{\Lambda}$O	-	-118(1)	-104.5(1)	-	13(1)	?
$^{41}_{\Lambda}$Ca	-326(2)	-293(1)	-279.4(5)	47(1)	13(2)	?

exchange. This term contributes to the tensor components, but it also includes a space exchange term between the nucleon and hyperon degrees of freedom. As assumed, and partially justified, in other works like [25, 26], we assume that the exchange term is quite negligible. The contribution can be considered as effectively included in the hyperon separation energy by the fitted value of the strength of the two–pion exchange term.

The three-body YNN interaction is instead of the general form:

$$V_{YNN} = V^D + V^{2\pi}, \tag{39}$$

where V^D is a dispersive term, an $V^{2\pi}$ is once more a standard two-pion exchange contribution. It does not present any special difficulty in the AFDMC scheme, due to the assumed distinguishability of the Λ with respect to the nucleons. Particular care has to be taken in treating the center of mass contributions.

The separation energy of the Λ particle is defined starting from the difference between the energies of the nuclear systems with and without the Λ hyperon:

$$-B(\Lambda) = \langle H_{N+\Lambda}\rangle_{A+\Lambda} - \langle H_N\rangle_A\,, \tag{40}$$

The hypernucleus wave function is built starting from single particle orbitals computed by HF with a Skyrme I force. A general expression of the mean-field part of the wave function is then:

$$|\Psi_{A+\Lambda}\rangle = \left(\prod_{i=1}^{A} f_\Lambda(r_{\Lambda i})\right)\phi^\Lambda_{nljm}(\mathbf{r}_\Lambda)|\Psi_A\rangle\,, \tag{41}$$

where $\phi^\Lambda_{nljm}(\mathbf{r}_\Lambda)$ is the orbital describing the hyperon. The function $f_\Lambda(r_{\Lambda i})$ is a two-body scalar (Jastrow) correlation between the hyperon and a single nucleon and $|\Psi_A\rangle$ is the correlated wave function describing the remaining A nucleons. In our calculation this function is defined as:

$$|\Psi_A\rangle = \left(\prod_{j>i=1}^{A} f_N(r_{ij})\right)\Phi_{N,Z}(1,\ldots,A)\,, \tag{42}$$

where $\Phi_{N,Z}$ is the Slater determinant of a set of single particle wave functions of N neutrons and Z protons. Obviously, $A = Z + N$.

The AFDMC calculations have been performed for different values of the parameters characterizing the hyperon-nucleon interaction. Despite the estimates of the Λ separation energies are still rather noisy, it was possible to determine a set of values of the potential parameters by which the experimental data are roughly reproduced at the same time for ${}^{5}_{\Lambda}$He and ${}^{17}_{\Lambda}$O. With this set of parameters we gave a preliminary estimate of the separation energies by both including and excluding the YNN term in the Hamiltonian. Notice that we are making the strong assumption that the separation energy, being the difference of two terms, is not strongly influenced by the quality of the nucleon-nucleon Hamiltonian that we employ in our calculations.

In table 1 we report the results obtained in our simulations. As it can be seen, the first indication is that using a YN Hamiltonian only, the separation energy tends to be overestimated. On the other hand, the YNN term seems to give an overall repulsive contribution. This result would confirm the fact that the inclusion of three-body forces is necessary to correctly discuss the onset of hyperons in infinite matter, and their contribution to correcting quantities like the mass-radius relation, which are presently the subject of an open discussion.

Acknowledgments

The authors from University of Trento are also at LISC, Interdisciplinary Laboratory for Computational Science, a joint venture with the Bruno Kessler Foundation. Part of this work was funded by the PAT/INFN *Aurora* project, and by a ISCRA cat. B computing grant. We acknowledge J.C. Miller, B. Gibson and J. Carlson for stimulating discussions and contributions to this work. The work of S.G. is supported by the U.S. Department of Energy, Office of Nuclear Physics, under contracts DE-FC02-07ER41457 (UNEDF SciDAC), and DE-AC52-06NA25396.

References

[1] Demorest P B, Pennucci T, Ransom S M, Roberts M S E and Hessels J W T 2010 *Nature* **467** 1081
[2] Vidaña I, Logoteta D, Providência C, Polls A and Bombaci I 2011 *Europhysics Letters* **941** 11002
[3] Schulze H J and Rijken T 2011 *Phys. Rev. C* **84** 035801
[4] Gandolfi S, Illarionov A Y, Fantoni S, Miller J, Pederiva F and Schmidt K 2010 *Mon. Not. R. Astron. Soc.* **404** L35
[5] Lovato A, Benhar O, Fantoni S, Illarionov A Y and Schmidt K E 2011 *Phys. Rev. C* **83** 054003
[6] Pieper S C 2007 *Preprint* arXiv:0711.1500
[7] Sarsa A, Fantoni S, Schmidt K E and Pederiva F 2003 *Phys. Rev. C* **68** 024308
[8] Gandolfi S, Illarionov A Y, Schmidt K E, Pederiva F and Fantoni S 2009 *Phys. Rev. C* **79** 054005
[9] Pederiva F, Sarsa A, Schmidt K E and Fantoni S 2004 *Nucl. Phys. A* **742** 255
[10] Gandolfi S, Carlson J and Pieper S C 2011 *Phys. Rev. Lett.* **106** 012501
[11] Gandolfi S, Carlson J and Reddy S 2011 *Preprint* arXiv:1101.1921
[12] Gandolfi S, Pederiva F, Fantoni S and Schmidt K E 2007 *Phys. Rev. Lett.* **98** 102503
[13] Shetty D V, Yennello S J and Souliotis G A 2007 *Phys. Rev. C* **76** 024606
[14] Worley A, Krastev P G and Li B A 2008 *Astrophys. J.* **685** 390
[15] Bombaci I, Kuo T T S and Lombardo U 1993 *Phys. Lett. B* **311** 9
[16] Nicotra O E, Baldo M, Burgio G F and Schulze H J 2006 *Phys. Rev. D* **74** 123001
[17] Schmidt K E and Pandharipande V R 1979 *Phys. Lett. B* **87** 11
[18] Fantoni S and Pandharipande V R 1988 *Phys. Rev. C* **37** 1697
[19] Mukherjee A and Pandharipande V R 2007 *Phys. Rev. C* **75** 035802
[20] Shen G, Horowitz C J and Teige S 2010 *Phys. Rev. C* **82** 045802
[21] Horowitz C J and Schwenk A 2006 *Phys. Lett. B* **638** 153
[22] Huang K 1963 *Statistical Mechanics*
[23] Friedman B and Pandharipande V R 1981 *Nucl. Phys. A* **361** 502
[24] Usmani A A 1995 *Phys. Rev. C* **52** 1773
[25] Usmani A A, Pieper S C and Usmani Q N 1995 *Phys. Rev. C* **51** 2347
[26] Usmani Q N and Bodmer A R 1999 *Phys. Rev. C* **60** 055215
[27] Pile P H, Bart S, Chrien R E, Millener D J, Sutter R J, Tsoupas N, Peng J C, Mishra C S, Hungerford E V, Kishimoto T, Tang L G, von Witsch W, Xu Z, Maeda K, Gill D, McCrady R, Quinn B, Seydoux J, Sleight J W, Stearns R L, Plendl H, Rafatian A and Reidy J 1991 *Phys. Rev. Lett.* **66** 2585

Journal of Physics: Conference Series **336** (2011) 012015 doi:10.1088/1742-6596/336/1/012015

Transport properties of β-stable nuclear matter

Arianna Carbone[1] **and Omar Benhar**[2]

[1] Departament d'Estructura i Constituents de la Matèria, Universitat de Barcelona, E-08028 Barcelona, Spain

[2] INFN, Sezione di Roma & Dipartimento di Fisica, "Sapienza" Università di Roma, I-00185 Roma, Italy

E-mail: `ariannac@ecm.ub.es; Omar.Benhar@roma1.infn.it`

Abstract. The transport properties of matter in the interior of rotating neutron stars play a critical role in determining the evolution of these compact objects. In this brief report we discuss a study of the shear viscosity of stellar matter composed by neutrons, protons and electrons in equilibrium with respect to β decay and electronic capture. The aim of this work is calculate in a fully consistent manner the equation of state and the transport properties of nuclear matter, in particular the shear viscosity coefficient, using the same dynamical model.

1. Introduction

Transport properties of neutron star matter characterize in a direct way its evolution. Especially viscosity plays an important role, because it measures the damping effect on the instability caused by the emission of gravitational waves by neutron stars. Hence, the knowledge of this quantity is an essential element to predict the stability of a rotating neutron star.

The CFS mechanism [1, 2, 3] describes this kind of instability. First pointed out by Chandrasekhar in 1970 [1, 2], gravitational radiation is emitted when a non-radial oscillation mode of the star is excited by an internal or external perturbation; this perturbation tends to dissipate in non rotating stars, while, as shown by Friedman and Schutz [3] it drives all perfect fluid rotating stars to instability. Furthermore, the r-modes, modes of rotating stars whose restoring force is the Coriolis force, were defined always unstable due to the emission of gravitational radiation in all perfect fluid rotating stars [4, 5].

If, on the other hand, we consider matter which does not behave as a perfect fluid, viscosity is that dissipative process which opposes the onset of the CFS instability; it appears that viscosity can damp or actually suppress totally the oscillating modes of rotating neutron stars which are driven unstable due to the emission of gravtiational radiation [6].

With this work we calculate the shear viscosity of a three-component, neutrons, protons and electrons, β-stable system [7], combining the Landau-Abrikosov-Khalatnikov formalism [8, 9, 10], to obtain the transport coefficients, with the many-body Correlated Basis Function (CBF) theory [11], necessary to obtain the nucleon-nucleon scattering probability, the proton fraction and the neutron and proton effective masses.

2. Landau theory for normal Fermi liquids

The Landau theory for normal Fermi liquids [8] applies to strongly interacting non superfluid fermionic systems at very low temperatures; it is therefore appropriate for the calculation of the

Published under licence by IOP Publishing Ltd

transport properties of matter interior to neutron stars.

Following the Landau-Abrikosov-Khalatnikov formalism [8, 9, 10], successively expanded by Flowers and Itoh to the case of a fermionic system composed by more then one kind of quasiparticle family [12, 13], the multicomponent Boltzmann-Landau transport equation which has to be solved takes the form

$$\frac{\partial n_\alpha}{\partial t} + \frac{\partial n_\alpha}{\partial \mathbf{r}} \cdot \frac{\partial \epsilon_{\mathbf{p}\alpha}}{\partial \mathbf{p}} - \frac{\partial n_\alpha}{\partial \mathbf{p}} \cdot \frac{\partial \epsilon_{\mathbf{p}\alpha}}{\partial \mathbf{r}} = \sum_\beta I_{\alpha\beta} , \tag{1}$$

where $n_\alpha = n_{\mathbf{p}\alpha}(\mathbf{r}, t)$ denotes the distribution of quasiparticles of type α ($\alpha = n, p, e$), carrying momentum $\mathbf{p}$ and energy $\epsilon_{\mathbf{p}\alpha}$. The term on the right hand side represents the collisions integral involving both like and unlike quasiparticles; in the low temperature limit, the collision term only takes into account processes involving quasipaticles carrying momenta close to the Fermi momenta [12, 13].

Solving the transport equation (1), the total shear viscosity results as the sum of the contributions given by all the quasiparticle families; each contribution of type α reads [12, 13]

$$\eta_\alpha = \frac{1}{5}\rho_\alpha m^\star_\alpha v^2_{F\alpha}\tau_\alpha \frac{2}{\pi^2(1-\ell_{\alpha\alpha})} C(\ell_{\alpha\alpha}) , \tag{2}$$

where ρ_α, $m^\star_\alpha$ and $v_{F\alpha}$ denote the density, effective mass and Fermi velocity, respectively; $\ell_{\alpha\alpha}$ and $C(\ell_{\alpha\alpha})$ are factors given in Ref.[13]; τ_α is the quasiparticle lifetime at temperature T and it is given by the expression [13]

$$\tau_\alpha = \frac{4\pi^4}{m^\star_\alpha T^2 \sum_\beta m^{\star 2}_\beta \langle W_{\alpha\beta}\rangle} . \tag{3}$$

In the above equation, the quantity $W_{\alpha\beta}$ represents the scattering probability between quasiparticles of type α and β. These quantities take into account both strong and electromagnetic interaction. Nuclear interactions contributing to W_{nn}, W_{pp} and W_{np}=W_{pn} have been described within the dynamical model of Ref.[11] using the CBF effective interaction.

The effective interaction $v_{\rm eff}$ described in Ref.[11] is based on a truncated version of the NN potential referred to as Argonne v_{18} [14], providing an excellent fit of deuteron properties and the full Nijmegen phase shift database. It also includes the effects of interactions involving three- and many-nucleon forces, described according to the approach originally proposed in Ref.[15]. The EOS of symmetric nuclear matter and pure neutron matter obtained using the dynamical model of Ref.[11] turn out to be in fairly good agreement with the results of the state-of-the-art calculations of Ref.[16], carried out using the full Argonne v_{18} potential.

The calculations of W_{ep}=W_{pe}, W_{en}=W_{ne} and the electromagnetic part of W_{pp} have been calculated following Refs.[12, 13].

3. Shear viscosity of β-stable nuclear matter

The calculation of the shear viscosity of β-stable nuclear matter requires the proton and electron fraction, which is obtained solving the coupled equations of β-equilibrium and charge neutrality [7]

$$\mu_n - \mu_p = \mu_e \quad , \quad \rho_p = \rho_e , \tag{4}$$

where μ_α denotes the chemical potential of quasiparticles of type α. The proton and neutron chemical potential have been calculated using the single particle spectrum obtained from the CBF effective interaction of Ref.[11].

Figure 1 shows the baryon density-dependence of the proton fraction resulting from the numerical solution of the coupled equations (4). We clearly see that the presence of protons and

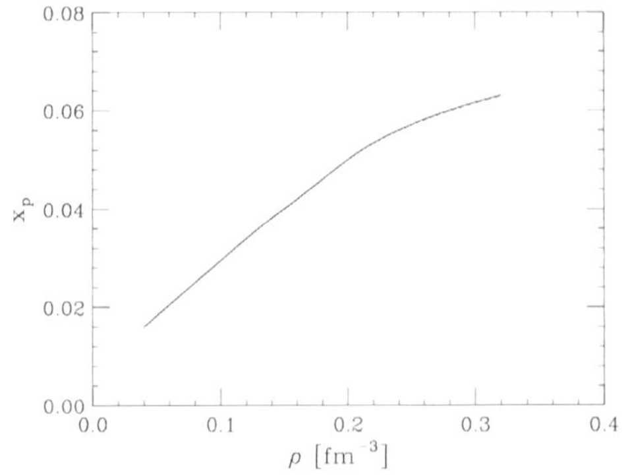

Figure 1. (Color online) Proton fraction as a fucntion of the baryon density of nuclear matter.

electrons is very low and it reaches a maximum value of 6% at double the value of equilibrium density of nuclear matter.

Figure 2 shows the results obtained for the temperature-independent quantities $\eta_\alpha T^2$, as well as ηT^2, plotted as a function of baryon density. It appears that the proton viscosity η_p is always very small, due to their low density and mobility. On the other hand, in spite of the fact that $\rho_p{=}\rho_e$, the electron viscosity η_e turns out to be much higher, as electrons are ultra-relativistic. As expected, the dominant contribution is η_n, as the neutron fraction is larger than 90% over the whole range of baryon density. For $\rho \leq \rho_0$, $\rho_0 = 0.16$ fm^{-3} being the equilibrium density of symmetric nuclear matter, the total viscosity can be identified with the neutron contribution. The electron contribution becomes barely visible only at larger density.

Furthermore, for comparison, we plot in figure 2 the shear viscosity of pure neutron matter calculated using both free-space and in medium collision probabilites [11], which shows how the presence of the medium strongly suppresses quasiparticle scattering probabilities, enhancing their mean lifetime and consequently their contribution to the shear viscosity.

The difference between the shear viscosity of pure neutron matter and β-stable nuclear matter lies in the larger number of channels available in collision processes, in the second case, which in turn lowers the quasiparticle lifetime and furthermore the total shear viscosity.

To underline the importance of the composition of matter in determining the different contributions of the quasiparticle families to the total shear viscosity, we plot in figure 3 the quantities $\eta_\alpha T^2$ and ηT^2 obtained using a proton fraction double the one plotted in figure 1. Note that changing the proton fraction amounts to using a different model of nuclear dynamics to determine the EOS. In this case the electron viscosity exceeds neutron viscosity at values of baryon density just above nuclear matter saturation density.

Table 1. Explanation of data in figure 2 and figure 3.

η of β-stable NM	——
η_n of β-stable NM	- - - -
η_e of β-stable NM	— · —
η_p of β-stable NM	· · · · · ·
η of pure NM with screening effects	◇
η of pure NM without screening effects	x

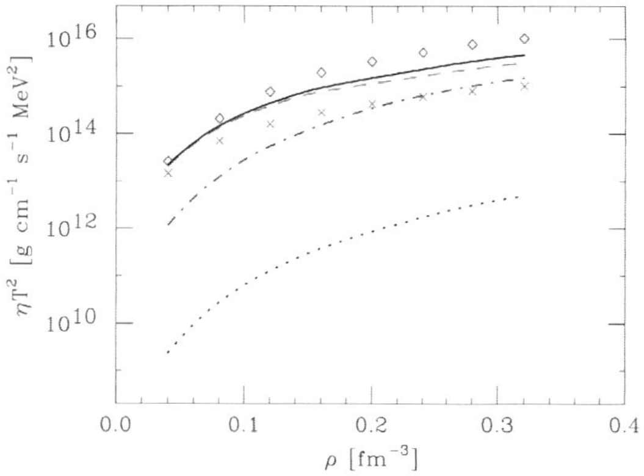

Figure 2. (Color online) Shear viscosity using proton fraction x_p given in figure 1.

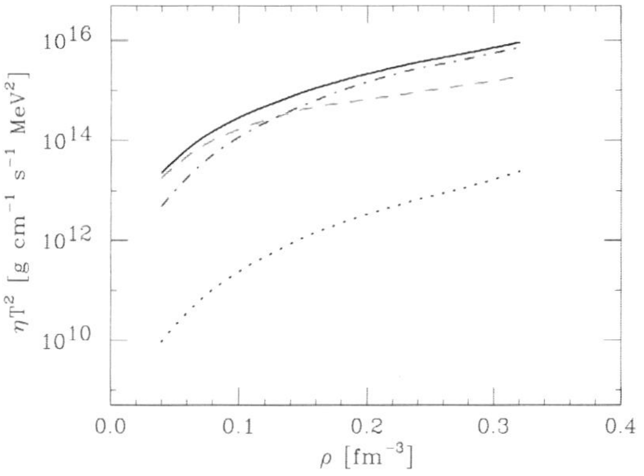

Figure 3. (Color online) Shear viscosity using double the proton fraction x_p given in figure 1.

4. Conclusions

We reported the results for the shear viscosity of β-stable nuclear matter composed by neutrons, protons and electrons, obtained combinig in a consistent manner the Landau-Abrikosov-Khalatnikov formalism with the CBF many-body approach. The effective strong interaction derived in the frame of this approach takes into account the presence of the medium which results in screening effects.

The role played by screening of the bare NN interaction, mainly due to short range correlations, was already pointed out in Ref.[11] for the case of pure neutron matter. In β-stable matter the inclusion of this effect turns out to be critical. The results displayed in figure 2 show that, in the absence of screening, the electron viscosity would become dominant at $\rho \geq 0.2$ fm^{-3}. Overall, due to the availability of a larger number of reaction channels, the viscosity of β-stable matter turns out to be lower than that of pure neutron matter. Furthermore electron viscosity would dominate at $\rho \geq \rho_0$ in the case of dynamical models predicting a larger proton fraction, typically $x_p \sim 8\%$ at nuclear matter equilibrium density.

References

[1] Chandrasekhar S 1970 *Phys. Rev. Lett.* **24** 611
[2] Chandrasekhar S 1970 *Astrophys. J.* **161** 561
[3] Friedman J L and Schutz B F 1978 *Astrophys. J.* **222** 281
[4] Andersson N 1998 *Astrophys. J.* **502** 708
[5] Friedman J L and Morsink S M 1998 *Astrophys. J.* **502** 714
[6] Cutler C and Lindblom L 1987 *Astrophys. J.* **314** 234
[7] Benhar O and Carbone A 2009 (*Preprint* `arXiv:0912.0129v1 [nucl-th]`)
[8] Baym G and Pethick C 1991 *Landau Fermi-Liquid Theory* (New York: John Wiley & Sons)
[9] Abrikosov A A and Khalatnikov I M 1957 *Soviet Phys.* JETP **5** 887
[10] Abrikosov A A and Khalatnikov I M 1959 *Rep. Prog. Phys.* **22** 329
[11] Benhar O and Valli M 2007 *Phys. Rev. Lett.* **99** 231501
[12] Flowers E and Itoh N 1976 *Astrophys. J.* **206** 218
[13] Flowers E and Itoh N 1979 *Astrophys. J.* **230** 847
[14] Wiringa R B, Stoks V G J and Schiavilla R 1995 *Phys. Rev.* C **51** 38
[15] Lagaris I and Pandharipande V R 1981 *Nucl. Phys.* A **359** 349
[16] Akmal A, Pandharipande V R and Ravenhall D G 1998 *Phys. Rev.* C **58** 1804

Density-dependent nucleon-nucleon interaction from Urbana UIX three-nucleon force

Alessandro Lovato[1], Omar Benhar[2,3], Stefano Fantoni[4], Alexey Yu Illarionov[5] and Kevin E Schmidt[6]

[1] SISSA and INFN, Sezione di Trieste, I-34014 Trieste, Italy
[2] INFN, Sezione di Roma, I-00185 Roma, Italy
[3] Dipartimento di Fisica, Università "La Sapienza", I-00185 Roma, Italy
[4] ANVUR, National Agency for the evaluation of Universities and Research Institutes, 20 P.za Kennedy, I-00144 Roma, Italy
[5] Dipartimento di Fisica, Università di Trento, I-38123 Povo, Trento, Italy
[6] Department of Physics, Arizona State University, Tempe, AZ 85287

E-mail: `lovato@sissa.it`

Abstract. A density-dependent two-nucleon potential has been derived within the formalism of correlated basis function. The effects of three-particle interactions has been included by integrating out the degrees of freedom of the third nucleon. The potential can be easily employed in nuclear matter calculations. It yields results in agreement with those obtained from the underlying three-body potential. Using the density dependent potential, we have carried out a study of the effects of three-nucleon interactions in symmetric nuclear matter within the Auxiliary Field Diffusion Monte Carlo (AFDMC) computational scheme.

1. Introduction

The Auxiliary Field Diffusion Monte Carlo (AFDMC) [1] calculations of Ref. [2] yield a binding energy of symmetric nuclear matter (SNM) lower than that predicted by both Fermi-Hyper-Netted-Chain (FHNC) and Brueckner-Hartree-Fock (BHF) calculations [3]. This result suggested that the underbinding of SNM resulting from FHNC calculation with the UIX three body potential [4] may not be avoided with a better choice of the variational wave function.

More refined three body potentials, while providing a quantitative account of the energies of the ground and low-lying excited states of nuclei with $A \leq 12$ [5], give very different results in pure neutron matter (PNM) [6].

Lagaris and Pandharipande [7] and Friedman and Pandharipande [8] proposed an effective, density-dependent, two-nucleon potential (TNI), suitable to take into account three- and many-nucleon forces. However, they adopted a purely phenomenological procedure, as the parameters of the density dependent functions appearing in this potential were determined through a fit of the saturation density, the binding energy per nucleon and the compressibility of symmetric nuclear matter (SNM). In general, fitting the potential to results obtained within many-body techniques makes the potential itself affected by the approximations implied in the many body calculations.

Our potential model [9] improves on the TNI model, since it has been derived from a realistic microscopic three-nucleon force, the UIX potential, which provides a good description of the

Published under licence by IOP Publishing Ltd

properties of light nuclei. It has to be mentioned that, as recently shown [10], the UIX does not reproduce the nd scattering lenght and the vector polarization observables A_y and iT_{11}. However, our approach can be applied to more refined three nucleon interactions, like the chiral NNLOL described in Ref. [10].

To obtain the density dependent potential, the average on the degrees of freedom of the third particle has been carried out using a formalism suitable to account for the full complexity of nuclear dynamics. Our results show that, in doing such reduction, of great importance is the proper inclusion of both dynamical and statistical NN correlations, whose effects on many nuclear observables have been found to be large [11, 12].

We have used Correlated Basis Function (CBF) theory and the Fantoni-Rosati (FR) cluster expansion formalism[13] to perform the calculation of the lowest order (linear-in-density) terms of the effective potential, arising from the irreducible three-nucleon interactions based on the UIX potential.

The effective potential has been implemented in the AFDMC computational scheme to obtain the equation of state (EoS) of SNM. A similar calculation using the UIX potential is not yet feasible, due to the complexities arising from the commutator term. In addition, the density-dependent potential can be used to include the effects of three-nucleon interactions in the calculation of the nucleon-nucleon scattering cross section in the nuclear medium. The knowledge of this quantity is required to obtain a number of nuclear matter properties of astrophysical interest, ranging from the transport coefficients to the neutrino emission rates [14, 15].

2. Two- and three- nucleon forces

The Argonne v_8 [16] two-body potential model is given by

$$\hat{v}_{ij} = \sum_{p=1}^{8} v^p(r_{ij})O^p_{ij} \tag{1}$$

where

$$O^{p=1-8}_{ij} = (1, \sigma_{ij}, S_{ij}, \mathbf{L}_{ij} \cdot \mathbf{S}_{ij}) \otimes (1, \tau_{ij})\,. \tag{2}$$

In the above equation, $\sigma_{ij} = \boldsymbol{\sigma}_i \cdot \boldsymbol{\sigma}_j$ and $\tau_{ij} = \boldsymbol{\tau}_i \cdot \boldsymbol{\tau}_j$, where $\boldsymbol{\sigma}_i$ and $\boldsymbol{\tau}_i$ are Pauli matrices acting on the spin or isospin of the i-th particle, while S_{ij} is the tensor operator, $\mathbf{L}_{ij}$ is the relative angular momentum and $\mathbf{S}_{ij}$ is the total spin of the pair.

We used the so called Argonne v'_8 and Argonne v'_6 potentials, which are not simple truncations of the Argonne v_{18} potential [17], but rather reprojections [18] obtained by refitting the scattering data and the deuteron binding energies. In all light nuclei and nuclear matter calculations the results obtained with the v'_8 are very close to those obtained with the full v_{18}.

Using a nuclear Hamiltonian including only two-nucleon interactions leads to the underbinding of light nuclei and overestimating the equilibrium density of nuclear matter. Hence, the contribution of three-nucleon interactions must be taken into account, by adding to the Hamiltonian the corresponding potential. One of the most widely used is Urbana IX (UIX) [4] potential, consisting of two terms: the Fujita and Miyazawa [19] attractive two-pion exchange interaction $V^{2\pi}$ and the purely phenomenological repulsive term V^R

$$\begin{aligned} \hat{V}^{2\pi} &= A^{2\pi} \sum_{cyclic} \left(\{\hat{X}_{ij}, \hat{X}_{jk}\}\{\tau_{ij}, \tau_{jk}\} + \frac{1}{4}[\hat{X}_{ij}, \hat{X}_{jk}][\tau_{ij}, \tau_{jk}] \right) \\ V^R &= U_0 \sum_{cyclic} T^2(m_\pi r_{ij}) T^2(m_\pi r_{jk})\,. \end{aligned} \tag{3}$$

The spin structure of $V^{2\pi}$ is given by

$$\hat{X}_{ij} = Y(m_\pi r)\sigma_{ij} + T(m_\pi r)S_{ij} \,. \tag{4}$$

The radial functions read

$$Y(x) = \frac{e^{-x}}{x}\xi_Y(x) \quad , \quad T(x) = \Big(1 + \frac{3}{x} + \frac{3}{x^2}\Big)Y(x)\xi_T(x) \tag{5}$$

where $\xi_Y(x) = \xi_T(x) = 1 - \exp(-cx^2)$ are short-range cutoff functions. The parameters $A_{2\pi}$ and U_0 are varied to fit the observed binding energies of ^{3}H and ^{4}He and to reproduce the empirical nuclear matter saturation density, while the cutoff parameter is kept fixed at $c = 2.1$ fm^{-2}.

3. Correlated Basis Function theory and Cluster Expansion technique

In the correlated basis theories of Fermi liquids [20, 21], the expectation value of the two-body potential can be written in the form

$$\langle \hat{v} \rangle = \frac{1}{2}\rho \sum_p \int d\vec{r}_1 d\vec{r}_2 v^p_{12}\, g^p_{12} \,, \tag{6}$$

where

$$g^p_{12} = \frac{A(A-1)}{\rho^2} \frac{\mathrm{Tr}_{12} \int dx_3 \ldots dx_A \Phi^*_0 F^\dagger O^p_{12} F \Phi_0}{\int dX\ \Phi^*_0 F^\dagger F \Phi_0} \,, \tag{7}$$

are the operatorial components of the two–body distribution function. The uncorrelated wavefunction Φ_0 in nuclear matter is conveniently chosen to be a Slater determinant of plane waves.

The structure of the correlation operator $\hat{F}$ reflects the complexity of the Argonne v'_6 nucleon-nucleon potential [22]:

$$F = \mathcal{S} \prod_{j>i=1}^{A} F_{ij} \qquad \text{with} \qquad \hat{F}_{ij} = \sum_{p=1}^{6} f^p(r_{ij}) \hat{O}^p_{ij} \,. \tag{8}$$

The radial functions $f^p(r_{ij})$, appearing in the definition of the correlation operator are determined by the minimization of the energy expectation value $E_V = \langle \Psi_0 | H | \Psi_0 \rangle$, which provides an upper bound to the true ground state energy E_0. As explained in detail in [9], the calculation of E_V in CBF theories is carried out by i) expanding $\langle \Psi_0 | H | \Psi_0 \rangle$ in powers of dynamical correlations $h(r_{ij}) = f^c(r_{ij})^2 - 1$, $2f^c(r_{ij})f^p(r_{ij})$, $f^{p>1}(r_{ij})f^{q>1}(r_{ij})$ that vanish in uncorrelated matter and ii) summing up the main series of the resulting cluster terms by solving a set of coupled integral equations. To accomplish the first of these two steps, an extension of the FR cluster expansion [13], that is able to treat scalar correlations only, was introduced in [22] to deal with spin–isospin dependent correlation operators, like those of Eq. (8).

The cluster terms are most conveniently represented by diagrams [22] consisting of dots (vertices) connected by different kinds of correlation lines. In particular h_{ij} is usually represented by a dashed line, $2f^c_{ij}f^p_{ij}$ by a single wavy line, and $f^{p>1}_{ij} f^{q>1}_{ij}$ by a doubly wavy line. In addition to the dynamical correlation lines, there are also statistical correlation lines, represented by solid oriented lines forming closed loops that never touch each other. They arise from the Slater determinant and are associated with the Slater function ℓ_{ij}

$$\ell(k_F r_{ij}) = 3\Big[\frac{\sin(k_F r_{ij}) - k_F r_{ij} \cos(k_F r_{ij})}{(k_F r_{ij})^3}\Big] \,. \tag{9}$$

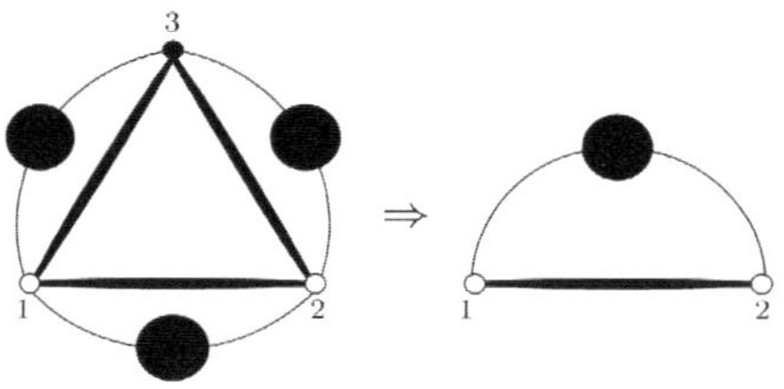

Figure 1. Diagrammatic representation of Eq. (13).

Open dots of the diagrams represent the active (or interacting) particles (1 and 2), while black dots are associated with passive particles, i.e. those in the medium. Integration over the coordinates of the passive particles leads to the appearance of a factor ρ.

It can be shown that all the diagrams contributing to E_V, hence to $\langle \hat{v} \rangle$, are linked. Those built with scalar passive bonds only, with the only exception of the so called *elementary* diagrams, can be summed up in closed form by solving the FHNC equations [13].

On the other hand, diagrams having one or more passive operatorial bonds are calculated at leading order only. This implies that at most two operatorial passive bonds can be attached to any internal point, thus only diagrams with Single Operator Chain (SOC) are considered. Such an approximation is justified by the observation that operatorial correlations are much weaker than the scalar ones.

4. Density dependent potential

As for the two-body potential v_8, it is possible to write the UIX three body potential in the following way

$$V_{123} \equiv \sum_p V^p_{123} O^p_{123} \,. \tag{10}$$

The expectation value of $\hat{V}_{123}$ then reads

$$\frac{\langle V \rangle}{A} = \frac{1}{3!} \rho^2 \sum_P \int dr_{12} dr_{13} V^p_{123}\, g^p_{123} \,, \tag{11}$$

with

$$g^p_{123} = \frac{A!}{(A-3)!} \frac{\mathrm{Tr}_{123} \int dx_4 \dots dx_A \Phi_0^\dagger F^\dagger O^p_{123} F \Phi_0}{\rho^3 \int dX \; \Phi_0^\dagger F^\dagger F \Phi_0} \,. \tag{12}$$

We require that the expectation values of V_{123} and of $v_{12}(\rho)$ be the same, implying in turn

$$\sum_P \frac{\rho}{3} \int d\vec{r}_3 V^P_{123}\, g^P_{123} = \sum_p v^p_{12}(\rho)\, g^p_{12} \,. \tag{13}$$

A diagrammatic representation of the above equation, which should be regarded as the definition of the $v_{12}(\rho)$, is shown in figure 1. The graph on the left-hand side represents the three-body potential times the three-body correlation function, integrated over the coordinates of particle 3. Correlation and exchange lines are schematically depicted with a line having a bubble in the middle, while the thick solid lines represent the three-body potential. The diagram in the right-hand side represents the density-dependent two-body potential, dressed with the two-body distribution function. Obviously, v^ρ_{12} has to include not only the three-body potential, but also the effects of correlation and exchange lines.

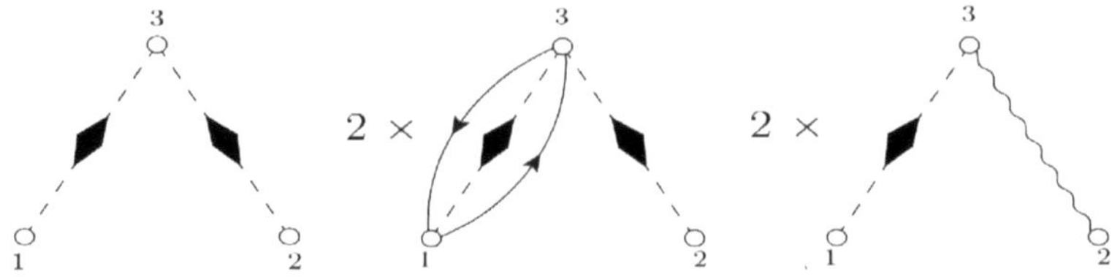

Figure 2. Diagrams contributing to the density-dependent potential. The dashed lines with diamonds represent the first order approximation to $g^{NLO}_{bose}(r_{ij})$, discussed in the text.

In the construction of the density dependent interaction we have found that the most relevant diagrams are those depicted in figure 2 that indeed include both statistical and dynamical correlations. To simplify the pictures, the three-body potential acting on particles 1, 2 and 3 is not explicitely depicted. In order to include higher order cluster terms, we have replaced the scalar correlation line $f^{c\,2}_{ij}$ with the Next to Leading Order (NLO) approximation to the bosonic two-body correlation function:

$$f^{c\,2}_{ij} \rightarrow g^{NLO}_{bose}(r_{ij}) = f^{c\,2}_{ij}\Big(1+\rho\int d\vec{r}_3 h_{13}h_{23}\Big)\,. \tag{14}$$

The inclusion of both statistical and dynamical correlations plays a fondamental role in the determination of a realistic density dependent interaction, as can be seen from figure 3, where different density dependent potentials contributions are plotted and compared with the full calculation with the genuine three-nucleon interaction. In particular $v^{I}_{12}(\rho)$, obtained by averaging over the third particle without statistical and dynamical correlations is far from the solid curve of UIX. When only statistical correlations are considered, the relative potential, $v^{II}_{12}(\rho)$, comes closer to the UIX curve. Only when the full set of diagrams of figure 2 is included in the calculation

$$v^{(III)}_{12}(\rho) = \frac{\rho}{3}\int dx_3\, V_{123}\Big[g^{NLO}_{bose}(r_{13})g^{NLO}_{bose}(r_{23})(1-2P_{13}\ell^2_{13}) + 4g^{NLO}_{bose}(r_{13})f_c(r_{23})\hat{f}(r_{23})\Big]\,, \tag{15}$$

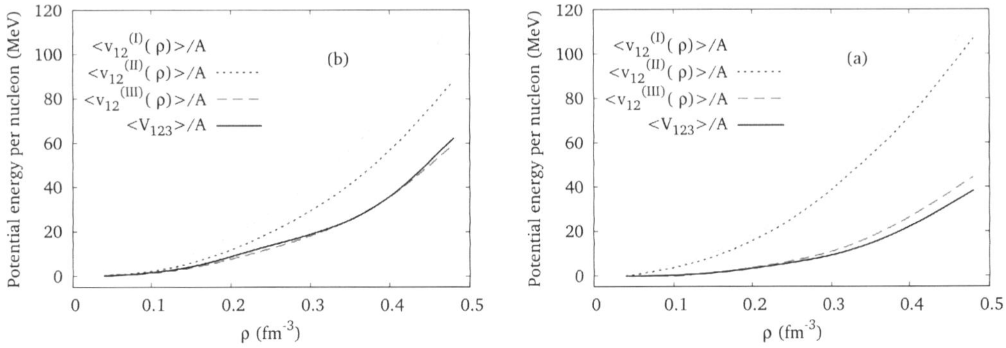

Figure 3. (Color online) Contributions of the density-dependent potential to the energy per particle of SNM (a) and PNM (b), compared to the expectation value of the three-body potential UIX: $\langle V_{123}\rangle/A$.

where $\hat{f}(r_{23})$ denotes the sum of non central correlations, the contribution of the density dependent potential comes very close to that of the original UIX. A proper treatment of the statistical correlation is crucial for the construction of a density dependent potential to be treated as an additive term to the standard two-body potential. Some care is needed for the density dependent potential to reproduce the exchange loop involving particles 1, 2 and 3 with the appropriate symmetry factor (see [9] for a thorough discussion).

5. Numerical results: PNM and SNM equation of state

As explained in [9], the correlation functions are determined by minimizing the two-body cluster term of the energy expectation value for a set of parameters d_c, d_t, β_p and α_p. The energy expectation value E_V calculated in full FHNC/SOC approximation has then to be minimized with respect to the variation of these parameters. To this aim a Simulated annealing [23] optimization algorithm has been implemented [9].

Both SOC approximation and the fact that elementary diagrams are neglected may lead to a violation of the variational principle. To keep this effect under control, an additional constraint on the kinetic energy with respect to the optimization performed in [24] has been considered.

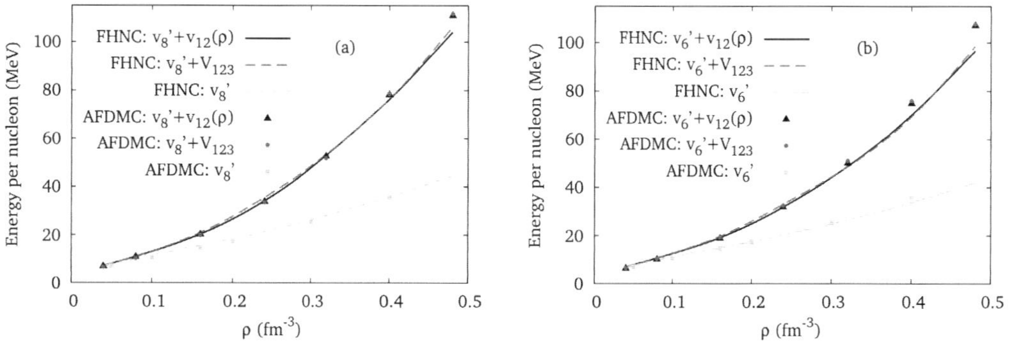

Figure 4. (Color online) Energy per particle for PNM, obtained using the density-dependent potential added to the Argonne v_8' (a) and to Argonne v_6' (b) potentials. The energies are compared to those obtained from the genuine three-body potential and from the two-body potentials alone.

We carried out AFDMC simulations [1] for PNM with $A = 66$ and SNM with $A = 28$ nucleons in a periodic box. The finite-size errors in PNM simulations have been investigated in [25] by comparing the Twist Averaged Boundary Conditions (TABC) with the Periodic Box Condition (PBC). The kinetic energy of 66 fermions approaches the thermodynamic limit very well, so that the energies of 66 neutrons computed using either TABC or PBC turn out to be almost the same. We can infer that the finite-size errors in the present AFDMC calculations for PNM do not exceed 2% of the asymptotic value of the energy. The finite-size effects of SNM calculations can be estimated from the difference of the energies of PNM obtained with 14 neutrons and the TABC asymptotic value, which is of the order of 7%.

FHNC/SOC and Monte Carlo calculations provide very close results, as shown in figures 4 and 5, to be compared with those of Ref. [2] where the agreement between FHNC and Monte Carlo methods were not nearly as good.

The EoS of PNM, displayed in figure 4, obtained with the three-body potential UIX and using the density-dependent two-body potential are very close to each other. In the same figure, for the sake of comparison, we also report the results of calculations carried out including the two–body potential only.

Despite cluster contributions proportional to ρ^2 have been neglected in the density dependent potential, with the exception of the line with diamonds of figure 2, even at high densities the EoS relative to $v_{12}(\rho)$ remain close to those obtained with UIX. Probably, in this case a compensation among second and higher order terms takes place.

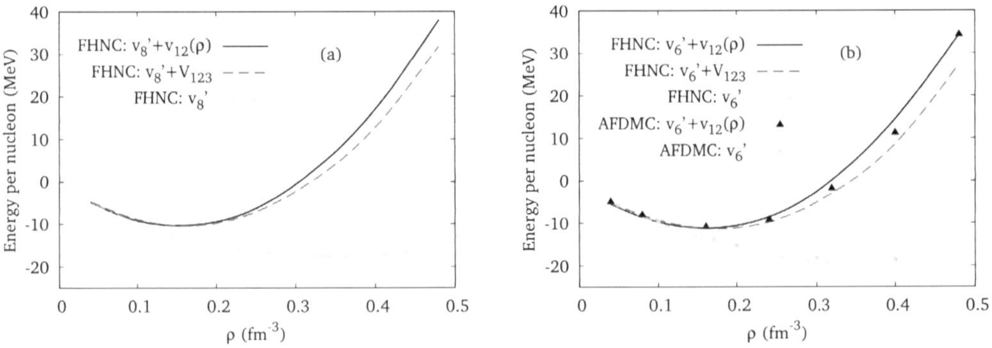

Figure 5. (Color online) Same as in figure 4, but for SNM.

The density-dependent potential has been also employed in AFDMC calculations. As can be seen in figure 4, the triangles representing the results of this calculation are very close, when not superimposed, to the circles corresponding to the UIX three-body potential AFDMC results.

For what concern the EoS of symmetric nuclear matter, see figure 5, at densities lower than $\rho = 0.32\,\mathrm{fm}^{-3}$, the curves resulting from UIX and the density-dependent potential are very close to one other, while for $\rho > 0.32\,\mathrm{fm}^{-3}$ a gap between them appears.

Table 1. Values for the saturation densities, the binding energy per particle, and the compressibility of SNM relative to the EoS of figure 5.

FHNC/SOC	$v_6' + V_{123}$	$v_6' + v(\rho)$	$v_8' + V_{123}$	$v_8' + v(\rho)$
ρ_0 (fm^{-3})	0.17	0.16	0.16	0.15
E_0 (MeV)	-11.3	-11.2	-10.3	-10.3
K (MeV)	205	192	189	198

AFDMC	$v_6' + v(\rho)$
ρ_0 (fm^{-3})	0.17
E_0 (MeV)	-10.9
K (MeV)	201

In table 1, the saturation density ρ_0, the binding energy per particle $E(\rho_0)$ and the compressibility $K = 9\rho_0(\partial E(\rho)/\partial\rho)^2$ for all the EoS of figure 5 are listed. It is remarkable that the values obtained with density dependent potential are very close to those resulting from the genuine UIX three body potential. While the saturation density is well reproduced, this being not surprising since the constant U_0 of the UIX potential has been fitted to this value, the values of the compressibility are slightly lower than the experimental ones. AFDMC calculations lower the variational results for E_0 by 0.6 MeV only, showing that the binding energy overstimation of the FHNC/SOC calculations performed with UIX potential can not be solved with a better choice of the variational wavefunction.

Journal of Physics: Conference Series **336** (2011) 012016 doi:10.1088/1742-6596/336/1/012016

6. Conclusions

We have developed a novel scheme, suitable to obtain an effective density-dependent NN potential taking into account the effects of three-nucleon interactions. Our approach is fully consistent with the treatment of correlations underlying the FHNC and AFDMC approaches.

The PNM and SNM equation of state resulting from the density-dependent potential turn out to be very close to those obtained with the UIX three-body potential. In this context, a critical role is played by the treatment of both dynamical and statistical correlations. This is a distinctive feature of our approach, as compared to different reduction schemes based on effective interactions [26, 27].

A AFDMC calculation of the equation of state of SNM consistently including the effects of three nucleon forces has been carried out for the first time,. The results of this calculation show that the v_6'+UIX hamiltonian, or equivalently the one including the effective potential, fails to reproduce the empirical data. The discrepancy has most likely to be ascribed either to deficiencies of the UIX model or to the effect of interactions involving more than three nucleons.

As a further development, we are studying the dependence on the specific model of three-nucleon force [28], as well as the inclusion of four- and many-nucleon interactions, whose effects are expected to be critical for the determination of the properties of high density neutron star matter .

As a final remark, we note that our density dependent potential could be easily employed in many-body approaches other than those based on the CBF formalism or quantum Monte Carlo simulations, such as the G-matrix and self-consistent Green function theories [29, 30, 31].

References

[1] Schmidt K E and Fantoni S 1999 *Phys. Lett. B* **446** 99
[2] Gandolfi S, Pederiva F, Fantoni S and Schmidt K E 2007 *Phys. Rev. Lett.* **98** 102503
[3] Bombaci I, Fabrocini A, Polls A and Vidanã I 2005 *Phys. Lett. B* **609** 232
[4] Pudliner B S , Pandharipande V R , Carlson J and Wiringa R B 1995 *Phys. Rev. Lett.* **74** 4396
[5] Pieper S C 2008 *Rivista del Nuovo Cimento* **31** 709
[6] Sarsa A, Fantoni S, Schmidt K E and Pederiva F 2003 *Phys. Rev. C* **68** 024308
[7] Lagaris I and Pandharipande V R 1981 *Nucl. Phys.* **A359** 349
[8] Friedman B and Pandharipande V R 1981 *Nucl. Phys.* **A361** 502
[9] Lovato A, Benhar O, Fantoni S, Illarionov A Yu and Schmidt K E 2011 *Phys. Rev. C***83** 054003
[10] Kievsky A, Viviani M, Girlanda L and Marcucci L E 2010 *Phys. Rev. C* **81** 044003
[11] Benhar O, Pandharipande V R and Pieper S C 1993 *Rev. Mod. Phys.* **65** 817
[12] Pandharipande V R, Sick I , DeWitt Huberts P K A 961 *Rev. Mod. Phys.* **69** 961
[13] Fantoni S and Rosati S 1974 *Nuovo Cimento* **A20** 179
[14] Benhar O and Valli M 2007 *Phys. Rev. Lett.* **99** 232501
[15] Benhar O, Farina N, Fiorilla S and Valli M 2008 *AIP Conf. Proc.* **1056** 248
[16] Pudliner B S, Pandharipande V R, Carlson J, Pieper S C and Wiringa R B 1997 *Phys. Rev. C* **56** 1720
[17] Wiringa R B, Stokes V G J, Schiavilla R 1995 *Phys. Rev. C* **51** 38
[18] Wiringa R B, Pieper S C 2002 *Phys. Rev. Lett.* **89** 182501
[19] Fujita J and Miyazawa H 1957 *Prog. Theor. Phys.* **17** 360
[20] Clark J W 1979 *Prog. Part. Nucl. Phys.* **2** 89
[21] Fantoni S and Pandharipande V R 1988 *Phys. Rev. C* **37** 37
[22] Pandharipande V R and Wiringa R B 1981 *Rev. Mod. Phys.* **51** 821
[23] Kirkpatrick S, Gelatt C D and Vecchi M P,*Science New Series* **220** 671
[24] Wiringa R B, Fiks V and Fabrocini A *Phys. Rev. C* **38** 1010
[25] Gandolfi S, Illarionov A Yu, Schmidt K E, Pederiva F and Fantoni S 2009, *Phys. Rev. C* **79** 054005
[26] Hebeler K and Schwenk A 2010 *Phys. Rev. C* **82** 014314
[27] Holt J W, Kaiser N and Weise W 2010 *Phys. Rev. C* **81** 024002
[28] Lovato A, Benhar O, Fantoni S and Schmidt K E, in preparation
[29] Baldo M and Burgio F 2001 *Physics of Neutron Stars Interior.* Eds. D Blaschke, N. K. Glendenning and A. Sedrakian *Lecture Notes in Physics* **578** 1
[30] Rios A, Polls A and Vidaña I 2009 *Phys. Rev. C* **79** 025802
[31] Dickhoff W H and Barbieri C 2004 *Prog. Part. Nucl. Phys.* **52** 377

Transport properties of the hot quark-gluon plasma

Vincenzo Greco

Department of Physics and Astronomy, V S Sofia 64, I-95125 Catania, IT
INFN Laboratori Nazionali del Sud, V S Sofia 62, I-95125 Catania, IT

E-mail: greco@lns.infn.it

Abstract. A phase where quark and gluon are the relevant degrees of freedom is expected for nuclear matter at energy density $\epsilon \geq 1$ GeV/fm^3 and a temperature $T > 160$ MeV. A transient state state of such a matter can be created by mean of ultra-relativistic heavy-ion collisions. We briefly overview some main results on the properties of the quark-gluon plasma emphasizing the necessity to develop a transport theory for quarks and gluons able to incorporate the main developments in lattice QCD and perturbative QCD. First results show that Boltzmann-Vlasov transport theory correctly predict the elliptic flow observed at both RHIC and LHC energies.

1. Introduction

The study of the fundamental theory of strong interactions, Quantum Chromo Dynamics (QCD), under extreme conditions of temperature and density has been one of the most challenging problems in physics during the last 20 years, capturing increasing experimental and theoretical attention. There are several reasons underlying such a vivid interest. QCD is a quantum field theory with an extremely rich dynamical content (asymptotic freedom, confinement, chiral symmetry, nontrivial vacuum ...). Besides Heavy-Ion Collisions at ultrarelativistic energies ($\sqrt{s} > 10$ AGeV) provide the unique possibility to create a transient state of matter at energy densities and temperatures similar to those of the Universe in the first $10^{-6} - 10^{-5}\, s$ after the Big Bang. Furthermore the most dramatic event during the first second was the quark-to-hadron phase transition associated to a reduction in the number of degrees of freedom by about a factor three [1]. Finally, in recent years the discovery of a duality between gauge and string theory has led to the development of a new field of intense research [2].

In 1965 Hagedorn conjectured the existence of a limiting temperature $T_c \sim 160$ MeV for the hadronic system due to an envisaged exponential increase of the density of hadronic states. The existence of a matter made by quark in a deconfined state was suggested for the first time in 1975 by Cabibbo and Parisi [3] soon after the Nobel Prize paper on the the asymptotic freedom of non-abelian gauge theory [4]. They pointed out that the so-called Hagedorn limiting temperature was associated to a divergency in the hadronic gas partition function and hence a sign of a phase transition to quark matter. However first evidences of the possibility to realize by mean of heavy-ion collision a transient state of such a matter came only in the $90's$ thanks to the SPS facility able to realize heavy-ion collisions up to $\sqrt{s} = 17$ AGeV. It was however only with the RHIC project conducted at Brookhaven National Laboratory (BNL) that it was possible to create a quark-gluon plasma (QGP) phase (expected at energy density $\epsilon_c > 1$ GeV/fm^3) lasting for about 4-5 fm/c at a maximum initial temperature $T \sim 2\,T_c$. A temperature and a duration time sufficiently long for strong interactions that have made possible to achieve several discoveries about the properties of the QGP and its hadronization [5, 6].

The present knowledge about the properties of the QGP is mainly based on three complementary sources: QCD calculations on the lattice (lQCD) and in the perturbative regime (pQCD), theoretical and phenomenological models, empirical information from heavy-ion experiments. The lQCD computations have clearly shown that a phase transition at energy density $\epsilon_c \sim 1$ GeV/fm^3 and a temperature $T_c \sim 160$ MeV occurs being most likely a cross over in the case of realistic quark masses. Furthermore the energy density and entropy density reach about 80% of their ideal gas limit value already relatively close to the critical temperature T_c, but then the full value is reached only for asymptotically large temperatures. Even more interestingly a large deviation from a non-interacting gas behavior is found in the large trace anomaly $T^\mu_\mu = \epsilon - 3P$ up to $T \sim 2T_c$ indicating a system far from a mere gas of quarks and gluons and hinting to a strongly interacting one. The development of pQCD calculations at high temperature has shown a slow convergency with both temperature and mass of the quarks. It is necessary to go to temperature $T > 3 - 4T_c$ [7] or at masses larger than $m_c \sim 1.3$ GeV [8] to have a higher order pQCD scheme applicable.

Our focus in these Proceedings is on the phenomenological models and in particular on the development of a tansport theory for quarks and gluons able to embed the information coming from lQCD and pQCD from one hand and to provide a tool for a direct comparison with the experimental observables on the other hand.

2. Main results at RHIC

The theorethical and experimental efforts around RHIC in the last decade has allowed a first important breakthrough in the knowledge of the properties of the QGP at least up to $T \sim 2T_c$. It became soon quite clear that a new state of matter has been created and there are several novel discoveries and results. We will discuss some of them focusing on two main type of observables that has allowed a first survey of the QGP. A first observable is called nuclear modification factor and provides a measure of the modification of the hadrons momentum spectra in ion-ion (AA) collisions respect to the pp collisions through the ratio of the respective spectra rescaled by the number of collisions N_{coll} according to a geometrical Glauber model:

$$R_{AA}(p_T) = \frac{d^2N^{AA}/d^2p_T}{N_{coll}\, d^2N^{NN}/d^2p_T} \tag{1}$$

It is clear that $R_{AA} = 1$ means that AA collisions are merely a superposition of nucleon-nucleon collisions. First observations at RHIC and more recently at LHC have shown $R_{AA}(p_T) \sim 0.2$ for most central collisions (see figure 2 for the case of heavy quars) corresponding to a strong interaction of the partons initially created as one could expect if a QGP plasma fireball has been really created.

The other main observables that allow to characterize several properties of the QGP medium is the elliptic flow. Its origin is the initial space eccentricity of the QGP coming from the non-central overlapping of the colliding ions, quantified by $\epsilon = \left\langle \frac{y^2-x^2}{y^2+x^2} \right\rangle$. Due to the pressure gradients of the QGP such a space eccentricity is converted into an unisotropic momentum distribution respect to the azimutal angle ϕ_p that can be expressed by mean of a Fourier expansion:

$$f_q(p_T, \phi_p) = f_q(p_t)(1 + 2\sum_n v_n\, cos(n\,\phi_p)) \tag{2}$$

where the first coefficient (v_1) is vanishing on the average to the space symmetry of the system while the second term v_2 namely the elliptic flow is the dominant one. Both at RHIC and LHC values of v_2 up to about 0.25 has been observed, see figure 1 which means that in the $\phi_p = 0$

direction the abundancy of hadrons can be about 3 times larger than at $\phi_p = \pi/2$. It is clear that it is a strong effect carying information on both the EoS and the shear viscosity of the QGP.

We now briefly focus on three main reslts of relevant to the study of the transport properties of QGP and its hadronization:

- *The QGP is a nearly perfect fluid with very low viscosity* - The plasma created at temperatures $T \sim 200 - 300$ MeV and small baryon chemical potential exhibits a nearly perfect fluid behavior in the bulk of the system, opposite to the asymptotic freedom expectations. Such a statement is mainly corroborated by the observation of large anisotropic flows that develop due to the initial anisotropy of the created fireball. Similarly to what has been observed in the same years for ultra-cold trapped atoms [9] the elliptic flow has values close to the ideal hydrodynamical predictions. A first estimate would suggest a $\eta/s \leq 0.4$ very close to the conjectured lower bound for supersymmetric gauge theories in the infinite coupling limit [2] and to that suggested by quite general Quantum Mechanics considerations [10]. It remains to be determined is the value of η/s of what could be the most ideal fluid ever observed and in particular its microscopic origin. As for the microscopic origin of the ideal fluid behavior, a possible explanation is the presence of quark-antiquark resonances that are reminiscent of hadronic-like or gluonic states [8] or could be a more subtle competition between electrically charged quasiparticles (quarks and gluons) and magnetically charged ones (magnetic monopoles) [11]. Lattice results trying to identify and isolate these objects and their contribution to thermodynamics are also becoming available [12]. In the next Section we will describe more in detail the issue of the QGP shear viscosity.
- *Hadronization is modified respect to the vacuum one* - The statement is justified by the fact that the ratio of baryons to mesons is up to a factor of 4 larger than the one in pp collision in an intermediate range of $p_T \sim 2 - 6$ GeV. The most convincing explanation of this phenomena is that most of the hadrons come from a coalescence of the quarks in the plasma [13, 14]. The basic idea is that instead of pop-up quarks from the vacuum as in the standard fragmentation picture one can hadronize combining the quark of the medium. In such a picture calling $f_q(p)$ the (anti-)quark distribution function, the spectra of the mesons is given by two different mechanisms:

$$\frac{d^2N_H}{d^2p_T} = \int_{\Sigma(1)} f_q(p) \otimes D_{q\to H}(z) + \int_{\Sigma(n_q)} f_q^{n_q}(p/n_q) \otimes \Phi_H(p_{rel}) \tag{3}$$

The first term represent the standard fragmentation contribution while the second is the coalescence one.The $\Sigma(n)$ is the n-particle phase space, $D(z)$ is the fragmentation function giving the probability that a parton q will give an hadron H of momentum $p_H = z\, p_q$ and Φ_H is the hadron wave function. It has been shown in several works that the fragmentation dominates parametrically at high p_T [15, 16, 13], but with the expected density and temperature of the quark plasma there is a dominance quark coalescence up to $p_T \sim 5 - 6$ GeV. It is easy to understand that in a coalescence process the baryon can be produced more abundantly respect to fragmentation, in fact the quarks are already present and in particular because the distribution function at low p_T are exponentials $f_q \sim e^{-p/T}$ therefore the integrand in Eq.(3) is the same for baryons and mesons:

$$f_q^{n_q} \sim \left[e^{-p/n_q T}\right]^{n_q} = e^{-p/T}$$

which means that baryons and mesons can be produced with the same probability apart from wave function and finite quark mass effect and differneces in the degeneracy of

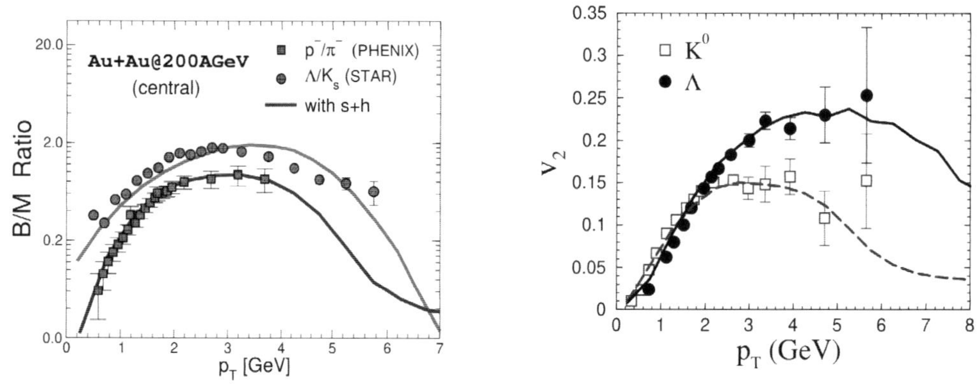

Figure 1. (Color online) Left Panel: The ratio of $\bar{p}/\pi^-$ (circles) and Λ/K_s (squares) in $Au+Au$ at 200 AGeV is shown as a function of p_T and compared to the prediction of a coalescence model (solid lines) [15]. Right Panel: elliptic flow in $Au + Au$ at 200 AGeV for K and Λ compared to a coalescence model prediction.

hadronic states. Therefore can account for the observed enhancement of the baryon over meson ratio observed in the data and shown in figure 1 for proton/pions (squares) and lambda/kaon (circles). The solid lines are the correspondent prediction of a coalescence plus fragmentation model [14, 15, 17].
A coalescence mechanism brings with it another and even more peculiar feature that is the scaling of the elliptic flow with the number of constituent quarks, a property firmly observed at RHIC and more recently also at the LHC energy. The understanding of QNS is straitfoward in the simplified version of collinear quarks coalescing with momenta p_q/n_q. Eeach quark distribution can be written as $f_q(p_T, \phi) = f_q(p_T)\,(1+2\,v_{2q}(p_T)\,cos(2\phi))$. Under this approximation substituting in the second term of Eq.3 it can be easily shown that

$$v_{2M}(p_T) = 2\,v_2(p_T/2) \qquad v_{2B}(p_T) = 3\,v_2(p_T/3) \rightarrow v_{2H}(p_T/n_q) \quad universal\ flow \tag{4}$$

In figure 1 (right) it is shown how the coalescing scaling is able to predict for example the Kaon and the Lambda $v_2(p_T)$ once the quark $v_{2q}(p_T)$ has been fixed fitting the pion v_2.
There several other observable that appear to be consistent with a quark coalescence mechanism like triggered angular correlation, charge fluctuations, R_{AA} and v_2 for heavy meson with charm and bottom quarks.
Of course what it has been briefly describe is a quite simplified version of a realistic coalescence models that has to include also the $r-$space coalescence, the possibility of quarks with different momenta, the radial flow, the feedown from resonance decays, the contribution from higher Fock states and so on. A review of this aspect of the QGP physics can be found in Ref. [13, 14] microscopic scale is important and the specific mechanism of hadronization can modify the observables from the partonic to the hadronic case.

- *Heavy Quarks strongly interact with the medium* - The trivial expectation was that due the large mass respect to the plasma temperature, $m_Q >> T$ and a presumable pQCD behavior due to the $m_Q >> \Lambda_{QCD}$ their in medium interaction relatively weak and the relaxation time of heavy quarks was much larger that the light quark one. Furthermore the main mechanism responsible for the in-medium energy loss, the gluon brehmsstrahlung, should be suppressed by a dead cone effect in the gluon radiation. Despite such pQCD expectation the experimental data [18] revealed a a strong suppression of the spectra, small R_{AA}, and a quite large elliptic flow both nearly comparable with the light quark ones, see figure 2.

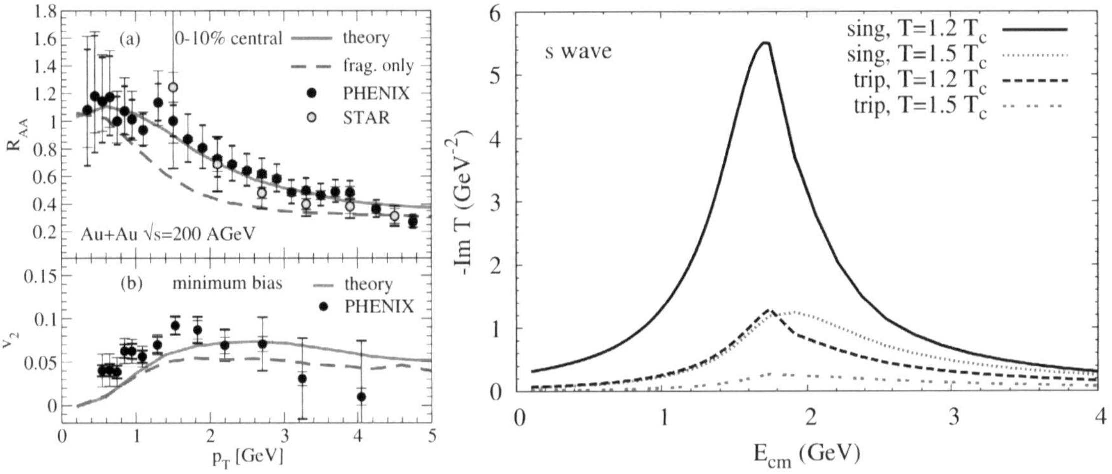

Figure 2. (Color online) Left Panel: the nuclear modification factor (upper panel) and elliptic flow (lower panel) of single electrons coming from D and B decay with/without (solid/dashed lines) quark coalescence in Au-Au collisions compared to RHIC data [23]. Right Panel: Imaginary part of the in-medium T-matrix for $S-$wave $c\bar{q}$ and cq scattering in color-singlet and antitriplet channels, respectively, at two different temperatures.

Models based on jet quenching or upscaled pQCD correction failed to explain the observed $R_{AA}(p_T)$ and the $v_2(p_T)$ measured. Only a non-perturbative approach to the heavy quark dynamics based on the solution of the T-matrix scattering under a potential derived from lQCD has been capable of account for the data [19, 20]. The main ingredients are the presence of a resonant scattering that leads to a peak in the imaginary part of the T-matrix especially in the color singlet state and again the presence of a coalscence mechanism for hadronization. In figure 2 it is shown by solid line the prediction of the T-matrix approach while the dashed line show the result if one discard the coalescence mechanism and assume an hadronization only by parton fragmentation. Predictions for at the LHC appears to be also quite succesfull showing a fairly good agreement with eralier results from the ALICE Collaboration [21, 22].

We have briefly discussed three main surprising and relevant discoveries at RHIC, of course there are other several aspects that could be discussed as the evidence of a strong jet quenching at high momenta, first possible signs of a Color Glass Condensate matter, the confirmation of the enhancement of strangeness. We note that the main findings about the properties of the Quark-Gluon Plasma both in the light and heavy quark sector ask for the development of a transport theory of quarks and gluons, as we discuss more in detail in the next section.

3. Transport Theory for the Quark-Gluon Plasma

A first comparison with preliminary data has shown an agreement of both the p_T spectra and the elliptic flow of different hadrons with the prediction of ideal hydrodynamics [24]. This has lead to the announcement of the creation of an almost ideal fluid. Nonetheless thanks to a more accurate comparison it has been found that dissipative effects cannot be neglected and even a small shear viscosity to entropy ratio η/s produce sizeable effect increasing with the transverse momentum p_T of the particles [25]. This has triggered a lot of activity in developing a relativistic theory of viscous hydrodynamics. The basic idea is to add a dissipative part in the energy momentum tensor by mean of a first order expansion in the space-momentum gradients:

$$T^{\mu\nu} = T^{\mu\nu}_{ideal} + \delta T^{\mu\nu}_{NS} \;=\; \epsilon\, u^\mu u^\nu - P\,\Delta^{\mu\nu} + \eta_s(\nabla^\mu u^\nu + \nabla^\nu u^\mu - \frac{2}{3}\Delta^{\mu\nu}\partial^\alpha u_\alpha) \qquad (5)$$

where $\Delta^{\mu\nu} \equiv g^{\mu\nu} - u^\mu u^\nu$, $\Delta^\mu \equiv \Delta^{\mu\nu}\partial_\nu$. However viscous corrections to ideal hydrodynamics are indeed large and a simple relativistic extension at first order, the so-called Navier-Stokes, is affected by causality and stability pathologies [26]. It is therefore necessary to go to second order gradient expansion, and in particular the Israel-Stewart theory has been implemented to simulate the RHIC collisions providing an upper bound for $\eta/s \leq 0.4$ [27].

Such an approach, apart from the present limitation to 2+1D simulations, has the more fundamental problem of a limited range of validity in η/s and in transverse momentum for $p_T > 1.5$ GeV. In this p_T region viscous hydrodynamics breaks its validity because the relative deviation of the equilibrium distribution function $\delta f/f_{eq}$ increases probably with p_T^2 becoming large already at $p_T \geq 3T \sim 1$ GeV. In fact viscous terms have two main effects one is the dissipative correction to the flow velocity $u^\mu(x)$ and to the density and tempeature evolution, the other is the non-equilibrium corrections to the distribution function $f \rightarrow f_{eq} + \delta f$. It has to be realize that however there is no biunivocal correspondence between the non-equilibrium distribution and the non-equilibrium energy momentum tensor. Therefore the theory of hydrodynamics cannot determine δf and an ansatz has to be choosen typically it used the Grad's ansatz:

$$\frac{\delta f}{f_{eq}} = \frac{\delta T^{\mu\nu}}{\epsilon + P}\frac{p_\mu p_\nu}{T^2}.$$

In this contest it appears important the development of a more complete transport theory for quarks and gluons that would have a more wide range of validity recovering the limiting case of hydrodynamics. This appears important not only for the issued of the viscosity of the QGP but more in general also because a transport theory has also microscopic scale that could result essential to treat consistenly the hadronization mechanism. Furthermore heavy quark significantly deviates from a full thermalization and cannot be described by viscous hydrodynamics while they can be self-constitently included in the transport theory and treated on equal footing as the light ones.

We are therefore developing a relativistic Boltzmann-Vlasov transport theory for on-shell particles [28, 29, 30]. Such a transport approach has the advantage to be a 3+1D approach not based on a gradient expansion in viscosity that is valid also for large η/s and for out of equilibrium momentum distribution allowing a reliable description also of the intermediate p_T range where the important property of quark number scaling (QNS) of $v_2(p_T)$ has been observed [13].

Furthermore Boltzmann-Vlasov transport theory distinguishes between the short range interaction associated to collisions and long range interaction associated to the field interaction, responsible for the change of the Equation of State (EoS) respect to that of a free gas. This last feature allows to unify two main ingredients that are relevant for the formation of collective flow. In ideal hydrodynamics the $v_2(p_T)$ depends only on the EoS namely on the sound velocity $c_s^2 = dP/d\epsilon$, while the mean free path λ is assumed to be vanishing. In the parton cascade approach the EoS is fixed to be the one of a free gas $c_s^2 = 1/3 = P/\epsilon$, while the mean free path $\lambda = 1/\rho\sigma$ is finite. In the first stage of RHIC the two different approaches were able to account for the large v_2 observed; in particular the parton cascade with large scattering cross section predicted the saturation of v_2 vs p_T [31]. Anyway once viscosity is finite both a finite λ and the EoS are important for the generation of the momentum anisotropies and this is naturally present in the Boltzmann-Vlasov transport approach.

The basic equation of transport for the (anti-) quark phase-space distribution function $f^{\pm}$ for the case of mean field interaction that generate massive quasi particles can be written as:

$$p^{\mu}\partial_{\mu}f^{\pm}(x,p) + M(x)\partial_{\mu}M(x)\partial_{p}^{\mu}f^{\pm}(x,p) = \mathcal{C}(x,p) \tag{6}$$

where the first term is related to the free streaming, the second term represents the effect of a scalar field modifying the $\epsilon = 3P$ relation (giving a finite interaction measure) and the last term is the effect of the collisions directly associated to a finite λ and therefore to a finite η/s. The collisions term if only two body collisions are considered can be written as:

$$\mathcal{C}(x,p) = \iiint_{21'2'} (f_{1'}f_{2'} - f_1 f_2)|\mathcal{M}_{1'2'\to 12}|^2 \delta^4(p_1 + p_2 - p_1' - p_2') \tag{7}$$

where $\int_j = \int_j d^3p_j/\left[(2\pi)^3\, 2E_j\right]$, $\mathcal{M}$ denotes the transition matrix for the elastic processes and f_j are the particle distribution functions.

The relevance of the transport equation for quasiparticles with a space-time dependent mass resides in the success of quasi particles in describing correctly the behavior of energy density and pressure of the QGP as computed in the lQCD approach.

3.1. Quasiparticle model

A successful way to account for non-perturbative dynamics is a quasi-particle approach, in which the interaction is encoded in the quasi-particle masses.

The model is usually completed by introducing a finite bag pressure that can account for further non-perturbative effects and could be directly linked to the gluon condensate at least in the pure gauge case [32]. It is already well known that, in order to be able to describe the main features of lattice QCD thermodynamics, a temperature-dependent mass has to be considered. This also implies that the bag constant has to be temperature-dependent, in order to ensure thermodynamic consistency.

The temperature-dependent effective mass for quarks and gluons can be evaluated in a perturbative approach that suggests the following relations [33]:

$$m_g^2 = \frac{1}{6}g^2\left(N_c + \frac{1}{2}n_f\right)T^2 \qquad m_{u,d}^2 = \frac{N_c^2 - 1}{8N_c}g^2T^2 \tag{8}$$

where n_f is the number of flavors considered, N_c is the number of colors, $m_{u,d}$ is the mass of the light quarks.

The coupling g is generally temperature-dependent. However, as mentioned in the introduction, the calculation of such a T-dependence by means of perturbation theory does not allow to have a good description of lattice QCD thermodynamics. Therefore, usually $g(T)$ is left as a function to be determined through the fit to lattice QCD data.

The pressure of the system can then be written as the sum of independent contributions coming from the different constituents, which have a T-dependent effective mass, plus a bag constant:

$$P_{qp}(m_u, m_d, ..., T) = \sum_{i=u,d,g} d_i \int \frac{d^3p}{(2\pi)^3}\frac{\vec{p}^2}{3E_i(p)}f_i(p) - B(T)\,, \tag{9}$$

where $f_i(p) = [1 \mp \exp(\beta E_i(p)]^{-1}$ are the Bose and Fermi distribution functions, with $E_i(p) = \sqrt{\vec{p}^2 + m_i^2}$; $d_i = 2 \times 2 \times N_C$ for quarks and $d_i = 2 \times (N_C^2 - 1)$ for gluons.

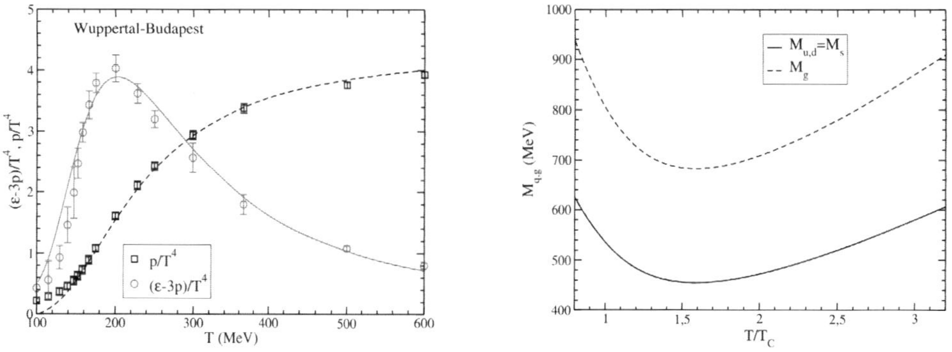

Figure 3. (Color online) Left panel : Lattice data [34] for the pressure and trace anomaly as functions of the temperature, together with the quasi-particle model curves. Right panel: quark and gluon quasi-particle masses as functions of T/T_c [34, 35].

In order to have thermodynamic consistency, the following relationship has to be satisfied:

$$\left(\frac{\partial P_{qp}}{\partial m_i}\right)_{T,\mu} = 0\ , \quad i = u, d, \ldots\ , \tag{10}$$

which gives rise to a set of equations of the form

$$\frac{\partial B}{\partial m_i} + d_i \int \frac{d^3p}{(2\pi)^3} \frac{m_i}{E_i} f_i(E_i) = 0\ . \tag{11}$$

Only one of the above equations is independent, since the masses of the constituents all depend on the coupling g through relationships of the form: $m_i(T, \mu = 0) = \alpha_i g(T) T$, where α_i are constants depending on N_c and N_f according to Eqs. (8). The energy density of the system is then obtained from the pressure through the thermodynamic relationship $\epsilon(T) = T dP(T)/dT - P(T)$ and will have the form

$$\epsilon_{qp}(T) = \sum_i d_i \int \frac{d^3p}{(2\pi)^3} E_i f_i(E_i) + B(m_i(T)) = \sum_i \epsilon^i_{kin}(m_i, T) + B(m_i(T)). \tag{12}$$

In the model there are therefore two unknown functions, $g(T)$ and $B(T)$, but they are not independent, they are related through the thermodynamic consistency relationship (10). Therefore, only one function needs to be determined, which we do by imposing the condition:

$$\epsilon_{qp}(T) = \epsilon_{lattice}(T). \tag{13}$$

We performed our fit to the lattice data for the energy density [34]; in figure 3 we show the good agreement between our curves and lattice results for other quantities like pressure and trace anomaly.

We notice that, at sufficiently high temperature, $m \sim T$, as we can expect because T remains the only scale of the problem. When we are approaching the phase transition, there is instead a tendency to increase the correlation length of the interaction: the quasiparticle model tells us that this can be described as a plasma of particles with larger masses. This essentially determines the fall and rise behavior of $m(T)$ seen in all quasiparticle model fits to lattice data in $SU(N)$ gauge theories including the $SU(3)$ case for QCD. We notice a quite smoother behavior of the masses when the Wuppertal-Budapest lQCD are considered respect to older data of the HotQCD collaboration [35], indicating that the strength of such correlation is significantly reduced when lattice simulations are performed at the physical quark masses and the continuum limit is taken.

The main point for our purposes here is the direct application of such a simple quasiparticle model in the transport theory supplies the possibility to include the correct equation of state of the QGP as computed in lattice QCD.

3.2. Transport at fixed shear viscosity

Our aim is to exploit the transport approach fixing the value of the η/s in order to make possible a direct comparison to viscous hydrodinamic approach and more generally to have a tool to directly estimate the viscosity of the plasma. To this end we do not calculate the cross section from a microscopic model (which could very well be an impossible task) but determine the local cross section σ in order to have the wanted local viscosity. Here we illustrate the procedure for the simplest case of a massless gas for simplicity, the extension to the finite mass case is easily achieved [36], In kinetic theory under ultra-relativistic conditions the shear viscosity can be expressed as [10]: $\eta = \frac{4}{15}\rho\langle p\rangle\lambda$, with ρ the parton density, λ the mean free path and $\langle p\rangle$ the average momentum. Therefore considering that the entropy density for a massless gas is $s = \rho(4 - \mu/T)$, μ being the chemical potential or fugacity, we get:

$$\eta/s = \frac{4}{15}\frac{\langle p\rangle}{\sigma_{tr}\rho(4 - \mu/T)} \tag{14}$$

where σ_{tr} is the transport cross section, i.e. the $\mathtt{sin}^2\theta$ weighted cross section.

From Eq. (14) we see that assuming locally the thermal equilibrium this can be obtained evaluating in each α cell the cross section according to:

$$\sigma_{tr,\alpha} = \frac{4}{15}\frac{\langle p\rangle_\alpha}{\rho_\alpha(4 - \mu_\alpha/T)}\frac{1}{\eta/s} \tag{15}$$

with $4\pi\eta/s$ set in the range $1-4$. This approach is equivalent to have a total cross section of the form $\sigma_{Tot} = K(\rho, T)\sigma_{pQCD} > \sigma_{pQCD}$ where K takes into account the non perturbative effects responsible for that value of viscosity. This approach have been shown to recover the viscous hydrodynamics evolution of the bulk system [26].

We notice that a guideline on the temperature and time dependence of the cross section can be obtained considering the simple case of a free massless gas for which $s = g\frac{2\pi^2}{45}T^3$, and therefore neglecting μ in Eq. (15) one gets $\sigma_{tr} \sim T^{-2}$ for $4\pi\eta/s = 1$. Furthermore a simple Bjorken expansion which means $T \sim \tau^{-1/3}$ gives $\sigma_{tr} \propto \tau^{2/3}$ which is the approximate prescription adopted in [21].In figure 4 it is shown $\sigma_{tr}(\tau)$ evaluated locally in space in a cylinder of radius 3 fm as a function of time for $Au + Au$ at $\sqrt{s} = 200$ AGeV, we see on the left the approximate $\tau^{2/3}$ and on the right the agreement with the estimated T^{-2} behavior.

4. First preliminary results

In our calculation the initial condition are longitudinal boost invariant with the initial parton density $dN/d\eta(b = 0) = 1250$ at RHIC and $dN/d\eta(b = 0) = 2250$ at LHC. The partons are initially distributed in coordinate space according to the Glauber model while in the momentum space at RHIC (LHC) the partons with $p_T \leq p_0 = 2$ GeV ($p_T \leq p_0 = 4$ GeV) are distributed according to a thermalized spectrum with a maximum temperature in the center of the fireball of $2T_c$ ($3.5T_c$), while for $p_T > p_0$ we take the spectrum of non-quenched minijets according to standard NLO-pQCD calculations. We also start our simulation at the time $t_0 = 0.6 fm/c$ at RHIC and $t_0 = 0.3 fm/c$ at LHC.

In order to study the effect of the kinetic freezeout on the generation of the elliptic flow we have performed two calculations one with a constant $4\pi\eta/s = 1$ during all the evolution

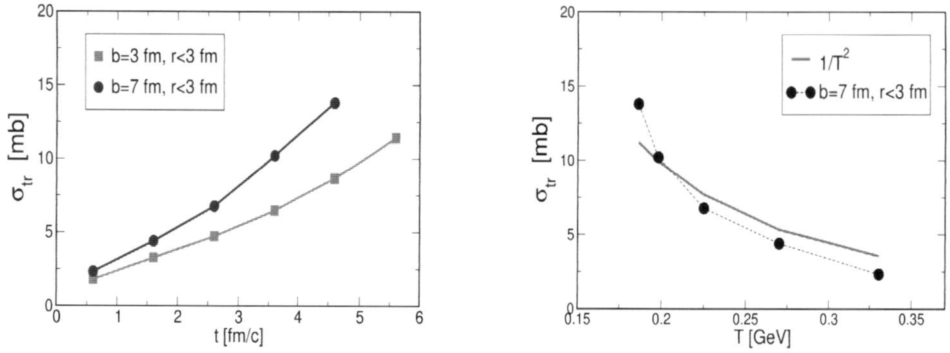

Figure 4. (Color online) Left: Time dependence of the cross section in a central region of rapidity ($|y| < 0.2$) and transverse radius $r < 3$ fm. Right: Temperature dependence of the cross section in the central region compared with the T^{-2} dependence.

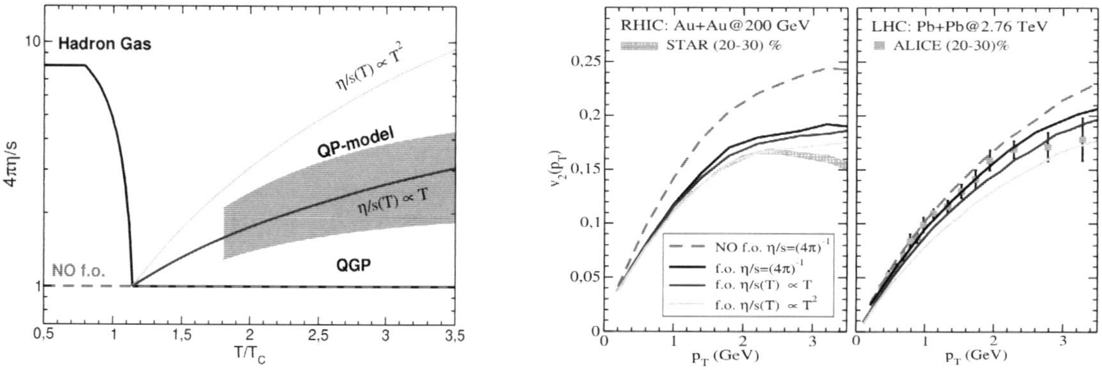

Figure 5. (Color online) Different temperature dependent parametrizations for η/s. The orange area take into account the quasi-particle model predictions for η/s [35]. Right: Differential elliptic flow $v_2(p_T)$ at mid rapidity for $20\% - 30\%$ collision centrality. The red dashed line is the calculation with $4\pi\eta/s = 1$ during all the evolution of the fireball and without the freeze out condition, while the black blue and green lines are calculations with the inclusion of the kinetic freeze out and with $4\pi\eta/s = 1$, $4\pi\eta/s \propto T$ and $4\pi\eta/s \propto T^2$ in the QGP phase respectively as shown in the left panel.

of the system (red dashed line of figure 5) the other (shown by black solid line in figure 5) with $4\pi\eta/s = 1$ in the QGP phase and an increasing η/s in the cross over region towards the estimated value for hadronic matter $4\pi\eta/s \sim 8$. Such an increase allows for a smooth realistic realization of the kinetic freeze-out. In figure 5 (right) it is shown the elliptic flow $v_2(p_T)$ at mid rapidity for $20\% - 30\%$ centrality for both RHIC Au+Au at $\sqrt{s} = 200$ GeV and LHC Pb+Pb at $\sqrt{s} = 2.76$ TeV. As we can see at RHIC energies, left panel of figure 5, the v_2 is sensitive to the hadronic phase and the effect of the freeze out is to reduce the v_2 of about of 20%, from red dashed line to black solid line in left panel of figure 5 (see also figure 6. For the p_T range shown we get a good agreement with the experimental data for a minimal viscosity $\eta/s \approx 1/(4\pi)$ once the f.o. condition is included. At LHC energies, right panel of figure 5, the scenario is different, we have that the v_2 is less sensitive to the increase of η/s at low temperature in the hadronic phase. The effect of large η/s in the hadronic phase is to reduce the v_2 by less than 5% as shown by the solid line for LHC in figure 6 while the RHIC case is shown by red dashed line and compared to the case at $4\pi\eta/s = 2$ but without the f.o. dynamics. This different behaviour of

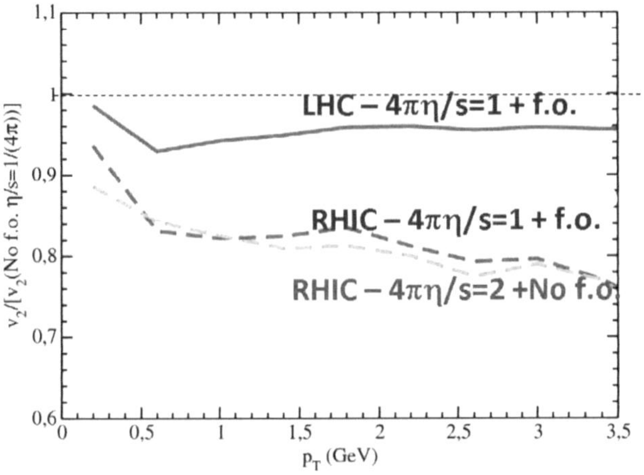

Figure 6. (Color online) The ratio between the elliptic flow $v_2(p_T)$ developed when the increase of η/s in the hadronic phase is included realizing a kinetic freeze-out (f.o.), see figure 5 and the maximal $v_2(p_T)$ obtained discarding the freeze-out. Results are shown for $Au + Au$ at $\sqrt{s} = 200$ AGeV (RHIC) and $Pb + Pb$ at $\sqrt{s} = 2.75$ ATeV (LHC) in minimum bias centrality.

v_2 between RHIC and LHC energies can be explained looking at the life time of the fireball. In fact at RHIC energies the life time of the fireball is smaller than that at LHC energies, $5fm/c$ at RHIC against the about $10fm/c$ at LHC. Therefore at RHIC the elliptic flow has not enough time to fully develop in the QGP phase. While at LHC we have that the v_2 can develop almost completely in the QGP phase.

Due to this large life time of the fireball at LHC and the larger initial temperature is interesting to study the effect of a temperature dependence in η/s. In the QGP phase η/s is expected to have a minimum of $\eta/s \approx (4\pi)^{-1}$ close to T_C as suggested by lQCD calculation. While at high temperature quasi-particle models (see previous Section) seems to suggest a temperature dependence of the form $\eta/s \sim T^{\alpha}$ with $\alpha \approx 1 - 1.5$ [35]. To analyze these possible scenarios for η/s in the QGP phase we have considered two different situation one with a linear dependence $4\pi\eta/s = T/T_0 = (\epsilon/\epsilon_0)^{1/4}$ (blue line) and the other one with a quadratic dependence $4\pi\eta/s = (T/T_0)^2 = (\epsilon/\epsilon_0)^{1/2}$ (green line) where $\epsilon_0 = 1.7$ GeV/fm^3 is the energy density at the beginning of the cross over regions where the η/s has its minimum, see figure 5.

At RHIC energies the v_2 is essentially not sensitive to the dependence of η/s on temperature in the QGP phase, see the blu and green lines in the left panel of figure 5. However the effect on average is to decrease the value of v_2 but at low $p_T < 1.5$ GeV the $v_2(p_T)$ appears to be insensitive to $\eta/s(T)$ while a quite mild dependence appears at higher p_T where however the transport approach tends always to overpredicted the elliptic flow observed experimentally. In any case still a strong temperature dependence in η/s has a small effect on the generation of v_2 we found that with a constant or at most linearly dependent $\eta/s(T)$ the transport approach can describe the data at both RHIC and LHC at least up to $p_T \sim 2$ GeV. It is quite likely that a more detailed analisis of all the anisotropic harmonics measurable up to $v_5 = \langle cos(5\,\phi_p\rangle$ will allow to better constraint the $\eta/s(T)$ as anticipated for the v_4/v_2^2 in Ref. [37].

5. Perspectives and conclusions

We have reviewed some of the main results of the QGP at high temperature created in ultrarelativisti heavy-ion collisions (HIC) at both RHIC and LHC energies. We are developing a transport approach to study the properties of the QGP and have the possibility of interpreting the rich fenomenology coming from ultrarelativisitc HIC. We have shown that a Boltzmann-

Vlasov transport approach has the potential to properly include the dynamics associated to an EoS as evaluated in the lattice QCD and the viscosity dissipation mechanisms. First results indicate that without any parameter tuning the approach is able to correctly predict the behavior of the elliptic flow going from the $Au + Au$ collisions at $\sqrt{s_{NN}} = 200$ GeV to the $Pb + Pb$ at $\sqrt{s_{NN}} = 2.75$ TeV showing its validity. A first important result is that at LHC a key observables like the elliptic flow is much less contaminated by the hadronic phase allowing a better study of the QGP properties.

In the next future the capability to naturally extend transport theory to self-consistenty include the developments in the quark-gluon quasiparticle models, the possibility extend the study of the collective flows to higher harmonics up to $v_5 = \langle cos(5\phi) \rangle$ and the extension of the transport approach to heavy quarks dynamics and the long-standing issue of the J/Ψ suppression-regeneration will potentially allow to obtain a deeper insight into the QGP properties and their microscopic origin.

6. Acknowledgments

This work was supported in part by funds provided by the MIUR under the Firb Research Grant RBFR0814TT and by the ERC-StG2010 under the QGPDyn Grant n.259684.

References

[1] Schwarz D J 2003 *Annalen Phys.* **12** 220 astro-ph/0303574)
[2] Kovtun P, Son D T and Starinets A O 2005 *Phys. Rev. Lett.* **94** 111601
[3] Cabibbo N and Parisi G 1975 *Phys. Lett. B* **59** 67
[4] Gross D and Wilczek F 1973 *Phys. Rev. Lett.* **30** 1343
[5] Adams J *et al.* (STAR Collaboration) 2005 *Nucl. Phys. A* **757** 102
[6] Adcox K *et al.* (PHENIX Collaboration) 2005 *Nucl. Phys. A* **757** 184
[7] Blaizot J P, Iancu E and Rebhan A 2003 (*Preprint* hep-ph/0303185)
[8] Caron-Huot S and Moore G D 2008 *Phys. Rev. Lett.* **100** 052301
[9] O'Hara K M, Hemmer S L, Gehm M E, Granade S R and Thomas J E 2002 *Science* **298** 2179
[10] Danielewicz P and Gyulassy M 1985 *Phys. Rev. D* **31** 53
[11] Ratti C and Shuryak E 2009 *Phys. Rev. D* **80** 034004
[12] Chernodub M, D'Alessandro A, D'Elia M and Zakharov V 2009 (*Preprint* 0909.5441)
[13] Fries R J, Greco V and Sorensen P 2008 *Ann. Rev. Nucl. Part. Sci.* **58** 177
[14] Greco V 2008 *Eur. Phys. J. ST* **155** 45
[15] Greco V, Ko C and Levai P 2003 *Phys. Rev. C* **68** 034904
[16] Fries R, Muller B, Nonaka C and Bass S 2003 *Phys. Rev. C* **68** 044902
[17] Greco V, Ko C and Levai P 2003 *Phys. Rev. Lett.* **90** 202302
[18] Adler S S *et al.* (PHENIX Collaboration) 2006 *Phys. Rev. Lett.* **96** 032301
[19] van Hees H, Mannarelli M, Greco V and Rapp R 2008 *Phys. Rev. Lett.* **100** 192301
[20] van Hees H, Greco V and Rapp R 2006 *Phys. Rev. C* **73** 034913
[21] Armesto N e, Borghini N e, Jeon S e, Wiedemann UA e, Abreu S *et al.* 2008 *J. Phys. G* **35** 054001
[22] Dainese A (ALICE Collaboration) 2011 (*Preprint* 1106.1341)
[23] Adare A *et al.* (PHENIX Collaboration) 2007 *Phys. Rev. Lett.* **98** 172301
[24] Heinz U W 2004 *J. Phys. G* **30** S251
[25] Romatschke P and Romatschke U 2007 *Phys. Rev. Lett.* **99** 172301
[26] Huovinen P and Molnar D 2009 *Phys. Rev. C* **79** 014906
[27] Song H and Heinz U W 2009 *J. Phys. G* **36** 064033
[28] Ferini G, Colonna M, Di Toro M and Greco V 2009 *Phys. Lett. B* **670** 325
[29] Greco V, Colonna M, Di Toro M and Ferini G 2008 *Prog. Part. Nucl. Phys.* **62** 562
[30] Xu Z and Greiner C 2005 *Phys. Rev. C* **71** 064901
[31] Molnar D and Gyulassy M 2002 *Nucl.Phys. A* **697** 495
[32] Castorina P, Miller D E and Satz H 2011 *Eur. Phys. J. C* **71** 1673
[33] Levai P and Heinz U W 1998 *Phys. Rev. C* **57** 1879
[34] Borsanyi S, Endrodi G, Fodor Z, Jakovac A, Katz S D *et al.* 2010 *JHEP* **1011** 077
[35] Plumari S, Alberico W M, Greco V and Ratti C 2011 *Phys. Rev. D* **84** 094004
[36] Plumari S, Baran V, Di Toro M, Ferini G and Greco V 2010 *Phys. Lett. B* **689** 18
[37] Plumari S, Baran V, Di Toro M and Greco V 2011 *J. Phys. Conf. Ser.* **270** 012061

Journal of Physics: Conference Series **336** (2011) 012018 doi:10.1088/1742-6596/336/1/012018

Dynamical probes of pairing correlations: two-particle transfer and two-particle break-up reactions

Ilyas Inci[1], **Francisco Perez-Bernal**[2] **and Andrea Vitturi**[3,4]

[1] Department of Physics, Akdeniz University, Antalya, Turkey
[2] Facultad de Ciencias Experimentales, Departamento de Fisica Aplicada, Universidad de Huelva, Huelva, Spain
[3] Dipartimento di Fisica G.Galilei, Via Marzolo 8, I-35131 Padova, Italy
[4] Istituto Nazionale Fisica Nucleare, Sezione di Padova, Via Marzolo 8, I-35131 Padova, Italy

E-mail: vitturi@pd.infn.it

Abstract. We investigate the role played by pairing correlations in two typical classes of nuclear reactions specifically involving pairs of particles: two-particle transfer reactions, where a pair of nucleons is exchanged between the reaction partners, and two-particle break-up reactions, where a pair of nucleons is emitted. In both cases we find that, although the process is mainly induced by the repeated action on each particle by the one-body field created by the reaction partner, a coherent enhancing effect is generated by the pair correlation present in the initial wave function, acting during the process and again present (in the case of two-particle transfer) in the final wave function. In the case of reactions involving weakly-bound nuclei the low-density features characterizing the vicinity of the drip line and the suggested consequent enhancement of the pairing correlations should show up as clear signature in both processes. From the point of view of their theoretical description, the vicinity of the drip lines will involve heavily the continuum part of the spectrum and we will test in simple models the validity of different discretization procedures used in the description of both structure and reaction.

1. Pairing correlations and two-particle trasfer proceeses

Two-particle transfer processes induced by light ions (reactions as (t,p), (p,t), (^{3}He,n), (α,d)) or heavy ions are considered the ideal tool to study the dynamical effects of pairing correlations. All these reactions do explore precisely the radial properties of pair correlations. Unfortunately, the situation is different, for example, from low-energy one-step Coulomb excitation, where the excitation probability is directly proportional to the B(Eλ) values. Here the reaction mechanism is much more complicated and not well established, so that the possibility of extracting spectroscopic information on the pairing field is not obvious.

It is often assumed that the cross section for two-particle transfer will scale with the square of the matrix element of the pair creation (or removal) operator. In this perspective in order to define and measure the collectivity of pairing modes one could compare with single-particle pair matrix elements to define some pairing single-particle units and therefore pairing enhancement factors. But the two-particle transfer process in not sensitive to just the pair matrix element. We have to look at the radial dependence of the pair transition density, which is relevant for the

Published under licence by IOP Publishing Ltd

reaction mechanism associated with pair transfer processes. As examples we show in figures 1 and 2 the comparison with pure single-particle configurations in the case of the addition mode around ^{16}O (i.e. the ground state of ^{18}O) and the removal mode around ^{208}Pb (i.e. the ground state of ^{206}Pb). The comparison is done in figure 1 directly on the local transition density ($\delta\rho_P(r,r)$). In figure 2 the comparison is done on the non-local transition density $\delta\rho_P(r_1,r_2)$ by fixing the position on one particle and plotting as contour plots in space the probability of finding the second one. One immediately sees that while for pure configuration one has equal probablity of finding the second particle on the same or on the opposite side with respect to the core, the pairing correlation induces a correlation in space, favouring the clustering of the pair.

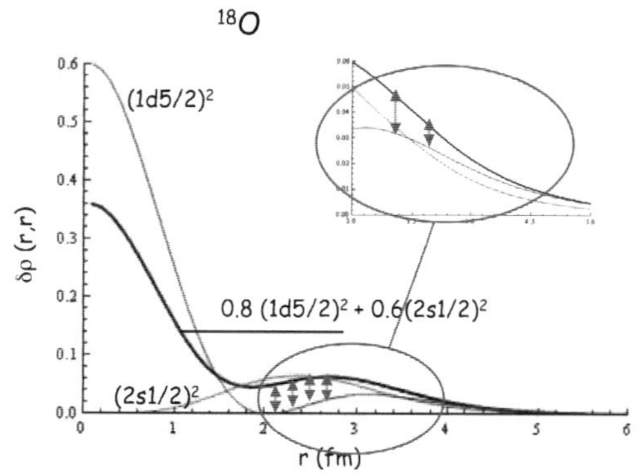

Figure 1. (Color online) Pair transition density in ^{18}O. The cases of pure $(1d5/2)^2$ and $(2s1/2)^2$ are compared with the case of correlated wave function.

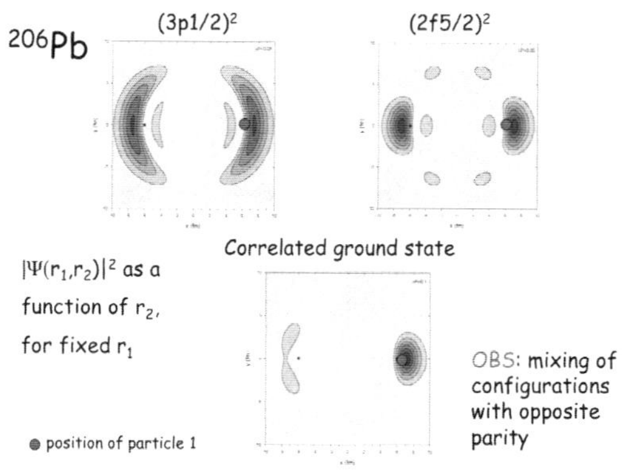

Figure 2. (Color online) Squared wave functions for the ground state of ^{206}Pb, plotted fixing the position of one particle as a function of the position of the second one. Pure configurations are compared with the case of pairing correlated wave function

One can see from these simple examples that the effect of the pairing on the spatial properties of the pair densities is clearly established. The difficult point is to clarify how the structure properties enter into the reaction mechanism. To this end one needs reliable models for the description of the reaction mechanism. There are actually in the literature a large number of different approaches, all trying to reduce the actual complexity of the problem, which is a genuine four-body scattering (the two cores plus the two transferred particles). Note also that different approaches are used in the case of heavy-ion induced reactions (where one may need detailed information on the wave functions in both target and projectile, but on the same time the reaction is mainly surface-peaked and semiclassical descriptions may apply) and in the case of light ions (as (p,t) reactions).

As an example, in a semi-microscopic approach the reaction mechanism is described as a one-step di-neutron (cluster) transfer. In this case the microscopy amounts to construct a formfactor obtained by double-folding the microscopic pair transition densities of initial and final states with some nucleon-nucleon interaction or by simple folding of microscopic pair density in the target with the one-body mean field of the projectile. Such an approach only involves the local transition densities and does not exploit the full non-local features of the pair correlation.

A fully microscopic approach is instead the one in which the two-particle transfer reaction is described by sequential two-step process (at each step one transfers one particle)[1, 2, 3]. In this case the pairing enhancement comes from the coherent interference of the different paths through the different intermediate states in (a-1) and (A+1) nuclei, due to the correlations in initial and final wave functions. Basis ingredients of the calculation are the individual formfactors for one-particle transfer and the microscopic pairing correlated wave functions. Since such fully quantal calculations are rather complex (taking into account full recoil), semiclassical approximation scheme are commonly used (although with approximate treatment of recoil).

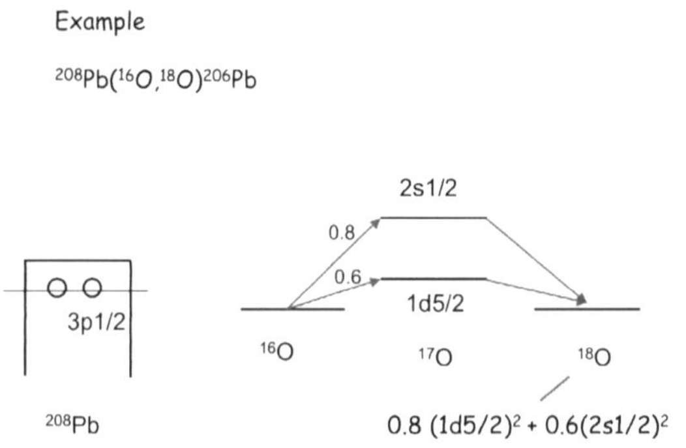

Figure 3. (Color online) Scheme of two-particle transfer etc in ^{18}O.

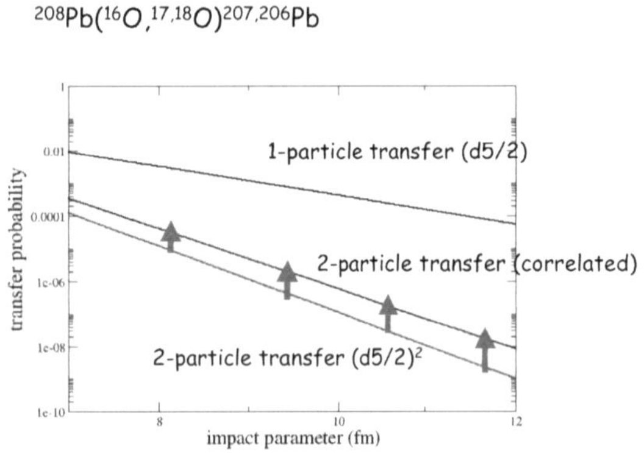

Figure 4. (Color online) Probabilities for one-, two- (uncorrelated) and two-particle (correlated) transfer as a function of the impact parameter

As an example, we consider the two-particle transfer reaction ^{208}Pb(^{16}O, ^{18}O$_{gs}$)^{206}Pb$_{gs}$ at a bombarding energy around the Coulomb barrier. We assume that the two particles are trasferred from the 3p1/2 level in ^{208}Pb into a correlated ground state of ^{18}O, taken as a linear combination of $(1d5/2)^2$ and $(2s1/2)^2$ (see figure 3). The effect of the correlation on the two-particle transfer probability is evident from figure 4, where the probabilities for one and two=particle transfer are shown as a function of the impact parameter. The results for the correlated case is compared with the corresponding values for the uncorrelated one in which the two particles are simply trasferred in pure $(1d5/2)^2$ orbits. Integrating over the impact parameter one obtains a two-particle transfer cross section which is approximately a factor two enhanced with respect to the uncorrelated case (3.90 mb with respect to 1.98 mb in the case of pure $(1d5/2)^2$ case).

As we move closer or even beyond the drip lines the previous picture is inevitably modified. As schematically shown in figures 5-6, we move from a situation in which all relevant intermediate states are bound (figure 5), to situations in which only part of these states are bound (figure 6), finally even reaching at the drip line situations in which no bound intermediate states are present and only due to the pairing interaction the (A+2) system becomes bound. Typical examples are given by the so-called Borromean nuclei such as ^{11}Li or ^{6}Li whose grond states are bound in spite of no available bound states in ^{10}Li or ^{5}He. For weakly-bound systems at the drip lines it is therefore mandatory to include in the models the positive energy part of the spectrum. If one wants to still use the same machinary used with bound states, one has to resort to the so-called discretization of the continuum. We will discuss this point in section 3.

2. Dineutron correlation and breakup reaction within a simple model: two-neutron halo nuclei in one dimension

Among neutron-rich nuclei, two-neutron halo nuclei are particularly intriguing systems to study. Their structure has often been described as a three-body system consisting of two valence neutrons interacting with each other and with the core nucleus. Some light neutron-rich nuclei,

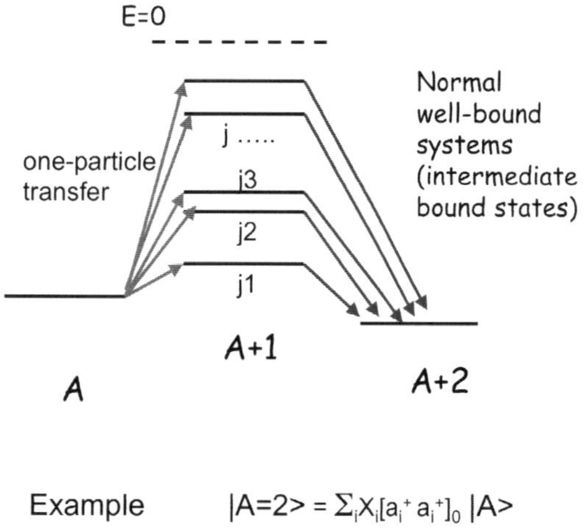

Figure 5. (Color online) Scheme of two-particle transfer for normal systems

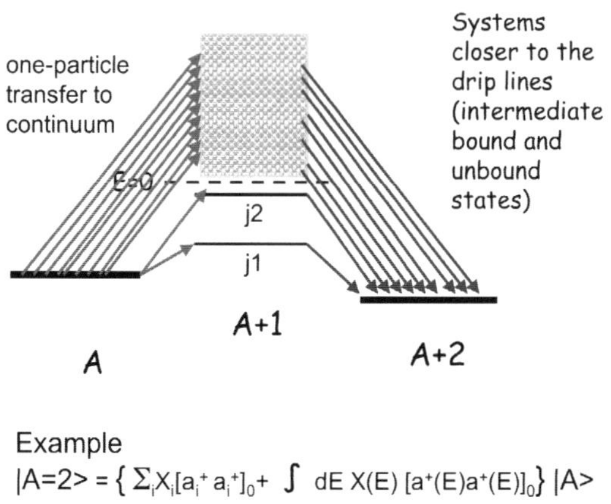

Figure 6. (Color online) Scheme of two-particle transfer for systems towards the drip line

such as ^{11}Li and ^{6}He, do not have a bound state in the two-body subsystem with a valence neutron and a core nucleus. These nuclei are referred to as Borromean nuclei, and their properties have been studied extensively both experimentally and theoretically . Nuclear breakup reactions of Borromean nuclei have also been investigated by the continuum-discretized-coupled-channels (CDCC) method and by the eikonal method

One of the most important current open questions concerning the Borromean nuclei is to clarify the characteristic nature of correlations between the two valence neutrons, which do not form a bound state in the vacuum. A strong dineutron correlation, where the two neutrons take a spatially compact configuration, has been theoretically predicted [4, 5, 6, 7, 8, 9, 10, 11] . Although the recent experimental observation of the strong low-lying dipole strength distribution in the ^{11}Li nucleus has provided an experimental signature of the existence of dineutron correlation in this nucleus, its direct evidence has not yet been obtained.

As discussed previously, the pair transfer reaction is another promising way to probe the dineutron correlation, as the cross section is known to be sensitive to the pairing correlation. However, as seen above, the reaction dynamics and the role of dineutron in the pair transfer reaction are rather complicated[12, 13, 14]. Apparently, it is urged to construct a theoretical framework for pair transfer which fully takes into account the pairing correlation and the dineutron correlation in its consequence.

We use here a simple toy model for two-neutron halo nuclei. Albeit its schematic nature, such model is rather useful as it allows detailed studies on the static and dynamical properties of two-neutron halo nuclei. A schematic model is also useful to deepen our understanding of two-neutron halo nuclei by providing intuitive pictures of several dynamical processes. To this end, we consider a three-body model in which the motion of two valence neutrons is confined within one-dimensional spatial space. The neutrons are assumed to move in a one-dimensional Woods-Saxon potential, while interacting with each other via a two-body interaction, which we take a density-dependent contact interaction [5, 15, 7, 16]. Denoting the coordinate of the two neutrons by x_1 and x_2, the three-body Hamiltonian reads,

$$H = -\frac{\hbar^2}{2m}\frac{d^2}{dx_1^2} + V(x_1) - \frac{\hbar^2}{2m}\frac{d^2}{dx_2^2} + V(x_2) + v_{nn}(x_1, x_2), \tag{1}$$

where m is the nucleon mass. Here, we neglect for simplicity the recoil kinetic energy of the

Journal of Physics: Conference Series **336** (2011) 012018 doi:10.1088/1742-6596/336/1/012018

core nucleus. $V(x)$ is the neutron-core potential, which we take the Woods-Saxon form[23],

$$V(x) = -\frac{V_0}{1 + e^{(|x|-R)/a}} \tag{2}$$

while v_{nn} is the neutron-neutron interaction. We take the density-dependent contact interaction for it, that is,

$$v_{nn}(x, x') = -g\left(1 - \frac{1}{1 + e^{(|x|-R)/a}}\right)\delta(x - x'), \tag{3}$$

where we assume that the density is given by the same functional form as the mean-field potential, Eq. (2). This is the so called surface type pairing interaction, which almost vanishes near the center of the core nucleus at $x \sim 0$. The density-dependent contact interaction has been successfully employed in the Hartree-Fock-Bogoliubov calculations for neutron-rich medium-heavy and heavy nuclei [9, 17] , as well as in describing the structure of ^{11}Li and ^{6}He nuclei with the three-body model [5, 15, 7] .

In order to obtain the ground state wave function of the two-neutron halo nucleus, we first solve the Schrödinger equation for the two-body subsystem,

$$\left[-\frac{\hbar^2}{2m}\frac{d^2}{dx^2} + V(x) - \epsilon_n\right]\phi_n(x) = 0. \tag{4}$$

Each eigenstate n is assumed to have a two-fold degeneracy, depending on the direction of the spin of neutron. The continuum states are discretized by putting the nucleus within a large box. Denoting the size of the box to be X_{box}, we impose the vanishing boundary condition $\phi_n(x) = 0$ at $x = \pm X_{\text{box}}$.

We expand the two-particle wave function $\Psi(x_1, x_2)$ with the single-particle wave functions $\phi_n(x)$ as

$$\Psi_{\text{gs}}(x_1, x_2) = \sum_{n \leq n'} \alpha_{nn'} \frac{1}{\sqrt{2(1 + \delta_{n,n'})}} [\phi_n(x_1)\phi_{n'}(x_2) + \phi_n(x_2)\phi_{n'}(x_1)]\,|S = 0\rangle. \tag{5}$$

Here we have assumed that the spin of the two neutrons form the total spin of S=0 configuration, so that the wave function is symmetric with respect to the interchange of x_1 and x_2. Notice that the Hamiltonian (1) conserves the parity. Since the ground state has positive parity, the expansion in Eq. (5) can therefore be restricted to the configurations where the states n and n' have the same parity. The expansion coefficients $\alpha_{nn'}$ are determined by diagonalizing the Hamiltonian matrix. Once the ground state wave function $\Psi_{\text{gs}}(x_1, x_2)$ is obtained, the two-particle density $\rho_2(x_1, x_2) = |\Psi_{\text{gs}}(x_1, x_2)|^2$ can be constructed. We now apply the one-dimensional model to a weakly-bound two-neutron halo nucleus. To this end, we take $R = 1.27 \times 9^{1/3}$ fm, $V_0 = -50.085$ MeV, and a=0.67 fm for the Woods-Saxon potential, Eq. (2), in order to mimic the ^{11}Li nucleus. This potential possesses four bound single-particle states. We assume that the lowest three bound states are already occupied by the neutrons in the core nucleus, and exclude those states explicitly from the expansion in Eq. (5). In this way, the Woods-Saxon potential has effectively only one bound state with odd parity at $\epsilon_{\text{WS}} = -0.15$ MeV. That is, the three-body system is bound by 0.3 MeV without the pairing interaction. We then include the pairing interaction. It is well known that a contact interaction has to be supplemented with an energy cutoff, E_{cut}. We arbitrarily take here the cutoff energy to be 30 MeV, and we have fixed the pairing strength g to the value -12.35 MeV$\cdot$ fm in order to obtain a correlated ground state energy $E_{\text{g.s.}} = -1$ MeV.

Figure 7 (a) shows the two-particle density of the ground state. The density distribution is largely concentrated along the line of $x_1 \sim x_2$, that is nothing but the manifestation of strong

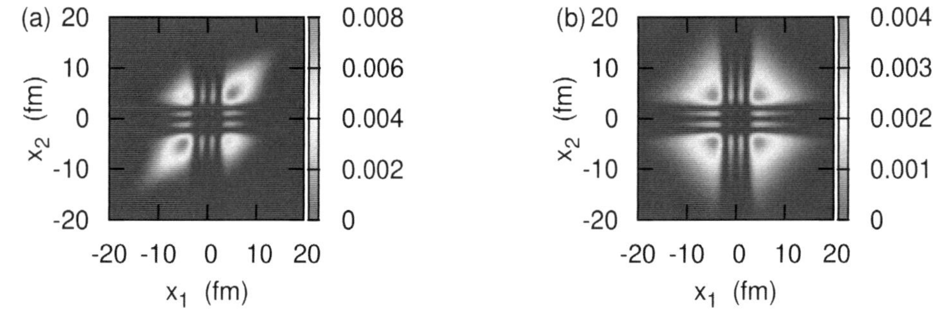

Figure 7. (Color online) The two-particle density for the correlated (left) and the non-correlated (right) ground states of a one-dimensional three-body model.

dineutron correlation discussed in Refs.[5, 6, 7, 8, 9, 10, 11]. The correlation largely hinders the density along the $x_1 = -x_2$ line, and only the peaks along $x_1 \sim x_2$ survive. When the pairing interaction is switched off, the two-particle density becomes that shown in figure 7 (b). The density for the non-correlated ground state is symmetric with respect to the transformation of $x_1 \to -x_1$ ($x_2 \to -x_2$) for a fixed value of x_2 (x_1), and it has therefore four symmetric peaks. See Refs. [18, 19, 20, 21, 22] for similar figures for ^{11}Li and other heavier nuclei.

Let us next discuss the time evolution of the two-particle wave function in the presence of an external field acting on each particle. As the external field, we take [23]

$$V_{\rm ext}(x_1, x_2, t) = \sum_{i=1,2} V_c\, e^{-t^2/2\sigma_t^2} e^{-(x_i-x_0)^2/2\sigma_x^2}, \tag{6}$$

with the parameters of V_c=3 MeV, $\sigma_t = 2.1\hbar$/MeV, and σ_x=2.2 fm. In order to investigate the time evolution, we solve the time-dependent two-particle Schrödinger equation,

$$i\hbar\frac{\partial}{\partial t}\Psi(x_1, x_2, t) = [H + V_{\rm ext}(x_1, x_2, t)]\Psi(x_1, x_2, t), \tag{7}$$

with the initial condition of $\Psi(x_1, x_2, t_0) = \Psi_{\rm gs}(x_1, x_2)$ at the initial time t_0. We orthonormalize the wave function to the occupied states using the projection procedure at each time step during the time evolution. In the calculations shown below, we take $ct_0 = -400$ fm, and use the implicit method [24] for the time propagation. We consider the symmetric perturbation, that is, $x_0 = 0$ in Eq. (6). We have confirmed that our conclusions remain qualitatively the same (except for the asymmetry in the two-particle density distribution along the $x_1 = x_2$ line) even if we choose an asymmetric perturbation [23], *e.g.*, with $x_0 = 2$ fm.

Figures 8(a), 8(b), 8(c), and 8(d) show the two-particle density at $ct = -400, 0, 300$, and 600 fm, respectively. As the time evolves, the extension of the peaks along the $x_1 = x_2$ line increases significantly. This is in marked contrast to the uncorrelated case shown in figures 9(a), 9(b), 9(c), and 9(d). In the uncorrelated case, the two-particle density expands democratically, indicating that there is the equal probability of emission of the two neutrons in the opposite directions to that in the same direction.

One can clearly see that due to the correlation the two neutrons fly apart from the core nucleus by sticking to each other due to the final state interaction. The effect of the correlation is clearly evidenced by the comparison with the uncorrelated case. We point out that both

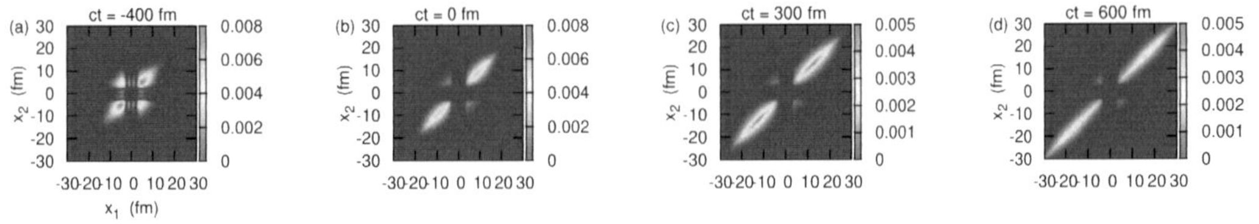

Figure 8. (Color online) The time evolution of the two-particle density. figures 8(a), 8(b), 8(c), and 8(d) show the two-particle density, $|\Psi(x_1, x_2)|^2$ at $ct = -400, 0, 300,$ and 600 fm, respectively. Notice the difference scales among figures 8(a), 8(b), 8(c), and 8(d).

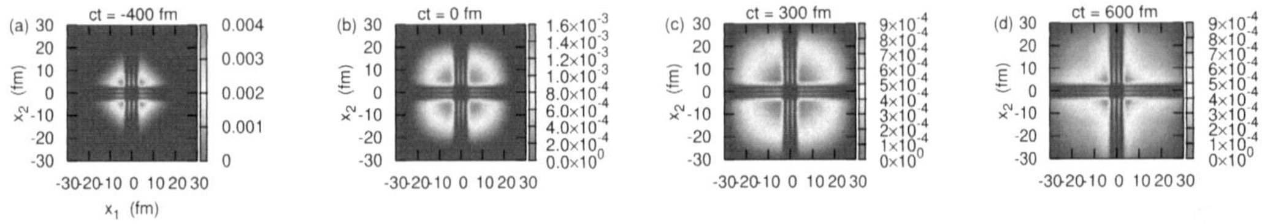

Figure 9. (Color online) Same as figure 8, but for the uncorrelated case with a vanishing neutron-neutron interaction. Notice the difference scales among figures 9(a), 9(b), 9(c), and 9(d).

the dineutron correlation in the ground state, as well as the neutron-neutron interaction acting during the time evolution (that is, the final state interaction) are important for the dineutron emission process.

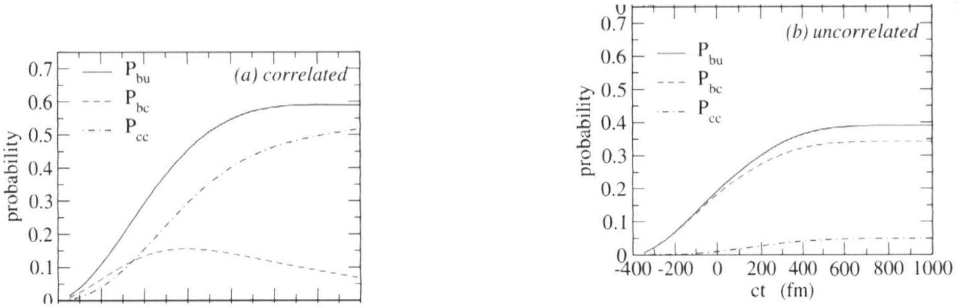

Figure 10. (Color online) The breakup probability as a function of time. The left and the right panels correspond to the correlated and the uncorrelated cases, respectively. The solid line shows the total breakup probability, while the dashed and the dot-dashed lines correspond to decompositions of the breakup probability into the (bc) (only one particle into the continuum) and (cc) (both particles into the continuum) processes, respectively.

The total breakup probability as well as the probability for each breakup process are shown in figure 10 as a function of time. The probabilities $P_{(bc)}$ and $P_{(cc)}$ correspond to the processes in which one or both particles are promoted to the continuum, respectively. The three probabilities (total, (cc) and (bc)) are denoted by the solid, the dashed, and the dot-dashed lines, respectively. The left and the right panels correspond to the correlated and the uncorrelated cases, respectively. By comparing the two panels of figure 10, one finds that the total breakup probability is increased due to the pairing correlation. This is the case especially for P_{cc}, that is, for the two-neutron emission process.

One also sees that the (bc) component first increases as a function of time while the increase of the (cc) component is somewhat delayed. For the correlated case, the (bc) component eventually decreases and the (cc) component takes over. This is a manifestation of the dominance of a sequential mechanism in the two-neutron breakup, if one intends to use a terminology of perturbation theory. Nevertheless, the strong final state interaction makes the dineutron-like emission the main breakup process as shown in figure 8.

In Ref. [25], it has been pointed out that the properties of the two-body subsystem with a neutron and the core nucleus play a decisive role in the Coulomb breakup of ^{11}Li and ^{6}He nuclei. This is because the external field is so weak for the Coulomb breakup that only one of the neutrons makes a transition to other (continuum) single-particle states. This corresponds to the (bc) process in our example. The external field is much stronger for the nuclear breakup process, and the two-step process, or even higher step processes, play an important role. This can be seen in a large probability for the (cc) process, P_{cc}, shown in the left panel of figure 10. The effect of dineutron correlation can therefore be much easily seen in the nuclear breakup process as compared to the Coulomb breakup. A similar conclusion has been reached also in Ref. [26] (see also Ref. [27]).

3. Continuum discretization procedures

As discussed in the previous sections in the description of systems close to the drip lines the inclusion of continuum states becomes mandatory. Quite a variety of approaches have been developed to include continuum states in existing models, both in structure and reaction. Some of them are intrinsically devised to only include in the many-body wave functions components with at maximum one particle in the continuum, others are based on the use of complex eigenstates of the hamiltonian [28]. Most of the approaches, on the contrary, introduce the so-called discretization procedure [29]. In this case the continuum part of the single-particle spectrum is replaced by an alternative discrete sum of states with positive real energies, chosen according to different prescriptions. Examples are given by the expansion in Harmonic Oscillator basis, or similar as the Transformed Harmonic Oscillator, or the use of a large box where to impose vanishing boundary conditions.

As a simple example of the discretization procedure we consider again pure single-particle states in ^{17}O. We can from one side construct the exact $d_{5/2}$ and $s_{1/2}$ states, both bound (discrete) and unbound (continuum), generated by a Woods-Saxon potential and from the other side diagonalize the same potential in a Harmonic Oscillator basis (including up to either 20 or 34 major shells) . In the latter approach we obtain only discrete states, for both negative and positive energies, the latter states being dependent on the number of included HO shells. To check the validity of the discretization procedure, we compare in figure 11 the $B(E2)$ matrix elements from the bound $s_{1/2}$ state to all $d_{5/2}$ states. As apparent from the figure, one has nonvanishing matrix elements not only to the bound $d_{5/2}$ state, but also to the continuum part. By increasing the number of HO shells (and therefore incresing the density of the discrete states) the discrete calculation seems to tend to converge and to approach the true continuum values both in shape and in the integrated value. To check the effect of the possible presence of a resonant state at the continuum threshold, in figure 12 the Woods-Saxon potential has been reduced in order to move the $d_{5/2}$ state into the continuum. As clear from the figure, also in this case the $B(E2)$ continuum distribution is rather well reproduced (again in both shape and integrated value) by the discretized basis, as long as a large number of shells are included. Note

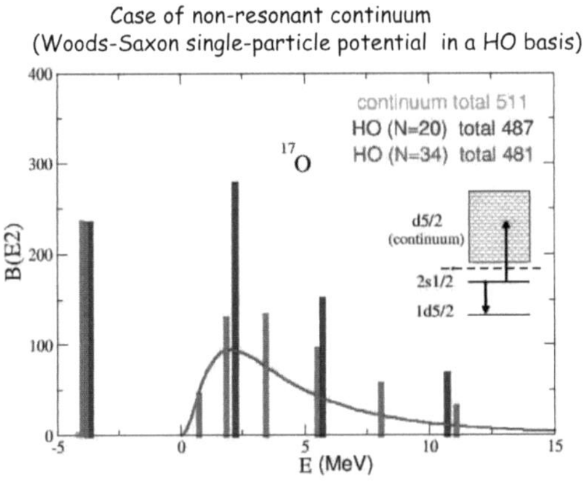

Figure 11. (Color online) $B(E2)$ matrix elements from the bound $s_{1/2}$ state to all $d_{5/2}$ states. Transitions to the true continuum states are compared with those obtained for the discretized states generated in HO basis.

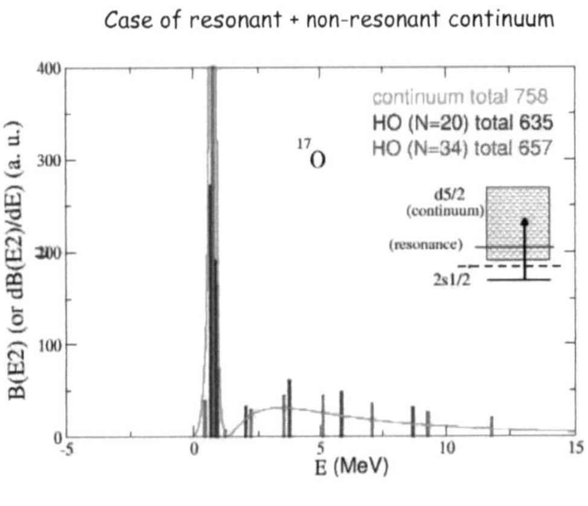

Figure 12. (Color online) Same as in figure 11, but with the one-body WS potential modified in order to have a resonant state close to the continuum threshold.

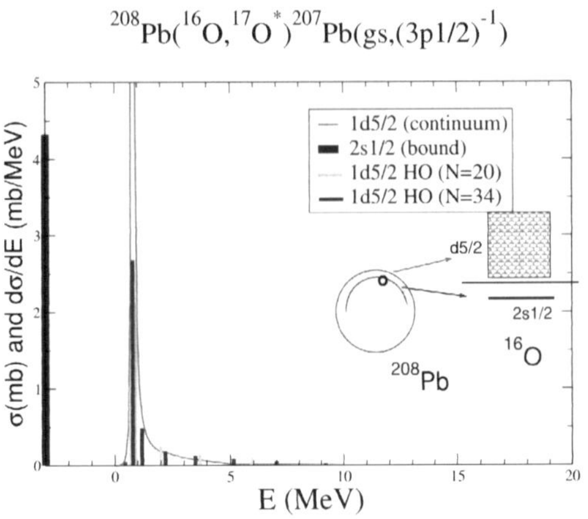

Figure 13. (Color online) Excitation function for one-particle transfer reaction ^{208}Pb(^{16}O, ^{17}O)^{207}Pb$_{gs}$ at a bombarding energy around the Coulomb barrier. We assume that the particle is trasferred from the 3p1/2 level in ^{207}Pb into the bound ($2s1/2$) and the continuum ($1d5/2$) states in ^{17}O. The total cross sections obtained with the true continuum states are compared with the results obtained with the discretized states

that in the $B(E2)$ matrix elements the presence of the initial bound $s_{1/2}$ wave function limits the relevant radial range for the integration to the nuclear volume, therefore cutting out the external region where the true continuum states display the characteristic oscillatory behaviour.

As a further test, we consider the one-particle transfer reaction ^{208}Pb(^{16}O, ^{17}O)^{207}Pb$_{gs}$ at a bombarding energy around the Coulomb barrier. We assume that the particle is trasferred from the 3p1/2 level in ^{207}Pb into the bound ($2s1/2$) and the continuum ($1d5/2$) states in ^{17}O. The total cross sections obtained with the true continuum states are compared in figure 13 with the results obtained with the discretized states. Again, for a large number of HO shells, results seem to converge to the exact ones both in shape and magnitude. Note that the one-particle transfer process involves the construction of one-particle transfer formfactors where the final (bound or unbound) final wave function is folded with the initial bound state in ^{208}Pb, that again gives the largest contribution in the volume region.

All discretization procedures are equivalent as long as a full complete basis is used. In practice each procedure involves a number of approximations and truncations in the actual computations. Computational constraints may in fact become a severe problem, in particular when a large number of particles are allowed to move in the continuum. Comparing the positive and negative aspects of the different approaches is not an easy task, in view of the different sets of parameters that characterize each procedure. In addition the problem used as a test case may dispay such a complexity not to allow for an exact solution to be used as a reference for checking convergence procedures.

For these reasons we consider here the simplest many-body problem: particles moving in the one-dimensional mean field and two particles outside the core interacting via an additional density-dependent short range residual interaction [30, 31]. We will take a case in which the particles of the core fully occupy all the bound states and chose an interaction strength such as to produce a bound two-particle correlate state. The situation can therefore model the case of a two-particle halo nucleus (such as ^{6}He or ^{11}Li), which is bound in spite of the fact that the neighbour core+particle is unbound (as ^{5}He or ^{10}Li in the considered examples).

The one-body potential is assumed to be of the usual simple Woods-Saxon form $V(x) = -V_0\,/\,(1+\exp((|x|-R)/a))$ with parameters $V_0 = 12$ MeV, R=5 fm and a=0.8 fm. This potential admits three bound states at energies $-10.22, -6.00$ and -1.39 MeV, all supposed to be fully occupied. We have used different techniques to produce the (discrete) single-particle basis associated with this one-body potential. We compare here the results obtained by calculating the single-particle Woods-Saxon states using either a harmonic oscillator basis or a finite box at which to impose boundary condition of infinite well. The energies of the resulting states are shown in figure 14 as a function of the number of shells used in the HO basis and in figure 15 as a function of the box radius. Note the variation, in density and in energy, of the unphysical positive energy states.

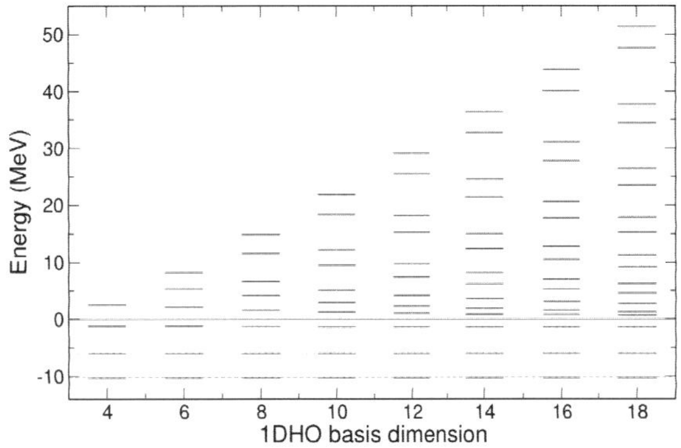

Figure 14. (Color online) Single particle energies obtained diagonalizing the WS one-body potential in a HO basis, for different values of the number of shells.

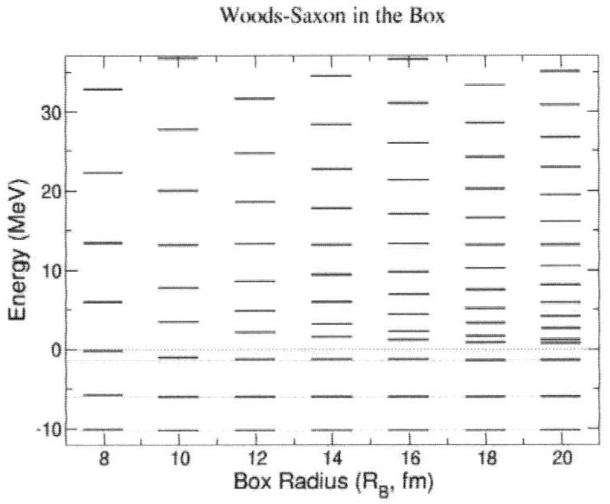

Figure 15. (Color online) Same as in figure 14, but diagonalizing the WS interaction in a finite box, for different values of the box radius.

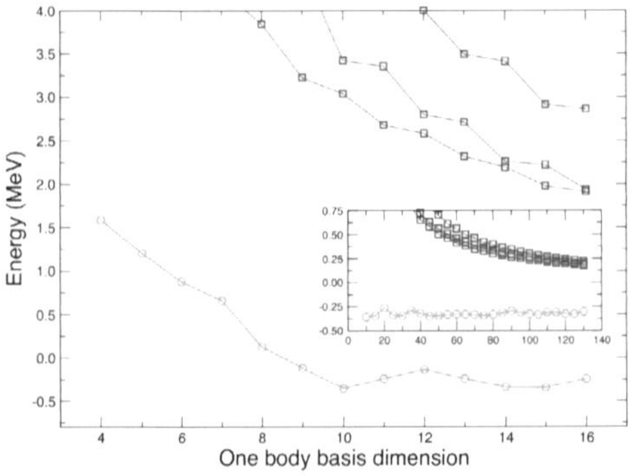

Figure 16. (Color online) Energies of the correlated two-particle states resulting from the diagonalization procedure. A Harmonic Oscillator has been as a basis to construct the WS single-particle states. In the abscissa is given the number of oscillator shells.

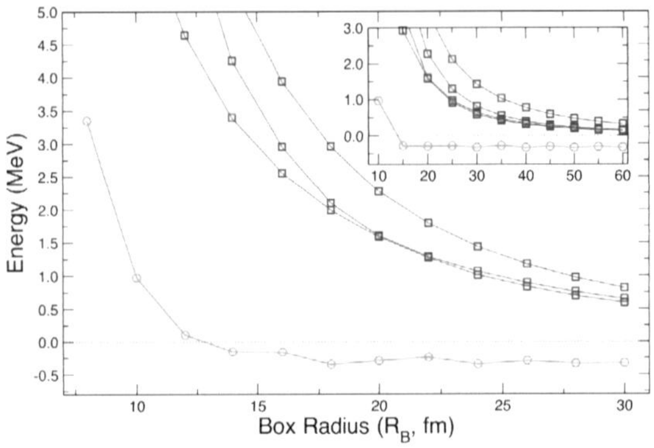

Figure 17. (Color online) Same as in figure 16, but with single-particle states obtained in a finite box, for different values of the box radius.

The residual interaction among the two additional particles is assumed of the form $V_{int}(x_1, x_2) = -V_{int}\ \delta(x_1 - x_2)\ \rho(x)$. We have diagonalized this residual interaction in the two-particle basis. Figures 16 and 17 show the energies of the resulting correlated states as a function of either the number of shells used in the HO basis or the box radius. Positive energy states have no physical meaning (dependent as they are from the chosen basis) but, as apparent from the figure, the residual interaction creates a (physical) bound state with a converging value of the energy (compare e.g. the inset in figure 16, where the number of shells oversizes one hundred or the inset in figure 17 where the box radius reaches 60 fm). A decent value of the energy can be achieved already with a dozen of HO shells or 15 fm for the box size. However, this is not sufficient to ensure a correct description of the resulting wave function. To check this point we display in figures 18 and 19 the correlated two-particle wave function (for $x_1 = x_2 = x$. As clear from the figure an extremely large number of shells (or a large box radius) has to be used to provide a proper behaviour in the tail region. Note that this behaviour is essential, for example, to describe properly the pairing field or the two-particle transfer processes in heavy-ion induced reactions.

The results of our analysis (only sketched here) shows how delicate is the problem of the discretization of the continuum in connection with reaction processes, where the tails of the wave functions may play a decisive role. A proper descrition of nuclear systems at the drip lines will therefore require further studies for both structure and reactions models.

Acknowledgments

This work presents results obtained in collaboration with Kouichi Hagino, Lorenzo Fortunato, Hiro Sagawa and Hugo Sofia. Important contributions on this subject have been obtained in these years by the Milano group, and the authors acknowledge fruitful discussions with Enrico Vigezzi, Francisco Barranco and Gregory Potel.

References

[1] Broglia RA and Winther A 1991 *Heavy-Ion Reactions: Lecture Notes*, Addison-Wesley, New York
[2] von Oertzen W and Vitturi A 2001 *Rep. Prog. Phys.* **64** 1247

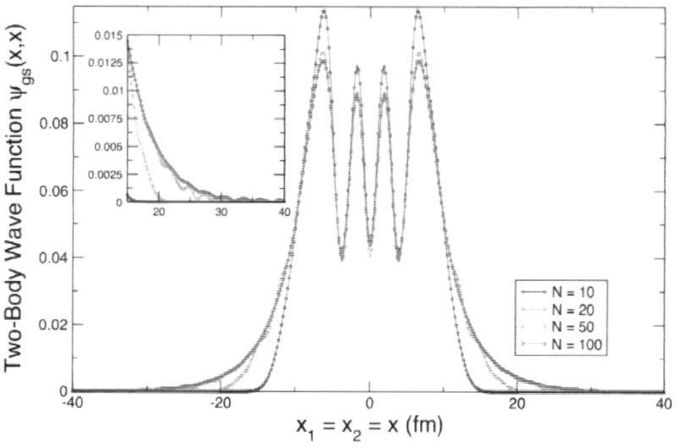

Figure 18. (Color online) Two-body bound state obtained after Hamiltonian diagonalization using a truncated one-body HO basis with number of shells 10, 20, 50, and 100.

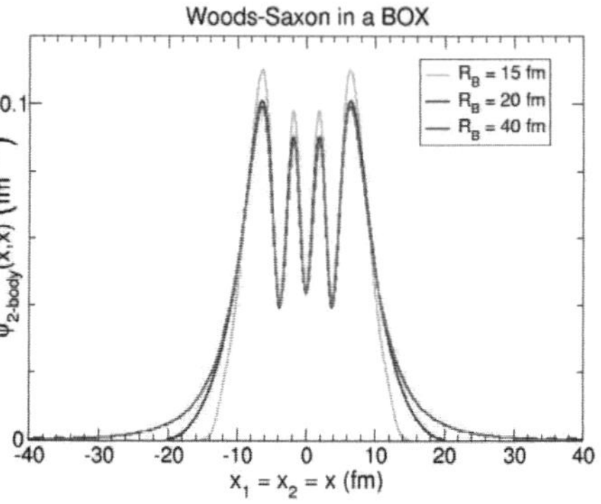

Figure 19. (Color online) Same as in figure 18, but with single-particle states obtained in a finite box, for values of the box radius 15, 20 and 40 fm.

[3] Potel G, Barranco F, Vigezzi E etal 2010 *Phys. Rev. Lett.* **105** 172502; Potel G, Marini F, Idini A, Barranco F, Vigezzi E, Broglia RA, http://arxiv.org/abs/1105.6250
[4] Hansen P G and Jonson B 1987 *Europhys. Lett.* **4** 409
[5] Bertsch G F and Esbensen H 1991 *Ann. Phys. (N.Y.)* **209** 327
[6] Zhukov M V, Danilin B V, Fedorov D V, Bang J M, Thompson I J, and Vaagen J S 1993 *Phys. Rep.* **231** 151
[7] Hagino K and Sagawa H 2005 *Phys. Rev. C* **72** 044321
[8] Hagino K, Sagawa H, Carbonell J, and Schuck P 2007 *Phys. Rev. Lett.* **99** 022506
[9] Matsuo M, Mizuyama K, and Serizawa Y 2005 *Phys. Rev. C* **71** 064326
[10] Pillet N, Sandulescu N, and Schuck P, 2007 *Phys. Rev. C* **76** 024310
[11] Matsuo M 2006 *Phys. Rev. C* **73** 044309
[12] Tanihata I *et al.* 2008 *Phys. Rev. Lett.* **100** 192502
[13] Chatterjee A *et al.* 2008 *Phys. Rev. Lett.* **101** 032701
[14] Lemasson A *et al.* 2009 *Phys. Rev. Lett.* **103** 232701
[15] Esbensen H, Bertsch G F and Hencken K 1997 *Phys. Rev. C* **56** 3054
[16] Hagino K and Sagawa H 2007 *Phys. Rev. C* **76** 047302
[17] Bender M, Heenen P.-H., and Reinhard P.-G. 2003 *Rev. Mod. Phys.* **75** 121
[18] Hagino K, Sagawa H, and Schuck P 2009 *Int. J. of Mod. Phys. E* **18** 2045
[19] Catara F, Insolia A, Maglione E, and Vitturi A 1984 *Phys. Rev. C* **29** 1091
[20] Ferreira L, Liotta R J, Dasso C H, Broglia R A and Winther A 1984 *Nucl. Phys. A* **426** 276
[21] Herzog M W, Civitarese O, Ferreira L, Liotta R J, Vertse T and Sibanda L J 1986 *Nucl. Phys. A* **448** 441
[22] Lotti P, Cazzola F, Bortignon P F, Broglia R A, and Vitturi A 1989 *Phys. Rev. C* **40** 1791
[23] Dasso C H and Vitturi A 2009 *Phys. Rev. C* **79** 064620
[24] Koonin S E, Davies K T R, Maruhn-Rezwani V, Feldmeier H, Krieger S J, and Negele J W 1977 *Phys. Rev. C* **15** 1359
[25] Hagino K, Sagawa H, Nakamura T, and Shimoura S 2009 *Phys. Rev. C* **80** 031301(R)
[26] Assié M and Lacroix D 2009 *Phys. Rev. Lett.* **102** 202501
[27] Assié M *et al.* 2009 *Eur. Phys. J. A* **42** 441
[28] Michel N, Nazarewicz W, Ploszajczak M, and Rotureau J, 2006 *Phys. Rev. C* **74** 54305; Hagen G, Jensen MH, and Michel N, 2006 *Phys. Rev. C* **73** 64307
[29] Tsukiyama K, Otsuka T and Fujimoto R 2009, *AIP Conf. Proc.* **1165** 59
[30] Pérez-Bernal F and Vitturi A 2009 *AIP Conf. Proc.* **1165** 305
[31] Vitturi A and Pérez-Bernal F 2010 *Nucl. Phys. A* **834** 428c

Journal of Physics: Conference Series **336** (2011) 012019 doi:10.1088/1742-6596/336/1/012019

QCD thermodynamics on the lattice and in the Hadron Resonance Gas model

Szabolcs Borsanyi[1], Gergely Endrodi[2], Zoltan Fodor[1,3], Christian Hoelbling[1], Sandor Katz[3], Stefan Krieg[1], Claudia Ratti[4,5] and Kalman Szabo[1]

[1] Dep of Physics, Wuppertal University, Gaussstr. 20, D-42119 Wuppertal, Germany
[2] Inst für Theoretische Physik, Regensburg University, D-93040 Regensburg, Germany
[3] Inst for Theoretical Physics, Eötvös University, Pázmány 1, H-1117 Budapest, Hungary
[4] Dip di Fisica Teorica, Università di Torino, via Giuria 1, I-10125 Torino, Italy
[5] INFN, Sezione di Torino

Abstract. We conclude our investigation on the QCD transition temperature and equation of state (EoS) with 2 + 1 staggered flavors and one-link stout improvement. We extend our previous studies by choosing even finer lattices (up to $N_t = 16$ for T_c, up to $N_t = 10$ and 12 for the EoS). These new results support our earlier findings. A Symanzik improved gauge and a stout-link improved staggered fermion action is taken; the light and strange quark masses are set to their physical values. Various observables are calculated in the temperature (T) interval of 100 to 1000 MeV. We compare our data to those obtained by the "hotQCD" collaboration.

1. Introduction

The study of QCD thermodynamics and that of the phase diagram are receiving increasing attention in recent years. A transition occurs in strongly interacting matter from a hadronic, confined system at small temperatures and densities to a phase dominated by colored degrees of freedom at large temperatures or densities. A systematic approach to determine the properties of this transition is through lattice QCD. Lattice simulations indicate that the transition at vanishing chemical potential is merely an analytic crossover [1]. One of the most interesting quantities that can be extracted from lattice simulations is the transition temperature T_c at which hadronic matter passes to a deconfined phase. We summarize here the final results on T_c by the Wuppertal-Budapest collaboration [2].

This field of physics is particularly appealing because the deconfined phase of QCD can be produced in the laboratory, in the ultrarelativistic heavy ion collision experiments at CERN SPS, RHIC at Brookhaven National Laboratory, ALICE at the LHC and the future FAIR at the GSI. The experimental results available so far show that the hot QCD matter produced experimentally exhibits robust collective flow phenomena, which are well and consistently described by near-ideal relativistic hydrodynamics. These hydrodynamical models need as an input an EoS which relates the local thermodynamic quantities.

Most of the results on the QCD EoS have been obtained using improved staggered fermions. This formulation does not preserve the flavor symmetry of continuum QCD; as a consequence, the spectrum of low lying hadron states is distorted. Recent analyses performed by various collaborations [2, 3] have pointed out that this distortion can have a dramatic impact on

Published under licence by IOP Publishing Ltd

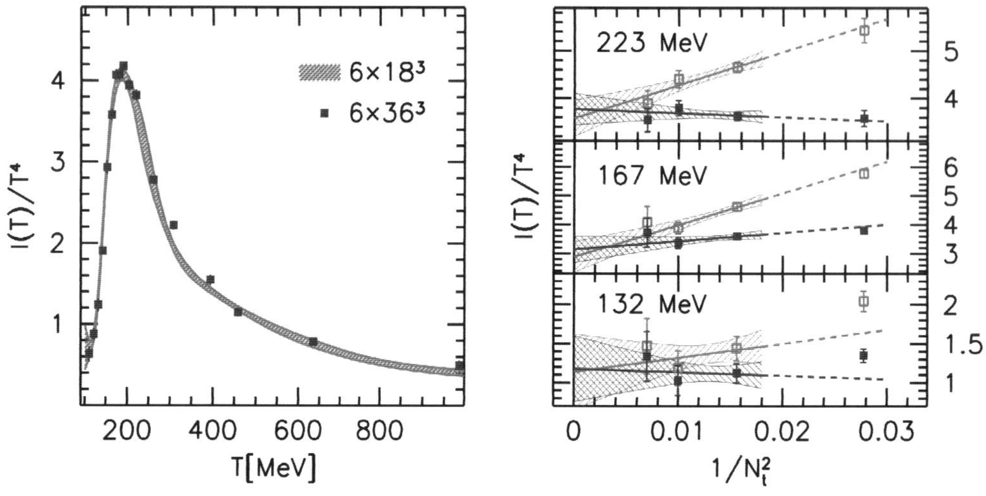

Figure 1. (Color online) Left: The trace anomaly on lattices with different spatial volumes: $N_s/N_t = 3$ (red band) and $N_s/N_t = 6$ (blue points). Right: I=ϵ-3p at three different Ts as a function of $1/N_t^2$. Filled/open symbols represent results with/without tree-level improvement.

the thermodynamic quantities. To quantify this effect, one can compare the low temperature behavior of the observables obtained on the lattice, to the predictions of the Hadron Resonance Gas (HRG) model.

2. Lattice framework

Here we present our results for the following observables: subtracted chiral condensate $\Delta_{l,s} = [\langle\bar{\psi}\psi\rangle_{l,T} - m_l/m_s\langle\bar{\psi}\psi\rangle_{s,T}]/[\langle\bar{\psi}\psi\rangle_{l,0} - m_l/m_s\langle\bar{\psi}\psi\rangle_{s,0}]$ for l=u,d, and with $\langle\bar{\psi}\psi\rangle_q = T\partial \ln Z/(\partial m_q V)$, pressure ($p$), trace anomaly ($I = \epsilon - 3p$) and speed of sound ($c_s$), for n_f=2+1 dynamical quarks. We improve our previous findings [4, 5, 6] by choosing finer lattices ($N_t = 16$ for T_c, $N_t = 8$, 10 and a few checkpoints at $N_t = 12$ for the equation of state). We work again with physical light and strange quark masses: we fix them by reproducing the physical ratios f_K/m_π and f_K/m_K and by this procedure [4, 5, 2] we get $m_s/m_{u,d}$=28.15. We checked that there were no significant finite size effects, by performing two sets of simulations in boxes with a size of 3.5 fm and 7 fm around T_c. The left panel of figure 1 shows the comparison between the two volumes for the normalized trace anomaly I/T^4. Let us note here, that the volume independence in the transition region is an unambiguous evidence for the crossover type of the transition.

To decrease lattice artefacts, we apply tree-level improvement for our thermodynamic observables: we divide the lattice results with the appropriate improvement coefficients. These factors can be calculated analytically for our action and in case of the pressure we have the following values on different N_t's: N_t=6 gives 1.517, N_t=8 gives 1.283, N_t=10 gives 1.159 and N_t=12 gives 1.099. Using thermodynamical relations one can obtain these improvement coefficients for the energy density, trace anomaly and entropy, too. The speed of sound receives no improvement factor at tree level. Note, that these improvement coefficients are exact only at tree-level, thus in the infinitely high temperature, non-interacting case. As we decrease the temperature, corrections to these improvement coefficients appear, which have the form $1 + b_2(T)/N_t^2 + \ldots$. Empirically one finds that the $b_2(T)$ coefficient, which describes the size of lattice artefacts of the tree-level improved quantities, is tiny not only at very high temperatures, but throughout the deconfined phase. The right panel of figure 1 illustrates at three temperature

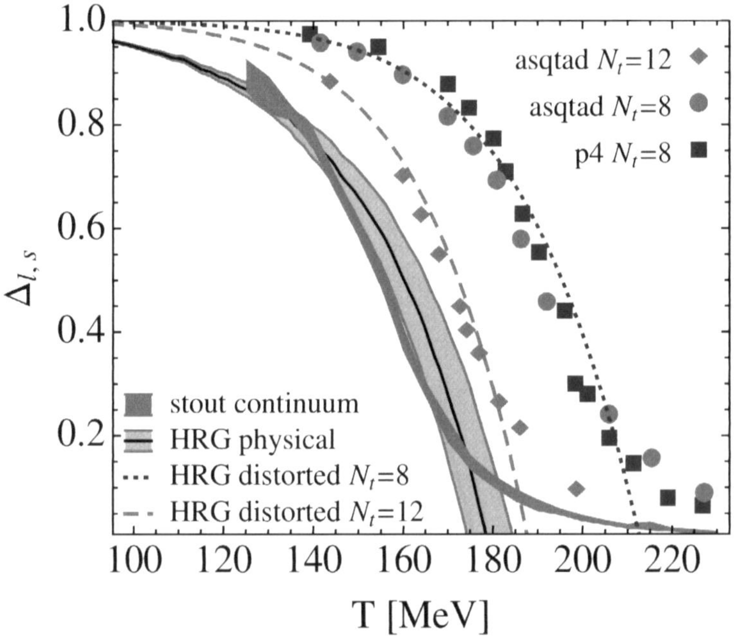

Figure 2. (Color online) $\Delta_{l,s}$ as a function of T. The gray band is our continuum result, obtained with the stout action. Full symbols are obtained with the asqtad and p4 actions [13, 14]. The solid line is the HRG model result with physical masses. The error band corresponds to the uncertainty in the quark mass-dependence of hadron masses. The dashed lines are the HRG+χPT model result with distorted masses of the hotQCD Collaboration [13, 14] for $N_t = 8$ and $N_t = 12$.

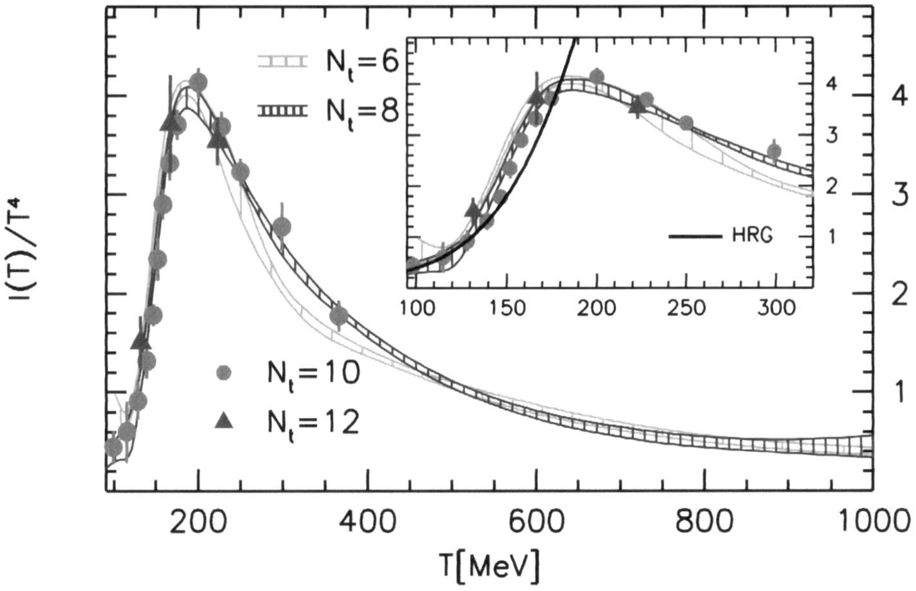

Figure 3. (Color online) The trace anomaly normalized by T^4 as a function of T on $N_t = 6, 8, 10$ and 12 lattices.

values ($T = 132$, 167 and 223 MeV) the effectiveness of this improvement procedure. We show both the unimproved/improved values of the trace anomaly for N_t=6,8,10 and 12 as a function of $1/N_t^2$. The lines are linear continuum extrapolations using the three smallest lattice spacings a. The $a \to 0$ limit of both the unimproved and the improved observables converge to the same value. The figure confirms the expectations, that lattice tree-level improvement effectively reduces the cutoff effects. At all three Ts the unimproved observables have larger cutoff effects than the improved ones. Actually, all the three values of $b_2(T)$, which indicate the remaining cutoff effects after tree-level improvement, differ from zero by less than one standard deviation.

The most popular technique to determine the EoS is the integral method. For large homogenous sytems p is proportional to the logarithm of the partition function. Its direct determination is difficult. Instead, one determines the partial derivatives with respect to the bare lattice parameters. Finally, p is rewritten as a multidimensional integral along a path in the space of bare parameters. To obtain the EoS for various m_π, we simulate for a wide range of bare parameters on the plane of $m_{u,d}$ and β (m_s is fixed to its physical value). Having obtained this large set of data we generalize the integral method and include all possible integration paths into the analysis [7, 8].

An additive divergence is present in p, which is independent of T. One removes it by subtracting the same observables measured on a lattice, with the same bare parameters but at a different T value. Here we use lattices with a large enough temporal extent, so it can be regarded as $T = 0$.

3. Hadron Resonance Gas Model

The Hadron Resonance Gas model has been widely used to study the hadronic phase of QCD in comparison with lattice data. The low temperature phase is dominated by pions. Goldstone's theorem implies weak interactions between pions at low energies. As the temperature T increases, heavier states become more relevant and need to be taken into account. The HRG model has its roots in the theorem of Ref. [9], which allows to calculate the microcanonical partition function of an interacting system, in the thermodynamic limit $V \to \infty$, to a good approximation, assuming that it is a gas of non-interacting free hadrons and resonances [10]. Staggered lattice discretization has a considerable impact on the hadron spectrum. In order to investigate these errors, we define a "lattice HRG" model, where in the hadron masses lattice discretization effects are taken into account.

In order to compare the HRG model results with our additional lattice simulations at larger-than-physical quark masses, we need the pion mass dependence (see e.g. [11] for a recent lattice result) of all hadrons and resonances included in the calculation. As we already did in [2], we assume that all resonances behave as their fundamental state hadrons as functions of the pion mass. For the fundamental hadrons, we use the pion mass dependence from Reference [12]. For larger-than-physical quark masses, the taste symmetry violation at finite a has a milder impact on the pressure. This also motivates to take a pion mass of about 700 MeV as the starting point of the integration of the pressure at $T = 100$ MeV.

4. Results

As we will see different sets of data corresponding to different N_t nicely agree with each other for all observables under study: thus, we expect that discretization effects are tiny.

In figure 2 we show the subtracted chiral condensate $\Delta_{l,s}$ as a function of the temperature. The gray band is our continuum result, obtained with the stout action. Full symbols are obtained with the asqtad and p4 actions [13, 14]. The solid line is the HRG model result with physical masses. The error band corresponds to the uncertainty in the quark mass-dependence of hadron masses. The dashed lines are the HRG+χPT model result with distorted masses of the hotQCD Collaboration [13, 14] for $N_t = 8$ and $N_t = 12$. As we can see, the HRG model predictions with physical resonance spectrum nicely agree with our results. The model also nicely reproduces the data obtained by the hotQCD collaboration, once the distorted resonance spectrum is taken into account. We find that the inflection point for the subtracted chiral condensate sits at $T = 157(3)(3)$ MeV, which can be identified as the transition temperature for chiral symmetry restoration given by this observable (since the transition is an analytic crossover, different observables can give rise to different transition temperatures).

In figure 3 we show the T dependence of ϵ-3p for $n_f = 2+1$. We have results at four different "a". Results show essentially no dependence on "a", they all lie on top of each other. Only the coarsest $N_t = 6$ lattice shows some deviation around $T \sim 300$ MeV. On the same figure, we zoom into the transition region. Here we also show the results from the HRG model: a good agreement with the lattice results is found up to $T \sim 140$ MeV. One characteristic temperature of the crossover transition can be defined as the inflection point of the trace anomaly. This and other characteristic features of the trace anomaly are the following: the inflection point of $I(T)/T^4$ is 152(4) MeV; the maximum value of $I(T)/T^4$ is 4.1(1), whereas T at the maximum of $I(T)/T^4$ is 191(5) MeV.

In figure 4 we show $p(T)$. We have results at three different values for "a". The $N_t = 6$ and $N_t = 8$ are in the T range from 100 up to 1000 MeV. In figure 5 we present the energy density. In figure 6 c_s^2(T), the speed of sound is shown. One can also read off the characteristic points of this curve: the minimum value of $c_s^2(T)$ is 0.133(5); T at the minimum of $c_s^2(T)$ is 145(5) MeV; whereas ϵ at the minimum of $c_s^2(T)$ is 0.20(4) GeV/fm^3.

As it was already discussed in Ref. [5], there is a disagreement between the results of current large scale thermodynamical calculations. The main difference can be described by a $\sim$ 20-30 MeV shift in the temperature. This means, that the transition temperatures are different: the temperature values that we obtain are smaller by this amount than the values of the "hotQCD" collaboration. Refs. [12] and [2] presented a possible explanation for this problem: the more severe discretization artefacts of the "asqtad" and "p4fat" actions used by the "hotQCD" collaboration lead to larger transition temperatures. This can be clearly seen in figure 2, where the results of the two collaborations for the subtracted chiral condensate are presented: the shift in temperature is evident, and can be reproduced by the HRG model once the correct resonance spectrum is taken into account.

It is interesting to look for this discrepancy in the equation of state as well. In figure 7 we compare the trace anomaly obtained in this study with the trace anomaly of the "hotQCD" collaboration. We plot the $N_t = 8$ data using the "p4fat" and "asqtad" actions, which we took from Refs. [13] and [15]. As it can be clearly seen, the upward going branch and the peak position are located at $\sim$ 20 MeV higher temperatures in the simulations of the "hotQCD" group. This is the same phenomenon as the one, which was already reported for many other quantities in [5]. We also see, that the peak height is about 50% larger in the "hotQCD" case. We determined the equation of state of QCD by means of lattice simulations. Results for the $n_f = 2+1$ flavor chiral condensate, pressure, trace anomaly and for the speed of sound were presented on figures. The results were obtained by carrying out lattice simulations at several "a" values, in the T range of $T = 100 \ldots 1000$ MeV. In order to reduce the lattice artefacts we applied tree-level improvement for all of the thermodynamical observables. We found that there is no difference in the results at the three finest lattice spacings. This shows that the lattice discretization errors are not significant and the continuum limit can be reliably taken. The details of the present work can be found in Refs. [2, 7]

Acknowledgements
Work supported in part by the EU grant (FP7/2007-2013)/ERC no 208740. The work of C R is supported by funds provided by the Italian Ministry of Education, Universities and Reserach under the Firb Research Grant RBFR0814TT.

Journal of Physics: Conference Series **336** (2011) 012019 doi:10.1088/1742-6596/336/1/012019

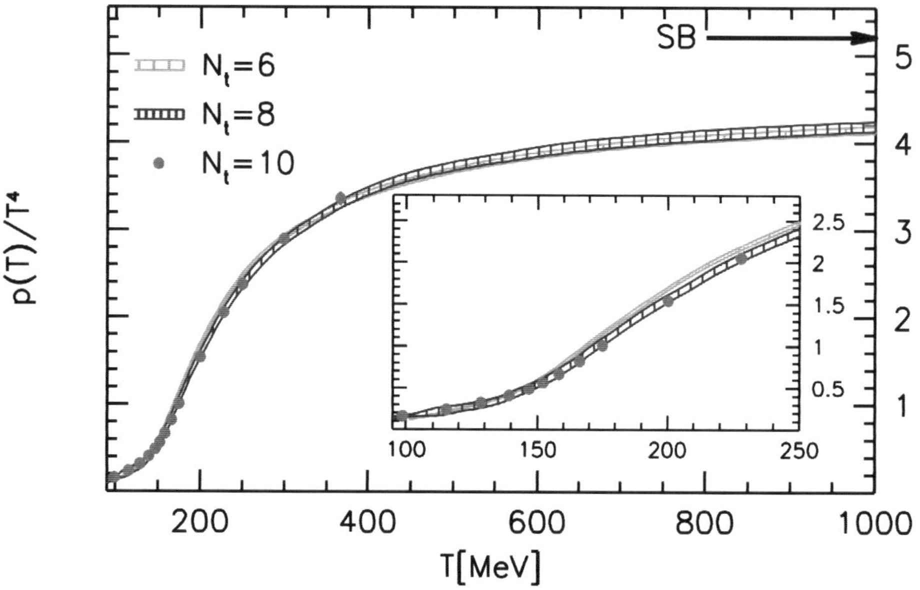

Figure 4. (Color online) $p(T)$ normalized by T^4 as a function of the temperature on $N_t = 6, 8$ and 10 lattices. The Stefan-Boltzmann limit $p_{SB}(T) \approx 5.209 \cdot T^4$ is indicated. At $T = 1000$ MeV p(T) is almost 20% below this limit.

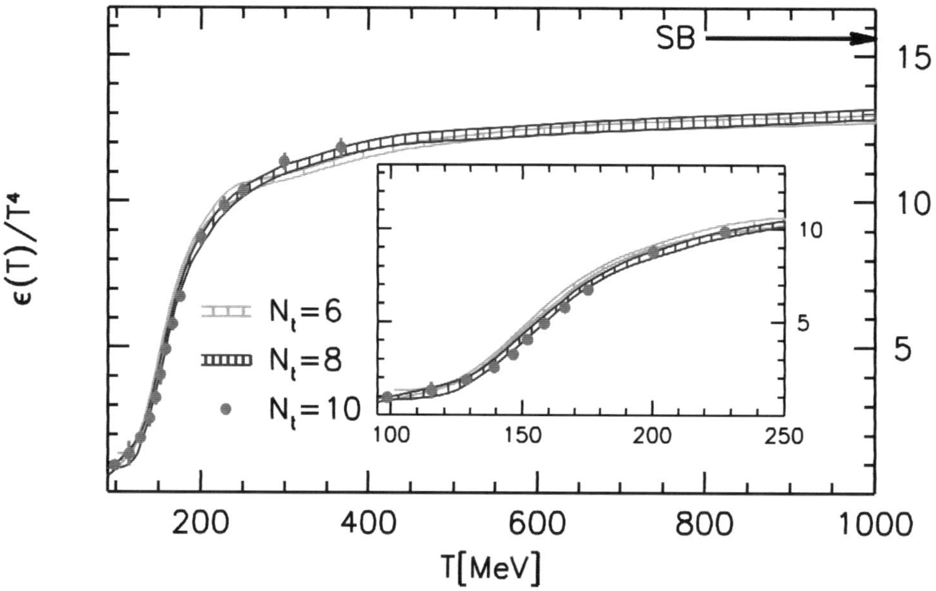

Figure 5. (Color online) The energy density normalized by T^4 as a function of the temperature on $N_t = 6, 8$ and 10 lattices. The Stefan-Boltzmann limit $\epsilon_{SB} = 3p_{SB}$ is indicated by an arrow.

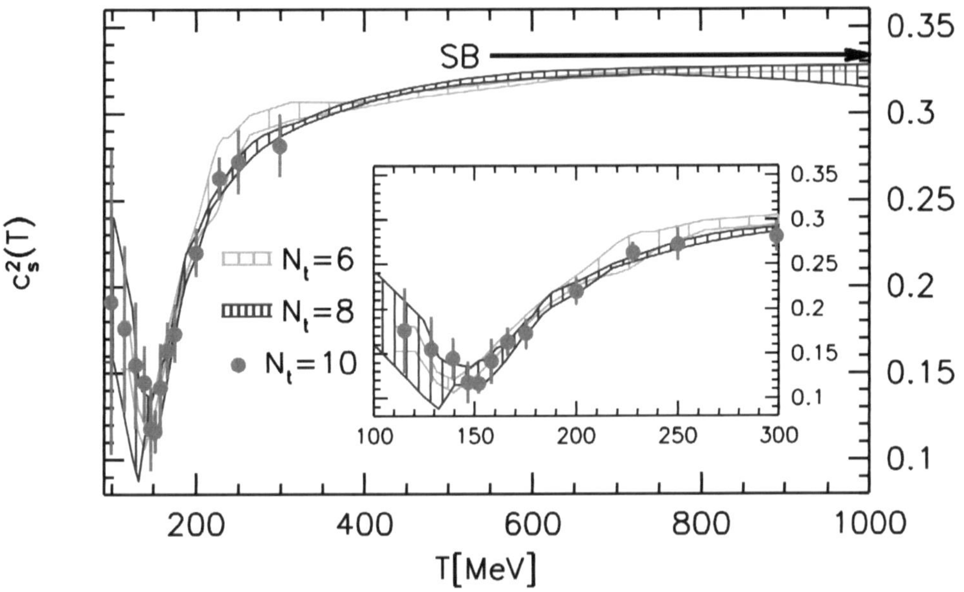

Figure 6. (Color online) The speed of sound squared as a function of the temperature on $N_t = 6, 8$ and 10 lattices. The Stefan-Boltzmann limit is $c^2_{s,SB} = 1/3$ indicated by an arrow.

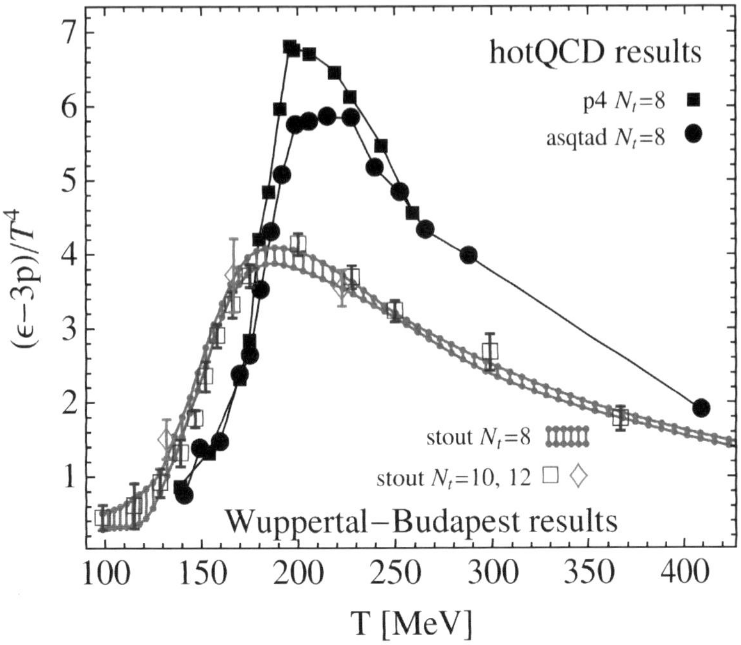

Figure 7. (Color online) The normalized trace anomaly obtained in our study is compared to recent results from the "hotQCD " collaboration [13, 15].

References

[1] Aoki Y, Endrodi G, Fodor Z, Katz S D and Szabo K K 2006 *Nature* **443**, 675
[2] Borsanyi S *et al.* [Wuppertal-Budapest Collaboration] 2010 *JHEP* **1009** 073
[3] Huovinen P and Petreczky P 2010 *J. Phys. Conf. Ser.* **230** 012012
[4] Aoki Y, Fodor Z, Katz S D and Szabo K K 2006 *Phys. Lett. B* **643** 46
[5] Aoki Y *et al.* 2009 *JHEP* **0906** 088
[6] Aoki Y, Fodor Z, Katz S D and Szabo K K 2006 *JHEP* **0601** 089
[7] S. Borsanyi *et al.* 2010 *JHEP* **1011**, 077
[8] Endrodi G 2010 *Preprint* 1010.2952 [physics.comp-ph].
[9] Dashen R, Ma S K and Bernstein H J 1969 *Phys. Rev.* **187** 345
[10] Venugopalan R and Prakash M 1992 *Nucl. Phys. A* **546** 718.
[11] Durr S *et al.* 2008 *Science* **322** 1224
[12] Huovinen P and Petreczky P 2010 *Nucl. Phys. A* **837** 26
[13] Bazavov A *et al.* 2009 *Phys. Rev. D* **80** 014504
[14] Bazavov A and Petreczky P 2010 *Preprint* 1005.1131 [hep-lat]
[15] Cheng M *et al.* 2010 *Phys. Rev. D* **81** 054504

Journal of Physics: Conference Series **336** (2011) 012020 doi:10.1088/1742-6596/336/1/012020

Relativistic fluids with spin

L Tinti[1]

[1] Università di Firenze and INFN Sezione di Firenze, Florence, Italy

E-mail: `tinti@fi.infn.it`

Abstract. We study the behavior of relativistic fluids with internal angular momentum, using the tools of quantum mechanics and statistical mechanics to find out the general properties of macroscopic systems. We found that, if we consider thermodynamics, different couples of stress-energy and spin tensors (previously considered equivalent) give different predictions on the momentum and angular momentum density. This difference remains even in the non-relativistic limit and can in principle be measured experimentally.

1. Introduction

The common relativistic generalization of a fluid involves the stress-energy tensor $T^{\mu\nu}$ which gives us the energy density T^{00}, the momentum density T^{0i} and the law of motion $\partial_\mu T^{\mu\nu} = 0$. The law of motion is enough to ensure the conservation of the total four momentum of the system, provided that the flux of the tensor on the boundary of the domain vanishes.

For the total angular momentum density we have the analog of the orbital angular momentum $x^\mu T^{0\nu} - x^\nu T^{0\mu}$ and another rank 3 tensor $\mathcal{S}^{0,\mu\nu}$ antisymmetric in the last two indices, which is the internal angular momentum of a fluid cell. So the total angular momentum density is $\mathcal{J}^{0,\mu\nu} = \mathcal{S}^{0,\mu\nu} + x^\mu T^{0\nu} - x^\nu T^{0\mu}$. In order to ensure the conservation of the angular momentum of the system, and the conservation of the four-momentum, we require:

$$
\begin{cases}
\partial_\mu T^{\mu\nu} = 0 \\
\partial_\lambda \mathcal{J}^{\lambda,\mu\nu} = \partial_\lambda \mathcal{S}^{\lambda,\mu\nu} + T^{\mu\nu} - T^{\nu\mu} = 0 \Rightarrow \partial_\lambda \mathcal{S}^{\lambda,\mu\nu} = T^{\nu\mu} - T^{\mu\nu}
\end{cases}
$$
$$
\begin{cases}
\int_{\partial V} dS\, T^{i\nu} n_i = 0 \\
\int_{\partial V} dS \mathcal{J}^{i,\mu\nu} n_i = 0.
\end{cases}
\tag{1}
$$

In the first phenomenological model describing relativistic fluids with spin [1] it was made an assumption on the form of the spin tensor:

$$\mathcal{S}^{\lambda,\mu\nu} = u^\lambda \sigma^{\mu\nu},$$

where u^μ is the four-velocity of the fluid cell and $\sigma^{\mu\nu}$, interpreted as a proper internal angular momentum density, *i.e.* the internal angular momentum in the comoving frame. A further condition, called Frenkel condition, was enforced:

$$t^\mu = \sigma^{\mu\nu} u_\nu = 0.$$

Published under licence by IOP Publishing Ltd

The physical validity of the last formula has been questioned [2], however the variational Lagrangian theory which is the base of many subsequent papers [3] still embodies the Frenkel condition.

In our work we found that even the simplest fluid, the ideal Boltzmann gas rotating at full thermodynamical equilibrium, doesn't fulfill the condition [4]. So we have been looking for a general method to study the behavior of relativistic fluids, and we decided to use the tools of quantum mechanics and statistical mechanics to find the general properties.

2. From quantum mechanics to hydrodynamics

From the correspondence principle we know that the classical quantity $\mathcal{O}_{\rm cl.}$, corresponding to the quantum operator $\widehat{O}$, is given by the mean value:

$$\mathcal{O}_{\rm cl.} = {\rm tr}\left(\widehat{\rho} : \widehat{O} :\right),$$

where $\widehat{\rho}$ is the density matrix, the state (pure or mixed) of the system, and the normal ordering $:\cdots:$ is introduced in the mean value in order to subtract the zero-point infinities[1].

So we would like to use as stress-energy and spin tensor:

$$T^{\mu\nu}(x) = {\rm tr}\left(\widehat{\rho} : \widehat{T}^{\mu\nu}(x) :\right) \qquad \mathcal{S}^{\lambda,\mu\nu}(x) = {\rm tr}\left(\widehat{\rho} : \widehat{\mathcal{S}}^{\lambda,\mu\nu}(x) :\right).$$

The Noether theorem give us a stress energy tensor (corresponding to the invariance of the action under space-time translation) and an angular momentum tensor (invariance under Lorentz boosts and rotations), and thus the spin tensor[2], but in quantum field theory (as long as we don't consider gravity) the stress-energy $\widehat{T}$ and spin tensor $\widehat{\mathcal{S}}$ are not uniquely defined. Once found a particular couple $(\widehat{T}, \widehat{\mathcal{S}})$, *e.g.* using the Noether's theorem to get the canonical tensors, we can always construct a new couple fulfilling the same continuity equations (the quantum analogue of (1)) and maintaining the same Poincaré generators:

$$\begin{aligned} \widehat{P}^{\mu} &= \int_V d^3\boldsymbol{x}\; \widehat{T}^{0\mu}(x) && \textit{total four-momentum} \\ \widehat{J}^{\mu\nu} &= \int_V d^3\boldsymbol{x} \left[x^{\mu}\widehat{T}^{0\nu}(x) - x^{\nu}\widehat{T}^{0\mu}(x) + \widehat{\mathcal{S}}^{0,\mu\nu}(x)\right], && \textit{total angular momentum} \end{aligned}$$

if we make this transformation [6]:

$$\begin{aligned} \widehat{T}'^{\mu\nu}(x) &= \widehat{T}^{\mu\nu}(x) + \tfrac{1}{2}\partial_{\alpha}\left(\widehat{\Phi}^{\alpha,\mu\nu}(x) - \widehat{\Phi}^{\mu,\alpha\nu}(x) - \widehat{\Phi}^{\nu,\alpha\mu}(x)\right) \\ \widehat{\mathcal{S}}'^{\lambda,\mu\nu}(x) &= \widehat{\mathcal{S}}^{\lambda,\mu\nu}(x) - \widehat{\Phi}^{\lambda,\mu\nu}(x), \end{aligned} \tag{2}$$

where $\widehat{\Phi}^{\lambda,\mu\nu}(x)$ is a rank 3 tensor antisymmetric in the last two indices that fulfills:

$$\int_{\partial V} dS \left(\widehat{\Phi}^{0,i\nu} - \widehat{\Phi}^{i,0\nu} - \widehat{\Phi}^{\nu,0i}\right) n_i = 0$$

$$\int_{\partial V} dS \left[x^{\mu}\left(\widehat{\Phi}^{0,i\nu} - \widehat{\Phi}^{i,0\nu} - \widehat{\Phi}^{\nu,0i}\right) - x^{\nu}\left(\widehat{\Phi}^{0,i\mu} - \widehat{\Phi}^{i,0\mu} - \widehat{\Phi}^{\mu,0i}\right)\right] n_i = 0.$$

[1] For a discussion of the meaning of normal ordering for interacting fields see, e.g., Ref. [5].
[2] Since $\widehat{\mathcal{J}}^{\lambda,\mu\nu} = \widehat{\mathcal{S}}^{\lambda,\mu\nu} + x^{\mu}\widehat{T}^{\lambda\nu} - x^{\nu}\widehat{T}^{\lambda\mu}$ having two of the operators $\widehat{\mathcal{J}}$, $\widehat{\mathcal{S}}$ or $\widehat{T}$ is enough to know all of them.

Journal of Physics: Conference Series **336** (2011) 012020 doi:10.1088/1742-6596/336/1/012020

This class of transformations leaves the Poincaré generators invariant, so the couples of tensors $(\widehat{T}', \widehat{\mathcal{S}}')$ we get through them are usually considered equivalent, but we found that, despite leaving the generators invariant, the new couples are not actually equivalent.

3. Thermodynamical inequivalence of quantum stress-energy and spin tensors[3]

Since the components T^{00}, T^{0i} and $\mathcal{J}^{0,ij} = \mathcal{S}^{0,ij} + x^i T^{0j} - x^j T^{0i}$ represent the energy density, the linear, and angular momentum density (measurable quantities) we would expect that the mean values of $(\widehat{T}, \widehat{\mathcal{S}})$ were the same for any couple of tensors $(\widehat{T}', \widehat{\mathcal{S}}')$ obtained with the previous class of transformation.

A minimal requirement would be:

$$T'^{0\mu}(x) = T^{0\mu}(x)$$
$$\mathcal{J}'^{0,ij}(x) = \mathcal{J}^{0,ij}(x),$$

but this is frame dependent; a Lorentz-boosted frame would measure a different energy-momentum density if only the first row of the stress-energy tensor was invariant in one particular frame, *e.g.* if Λ^μ_ν is the matrix of the reference change, in the new inertial frame $\tilde{\mathrm{R}}$ we would measure:

$$(T'^{0\mu})_{\tilde{\mathrm{R}}} = \Lambda^0_\alpha \Lambda^\mu_\beta T'^{\alpha\beta} = \Lambda^0_0 \Lambda^\mu_\beta T'^{0\beta} + \Lambda^0_i \Lambda^\mu_\beta T'^{i\beta} = \Lambda^0_0 \Lambda^\mu_\beta T^{0\beta} + \Lambda^0_i \Lambda^\mu_\beta T'^{i\beta},$$

which in general is different from $\Lambda^0_\alpha \Lambda^\mu_\beta T^{\alpha\beta}$ since we didn't enforce $T'^{i\beta} = T^{i\beta}$.We can make a similar argument for the angular momentum densities, but since we require an equivalence only on the spatial indices i, j of $\mathcal{J}'^{0,ij}$ in this case we can make a looser request than $\mathcal{J}'^{\lambda,\mu\nu} = \mathcal{J}^{\lambda,\mu\nu}$. We are thus to enforce, in a covariant form:

$$T'^{\mu\nu}(x) = T^{\mu\nu}(x)$$
$$\mathcal{J}'^{\lambda,\mu\nu}(x) = \mathcal{J}^{\lambda,\mu\nu}(x) + g^{\lambda\mu} K^\nu(x) - g^{\lambda\nu} K^\mu(x). \tag{3}$$

If we limit ourselves to spatial indices $\mu, \nu = 1, 2, 3$, the above equation is enough to ensure that the angular momentum densities, with $\lambda = 0$, are the same in any inertial frame, since in any of them $g^{0i} = g^{0j} = 0$ so the four-vector $K^\mu(x)$ doesn't change $\mathcal{J}^{0,ij}(x)$ by construction.

As long as we use (2) to make $(\widehat{T}', \widehat{\mathcal{S}}')$, and trough their average values T' and $\mathcal{J}'$, we will have some constrains on the possible form of K^μ since every difference between $(T', \mathcal{J}')$ and $(T, \mathcal{J})$ stems from a specific combination of the components of Φ (the average of the operator $\widehat{\Phi}$). In fact we proved that, taking the average of (2), when $\Phi^{\alpha,\mu\nu}$ is such that the new fields fulfill the last conditions (3), necessarily K must fulfill the following:

$$\partial^\nu K^\mu = 0, \tag{4}$$

and so K^μ has to be a constant vector field.

It is important to stress how these conditions are for the mean values, on the quantum level operators don't have to fulfill the same requirements. Even when the quantum analogue of the above equations are not enforced on the couples $(\widehat{T}, \widehat{\mathcal{S}})$ and $(\widehat{T}', \widehat{\mathcal{S}}')$ the density matrix $\widehat{\rho}$ could be such that the classical counterparts are still equivalent. The symmetry properties of $\widehat{\rho}$ are crucial, we found that in the grand-canonical ensamble (a very symmetric system) every couple of tensors gives the same average four-momentum and angular momentum density but in a less symmetric case the situation is very different.

[3] All the calculations of this section are found in Ref.[6]

3.1. A particular case: an axisymmetric system

We studied the case of a *rotational* grand-canonic system, where the density matrix, in the rest frame, is given by[4]:

$$\widehat{\rho} = \frac{1}{Z_\omega} \mathrm{P}_V \exp(-\widehat{H}/T + \boldsymbol{\omega} \cdot \widehat{\boldsymbol{J}}/T + \mu \widehat{Q}/T) \tag{5}$$

where Z_ω is the *rotational* partition function:

$$Z_\omega = \mathrm{tr}\left(\mathrm{P}_V \exp(-\widehat{H}/T + \boldsymbol{\omega} \cdot \widehat{\boldsymbol{J}}/T + \mu \widehat{Q}/T) \right),$$

and $\boldsymbol{\omega}$ has the physical meaning of an angular velocity; but, in order to consider the (tri-) vector $\boldsymbol{\omega} \times \mathbf{x}$ as the velocity of a fluid cell, we need $\|\boldsymbol{\omega} \times \mathbf{x}\| < 1$ (we are using here the natural units where the speed of light in the vacuum is $c = 1$). For that reason we put the projector P_V in the density matrix, to ensure that the tangential velocity never exceeds the speed of light. For simplicity we took as the volume V a cylindrical region centered on the axis z infinitely long and with a radius $R < 1/\|\boldsymbol{\omega}\|$.

Due to the symmetry of the system it's useful to decompose the average values of vectors and tensors along the tetrad (essentially four vector fields that act as a base of the tangent space, see figure 1):

$$\begin{aligned} u &= (\gamma, \gamma \mathbf{v}) \qquad & \tau &= (\gamma v, \gamma \hat{\mathbf{v}}) \\ n &= (0, \hat{\mathbf{r}}) \qquad & k &= (0, \hat{\mathbf{k}}), \end{aligned} \tag{6}$$

where $\mathbf{v} = \boldsymbol{\omega} \times \mathbf{x}$, $\gamma = 1/\sqrt{1 - v^2}$, $\hat{\mathbf{r}}$ is the radial versor in cylindrical coordinates, while $\hat{\mathbf{k}}$ is the versor of the z axis, that is, the axis of the cylinder.

If we decompose along the tetrad a rank three tensor antisymmetric in the last two indices like $\Phi^{\lambda\mu\nu}$, the mean value of the operator in (2), we obtain:

$$\begin{aligned} \Phi^{\lambda,\mu\nu} &= D(r)(n^\mu \tau^\nu - n^\nu \tau^\mu) u^\lambda + E(r)(\tau^\mu u^\nu - \tau^\nu u^\mu) u^\lambda + F(r)(n^\mu u^\nu - n^\nu u^\mu) u^\lambda + \\ &+ N(r)(n^\lambda \tau^\mu - n^\mu \tau^\lambda) u^\nu - N(r)(n^\lambda \tau^\nu - n^\nu \tau^\lambda) u^\mu + P(r) \tau^\lambda (\tau^\mu u^\nu - \tau^\nu u^\mu) + \\ &+ Q(r)(n^\lambda \tau^\mu + n^\mu \tau^\lambda) u^\nu - Q(r)(n^\lambda \tau^\nu + n^\nu \tau^\lambda) u^\mu + R(r) n^\lambda (n^\mu u^\nu - n^\nu u^\mu) + \\ &+ S(r) k^\lambda (k^\mu u^\nu - k^\nu u^\mu) + T(r)(\tau^\mu n^\nu - \tau^\nu n^\mu) n^\lambda + U(r) \tau^\lambda (\tau^\mu n^\nu - \tau^\nu n^\mu) + \\ &+ V(r) k^\lambda (k^\mu n^\nu - k^\nu n^\mu) + W(r) k^\lambda (k^\mu \tau^\nu - k^\nu \tau^\mu), \end{aligned}$$

with $E(0) = F(0) = Q(0) = T(0) = 0$.

Differently from the case of the grand-canonical ensamble, different couples of tensor $(\widehat{T}, \widehat{\mathcal{S}})$ don't give the same predictions, in fact the tensor Φ that we obtain from the *rotational* grand-canonic density matrix (5) in order to fulfill the conditions (3) must have:

$$\begin{aligned} V(r) &= U(r) = F(r) \qquad & P(r) &= R(r) = S(r) \\ E(r) &= W(r) = -T(r) \qquad & D(r) &= N(r) = Q(r) = 0, \end{aligned} \tag{7}$$

[4] It can be obtained in several fashions: by maximizing the entropy with the constraint of fixed mean value of angular momentum [7], generalizing to the quantum-relativistic case an argument used by Landau for classical systems [8], or as the limiting macroscopic case of a quantum statistical system with finite volume and fixed angular momentum in its rest frame in an exact quantum sense, i.e., belonging to a specific representation of the rotation group [9]

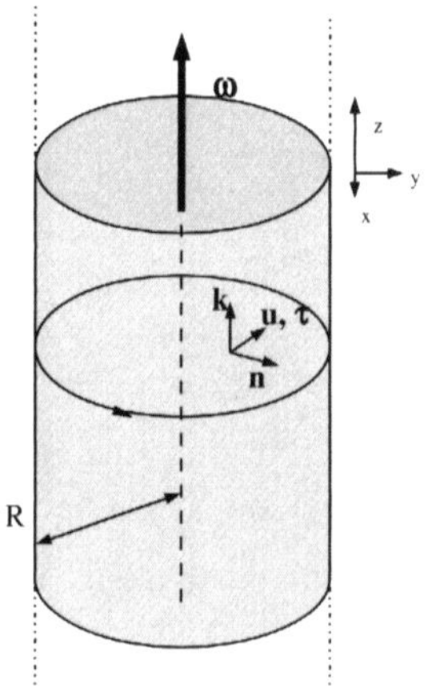

Figure 1. Rotating cylinder with finite radius R at temperature T. Also shown are the inertial frame axes and the spatial parts of the vectors of the tetrad (6).

which are not trivial relations. If these are fulfilled, then the mean value of $\widehat{\Phi}$ reduces to:

$$\begin{aligned}\Phi^{\lambda,\mu\nu} &= (F(r)n^\mu + E(r)\tau^\mu + P(r)u^\mu)g^{\lambda\nu} - (F(r)n^\nu + E(r)\tau^\nu + P(r)u^\nu)g^{\lambda\mu} \equiv \\ &\equiv K^\mu g^{\lambda\nu} - K^\nu g^{\lambda\mu}\end{aligned}$$

The field $K^\mu = F(r)n^\mu + E(r)\tau^\mu + P(r)u^\mu$ ought to be a constant one, according to (4). Since its divergence vanishes, then $F(r) = 0$ and, by using the definitions (6), we readily obtain the conditions:

$$F(r) = 0 \qquad P(r)/\gamma = \text{const} \qquad E(r) = -P(r)\omega r \tag{8}$$

In conclusion, only if $\widehat{\Phi}$ is such that its mean value, calculated with the density operator (5), fulfills conditions (7) and (8), is the corresponding transformation (3) possible. Otherwise, the original and transformed stress-energy and spin tensors are inequivalent because they imply different values of mean energy, momentum or angular momentum densities.

3.2. Example: Free rotating Dirac field in a cylindrical region

We studied the case of the free Dirac field in a cylindrical region to see if there are interesting cases in which the conditions (7) and (8) are not fulfilled.

From the Lagrangian density of the system:

$$\mathcal{L} = \frac{i}{2}\overline{\Psi}\gamma^\mu \overleftrightarrow{\partial}_\mu \Psi - m\overline{\Psi}\Psi,$$

we get the canonical tensors using the Noether's theorem [10]:

$$\widehat{T}^{\mu\nu} = \frac{\partial\mathcal{L}}{\partial(\partial_\mu\Psi)}\partial^\nu\Psi + \partial^\nu\overline{\Psi}\frac{\partial\mathcal{L}}{\partial(\partial_\mu\overline{\Psi})} - g^{\mu\nu}\mathcal{L} = \frac{i}{2}\overline{\Psi}\gamma^\mu\overleftrightarrow{\partial}^\nu\Psi$$

$$\widehat{\mathcal{S}}^{\lambda,\mu\nu} = -i\frac{\partial\mathcal{L}}{\partial(\partial_\lambda\Psi)}\Sigma^{\mu\nu}\Psi + i\overline{\Psi}\Sigma^{\mu\nu}\frac{\partial\mathcal{L}}{\partial(\partial_\lambda\overline{\Psi})} = \frac{1}{2}\overline{\Psi}\{\gamma^\lambda,\Sigma^{\mu\nu}\}\Psi = \frac{i}{8}\overline{\Psi}\{\gamma^\lambda,[\gamma^\mu,\gamma^\nu]\}\Psi,$$

where $\Sigma^{\mu\nu} = \frac{i}{4}[\gamma^\mu,\gamma^\nu]$ is related to the infinitesimal Lorentz transformation on the fields by[5]:

$$\delta\Psi = -\frac{i}{2}\omega_{\mu\nu}\Sigma^{\mu\nu}\Psi \qquad \delta\overline{\Psi} = -\frac{i}{2}\omega_{\mu\nu}\overline{\Psi}\left(-\Sigma^{\mu\nu}\right),$$

[5] $\delta\overline{\Psi} = \delta\Psi^\dagger\gamma^0 = \frac{i}{2}\omega_{\mu\nu}\Psi^\dagger(\Sigma^{\mu\nu})^\dagger\gamma^0$, but $(\Sigma^{\mu\nu})^\dagger = -\frac{i}{4}[(\gamma^\nu)^\dagger,(\gamma^\mu)^\dagger]$ and $(\gamma^\mu)^\dagger\gamma^0 = \gamma^0\gamma^\mu$, so $(\Sigma^{\mu\nu})^\dagger\gamma^0 = \gamma^0\Sigma^{\mu\nu}$ hence $\delta\overline{\Psi} = -\frac{i}{2}\omega_{\mu\nu}\overline{\Psi}\left(-\Sigma^{\mu\nu}\right)$ *q.e.d.*

$\omega_{\mu\nu}$ in this case is simply an infinitesimal number, it's not related at all with the angular velocity $\boldsymbol{\omega}$ of the system.

Because of the properties of the gamma matrices the canonical spin tensor is antisymmetric even in the first two indices, and thus the mean value, much more simple to handle:

$$\mathcal{S}^{\lambda,\mu\nu} = D(r)[(n^\mu\tau^\nu - n^\nu\tau^\mu)u^\lambda + (n^\lambda\tau^\mu - n^\mu\tau^\lambda)u^\nu - (n^\lambda\tau^\nu - n^\nu\tau^\lambda)u^\mu],$$

depends only on one function:

$$D(r) = \mathcal{S}^{0,12} = \mathrm{tr}\left(\widehat{\rho} : \widehat{\mathcal{S}}^{0,12} :\right).$$

It turns out that this function is the only one from which depends the difference between the mean momentum and angular momentum density of the canonical tensors and the Belinfante symmetrized tensors[6]:

$$\begin{aligned} T^{0i}_{\text{Belinfante}} - T^{0i}_{\text{canonical}} &= -\frac{1}{2}\frac{\mathrm{d}D(r)}{\mathrm{d}r}\hat{v}^i \\ \mathcal{J}_{\text{Belinfante}} - \mathcal{J}_{\text{canonical}} &= -\left(\frac{1}{2}r\frac{\mathrm{d}D(r)}{\mathrm{d}r} + D(r)\right)\hat{\mathbf{k}}. \end{aligned}$$

The problem of the Dirac field within a cylinder with finite radius has been tackled by several authors in the context of the MIT bag model [11, 12]. Using the complete solution of the free Dirac equation for a massive particle in a longitudinally unlimited cylinder with finite transverse radius (obtained by Bezerra de Mello *et al.* in Ref.[13]) we found:

$$\begin{aligned} D(r) &= \mathrm{tr}_V\left[\widehat{\rho} : \Psi^\dagger(0,\mathbf{x})\Sigma_z\Psi(0,\mathbf{x}) :\right] = \\ &= \sum_M \sum_{\xi=\pm1} \sum_{l=1}^{\infty} \int_{-\infty}^{\infty} \mathrm{d}p_z \left[\frac{1}{\mathrm{e}^{(\varepsilon - M\omega + \mu)/T} + 1} + \frac{1}{\mathrm{e}^{(\varepsilon - M\omega - \mu)/T} + 1}\right] \times \\ &\times \frac{p_{Tl}^2\left[J^2_{|M-(1/2)|}(p_{Tl}r) - b_\xi^{(+)^2} J^2_{|M+(1/2)|}(p_{Tl}r)\right]}{8\pi^2 R J^2_{|M-(1/2)|}(p_{Tl}R)\left(2Rm_{Tl}^2 + 2\xi M m_{Tl} + m\right)}, \end{aligned}$$

and we proved that, even being vanishing when the angular velocity is zero[7], $D(0)$ is non vanishing in a neighborhood of $\omega = 0$.

3.3. The non relativistic limit

It would be very important to evaluate the function $D(r)$ numerically. The difference between the Belinfante and the canonical tensors seen in the previous section should be measurable to have a practical meaning, if it was too small or a rapidly oscillating function on a microscopic scale, the macroscopic observation of the difference would be impossible. The numerical evaluation of this difference is very hard in the fully relativistic case, but turned out to be relatively easy in the non relativistic limit $m/T \gg 1$ where $D(r)$ is the sum of the particle and antiparticle contribution, $D^+(r)$ and $D^-(r)$:

[6] The ones obtained by the transformation in which $\widehat{\Phi} = \widehat{\mathcal{S}}$, i.e. $\widehat{T}'^{\mu\nu} = \frac{i}{4}\left[\overline{\Psi}\gamma^\mu\overleftrightarrow{\partial}^\nu\Psi + \overline{\Psi}\gamma^\nu\overleftrightarrow{\partial}^\mu\Psi\right]$, $\widehat{\mathcal{S}}'^{\lambda,\mu\nu} = 0$.

[7] Actually for $\omega = 0$ the whole function $D(r)$ must be vanishing because of symmetry reasons. In fact in this case the density operator enjoys an additional symmetry: the rotation of an angle π around any axis orthogonal to the cylinder axis. This transformation corresponds to flipping over the cylinder, which leaves the system invariant provided that $\omega = 0$, and has the consequence that any pseudovector field directed along the axis must vanish.

$$D(r)^{\pm} = \hbar \mathrm{tr}\left[\widehat{\rho}\left(:\Psi^{\dagger}\Sigma_z\Psi:\right)^{\pm}\right] \simeq \frac{1}{2}\frac{\hbar\omega}{KT}\hbar\mathrm{tr}\left[\widehat{\rho}\left(:\Psi^{\dagger}\Psi:\right)^{\pm}\right] = \hbar\frac{1}{2}\frac{\hbar\omega}{KT}\left(\frac{\mathrm{d}n}{\mathrm{d}^3\mathbf{x}}\right)^{\pm} \quad (9)$$

For the numerical computation to be accurate enough we have to make the series involved quickly convergent at any r. This leads to some requirements: first (in natural units) $\omega/T \ll 1$, secondly the radius R should be such that $R\sqrt{mT}$ is not too large, then $m/T \gg 1$ (the non relativistic approximation itself); we have to choose the chemical potential μ so as to keep far from the degenerate Fermi gas case.

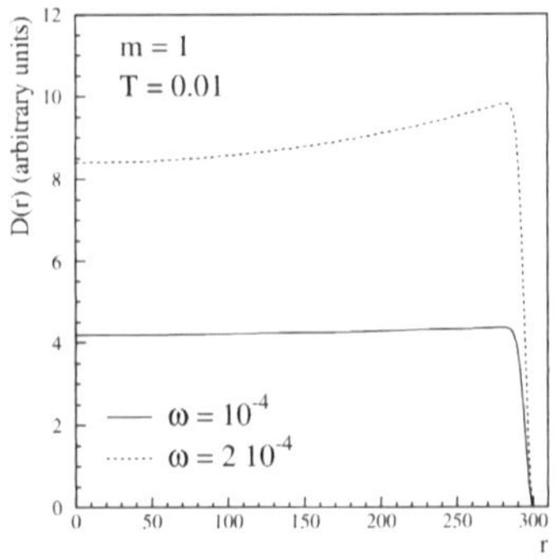

Figure 2. Function $D(r)$, corresponding to the mean value of the canonical spin tensor for the free Dirac field, in a rotating cylinder at thermodynamical equilibrium as a function of radius r, in the non-relativistic limit.

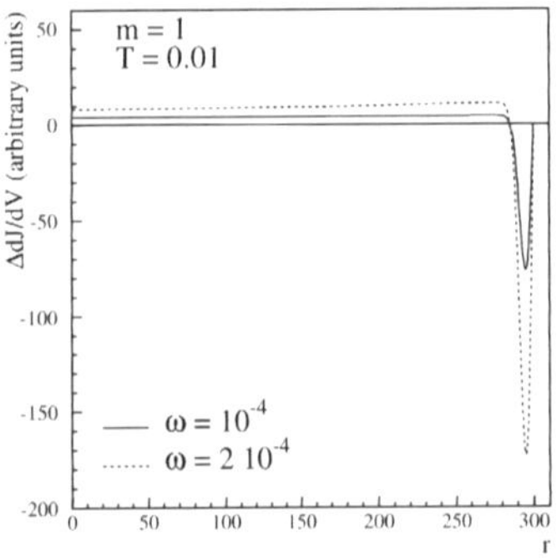

Figure 3. Difference between the mean value of the canonical angular momentum density and the Belinfante angular momentum density for the free Dirac field, in a rotating cylinder at thermodynamical equilibrium as a function of radius r, in the non-relativistic limit.

The function $D(r)$ as a function of r is shown in figure 2 for $\mu = 0$, $R = 300$, $T = 0.01$, $m = 1$ and two different values of ω, 10^{-4} and $2 \cdot 10^{-4}$; the function $(r/2)D'(r) + D(r)$, which is the difference between angular momentum densities for the canonical and Belinfante tensors, is shown in figure 3.

The plots in figures 2,3 show that the angular momentum density is larger in the canonical than in the Belinfante case almost everywhere, except for a narrow space near the boundary, whose thickness is plausibly determined by the microscopic scales of the problem (thermal wavelength or Compton wavelength). Thereby, the observable macroscopic value of the differences between angular momentum densities, for a rotating system of free fermions, is the slowly varying positive one in the bulk. While the boundary conditions are needed to ensure the invariance of the total angular momentum, the rapid drop to zero within a microscopic distance from the cylinder surface tells us that the chosen boundary conditions at a mascroscopic scale of observation correspond to a discontintuity or a surface effect. Any *macroscopic* coaxial sub-cylinder of the full cylinder with a radius $r < R$ will therefore have different total angular momenta whether one chooses the canonical or the Belinfante tensors. Such an ambiguity is physically unacceptable and can be solved only by admitting that these tensors are in fact inequivalent.

4. Conclusions and outlook

In conclusion, starting from the study of relativistic fluids with spin, we found that a previously assumed condition, the Frenkel condition, is not generally justified and that previously considered equivalent couples of spin and energy-momentum tensors are in fact thermodynamically inequivalent.

It is a very important issue what is the right couple of tensors; for instance, if it was found that the quantum energy-momentum tensor is not the Belinfante one, this would have major consequences in hydrodynamics and gravity, because a non-symmetric part could imply a torsion of the spacetime [8].

In principle we could decide if a specific stress-energy or a spin tensor is *wrong* by measuring with sufficient accuracy the angular momentum density of a rotating system at full thermodynamical equilibrium kept at fixed temperature T and angular velocity ω. This measurement would not seem an easy one since the difference in the average polarization of particles, proportional to $\hbar\omega/KT$, is extremely small for ordinary macroscopic systems[9]. But this is precisely the polarization responsible for the observed magneto-mechanical phenomena, the Barnett (magnetization induced by a rotation, see e.g.[15, 16]) and Einstein-De Haas (rotation induced by magnetization) effects. It is therefore possible that with some suitable experiment of this sort one can discriminate between spin tensors, for example this effect could be enhanced lowering the temperature so much to increase the ratio $\hbar\omega/KT$ with cold atom techniques.

We cannot, for the present, determine a thermodynamically"best" couple of stress-energy and spin tensor. We can argue, on the basis of a thermodynamical argument - which can in principle be used to assess any other couple of tensors - that the canonical spin tensor is favored over the Belinfante one[10]but, for the present, does not permit to single out the "best" couple of tensors for a given quantum field theory.

References

[1] Weyssenhoff J and Raabe A 1947 *Acta Phys. Pol.* **9** 7
[2] Bohm D and Vigier J P 1958, *Phys. Rev.* **109** 1882
[3] Halbwachs F 1960 *Theorie Relativiste des Fluids 'a Spin* (Paris: Gauthier Villars)
[4] Becattini F and Tinti L 2010 *Ann. Phys.* **325** 1566
[5] Van Weert C G 1980 *Physica A (Amsterdam)* **100** 641
[6] Becattini F and Tinti L 2011 *Phys. Rev. D* **84** 025013
[7] Balian R *From Microphysics to Macrophysics* 2007 (Heidelberg: Springer)
[8] Landau L and Lifshitz L *Statistical Physics* 1980 (New York: Pergamon Press)
[9] Becattini F and Ferroni L 2007 *Eur. Phys. J. C* **52** 597
[10] Itzykson C and Zuber J B 1985*Quantum field theory* (Mc-Graw Hill)
[11] Chodos A, Jaffe R L, Johnson K, Thorn C B and Weisskopf V F 1974 *Phys. Rev. D* **9** 3471
[12] Shanker R et al. 1983 *Z. Phys. C* **18** 327
[13] Bezerra de Mello E R, Bezerra V B, Saharian A A, and Tarloyan A S 2008 *Phys. Rev. D* **78** 105007
[14] Bjorken J D *arXiv:1008.0033*
[15] Landau L and Lifshitz L 1984*Electrodynamics of Continuous Media* (New York: Pergamon Press), Section 36
[16] Purcell E 1979 *Astrophys. J.* **231** 404

[8] For a recent discussion see e.g. Ref. [14].

[9] According to eq. (9) in the non-relativistic limit the difference between these two tensors is of the order of $\hbar\omega/KT$ times $\hbar$ times the particle density,*i.e.* particles have a polarization of the order of $\hbar\omega/KT$.

[10] The argument (written in the conclusions of[6]) basically states that, if we assume that the thermodynamical potential $\log Z_\omega$ has a physically objective density, we can consider a sub-cylinder with radius $r \ll R$ as "the system" and the rest as an angular momentum reservoir at full thermodynamical equilibrium. So, since the thermodynamical potential of the sub-cylinder is greater in the canonical case, if the system could choose between canonical and Belinfante tensors the former would be certainly favored.

Journal of Physics: Conference Series **336** (2011) 012021 doi:10.1088/1742-6596/336/1/012021

Evolution of newborn neutron stars: role of quark matter nucleation

Ignazio Bombaci

Dipartimento di Fisica Università di Pisa and Istituto Nazionale di Fisica Nucleare, sezione di Pisa, largo B. Pontecorvo, 3, 56126, Pisa, Italy

E-mail: bombaci@df.unipi.it

Domenico Logoteta, Constança Providencia, and Isaac Vidaña

Centro de Física Computacional, Department of Physics, University of Coimbra, 3004-516 Coimbra, Portugal

Abstract. A phase of strong interacting matter with deconfined quarks is expected in the core of massive neutron stars. We study the quark deconfinement phase transition in cold ($T = 0$) and hot β-stable hadronic matter. Assuming a first order phase transition, we calculate and compare the nucleation rate and the nucleation time due to thermal and quantum nucleation mechanisms. We show that above a threshold value of the central pressure a pure hadronic star (HS) is metastable to the conversion to a quark star (QS) (*i.e.* hybrid star or strange star). We introduce the concept of critical mass M_{cr} for cold HSs and proto-hadronic stars (PHSs), and the concept of limiting conversion temperature for PHSs. We show that PHSs with a mass $M < M_{cr}$ could survive the early stages of their evolution without decaying to QSs. Finally, we discuss the possible evolutionary paths of proto-hadronic stars.

1. Introduction

Neutron stars, the compact remnants of supernova explosions, are the densest macroscopic objects in the Universe. They represent the limit beyond which gravity overwhelm all the other forces of nature and lead to the formation of a black hole. In fact, neutron star structure calculations (see *e.g.* Refs. [1, 2, 3, 4, 5]) based on a large variety of modern equations of state (EOS) of hadronic matter, predict a maximum stellar central density (the one for the maximum mass star configuration) in the range of 4 – 8 times the saturation density ($\sim 2.8 \times 10^{14}$ g/cm^3) of nuclear matter. Thus the core of a neutron star is one of the best candidates in the universe where a phase of strong interacting matter with deconfined quarks could be found, and these compact stars can be viewed as natural laboratories to test the low temperature T and high baryon chemical potential region of the QCD phase diagram.

At high baryon chemical potential and low T, many QCD inspired models suggest the deconfinement transition to be a first order phase transition [6, 7]. As it is well known, first order phase transitions are triggered by the nucleation of a critical size drop of the new (stable) phase in a metastable mother phase. In the last few years, we have investigated [8]–[17] (see also Refs. [18, 19, 20, 21]) the astrophysical consequences of the nucleation process of quark matter (QM) in the core of massive pure hadronic compact stars (hadronic stars, HSs) *i.e.* neutron

Published under licence by IOP Publishing Ltd

stars in which no fraction of QM is present. In this contribution, we report some of the main findings of these studies.

2. Equation of state of dense matter

All the results we report in the present paper are relative to the finite temperature version of the following models for the EOS of dense matter. For the hadronic phase we use the Glendenning–Moszkowski model [22, 23], and particularly the GM1 parametrization [22, 23, 14]. The nucleon coupling constants are fitted to the bulk properties of nuclear matter. The inclusion of hyperons involves new couplings, which can be written in terms of the nucleonic ones: $g_{\sigma Y} = x_\sigma \, g_\sigma$, $g_{\omega Y} = x_\omega \, g_\omega$, $g_{\rho Y} = x_\rho \, g_\rho$. In this model [22, 23] it is assumed that all the hyperons in the baryonic octet have the same coupling and, in addition, it is assumed that $x_\rho = x_\sigma$. The binding energy of the Λ particle in symmetric nuclear matter $B_\Lambda/A = -28$ MeV $= x_\omega \, g_\omega \, \omega_0 - x_\sigma \, g_\sigma \sigma$ is used [22, 23] to determine x_ω in terms of x_σ. Measured neutron star masses can be used to put additional restrictions to the possible ranges of variability of the hyperon couplings. In this work we will consider the cases $x_\sigma = 0.6$ (hereafter $\mathrm{GM1}_{0.6}$) and $x_\sigma = 0.7$ ($\mathrm{GM1}_{0.7}$). Notice that the case with $x_\sigma = 0.6$ produces stars with a larger hyperon population (for a given stellar gravitational mass) with respect to the case $x_\sigma = 0.7$ [23, 14].

For the deconfined quark phase we have used: (i) the MIT bag model EOS [24] with $m_s = 150$ MeV, $m_u = m_d = 0$, $\alpha_s = 0$ and different values for the bag constant B; (ii) the Nambu–Jona-Lasinio (NJL) model [25], with the lagrangian density given in Ref. [26]; (iii) the Chromo Dielectric model [27, 28].

3. Phase equilibrium

For a first-order phase transition the conditions for phase equilibrium are given by the Gibbs' phase rule

$$T_H = T_Q \equiv T \,, \qquad P_H = P_Q \equiv P_0 \,, \qquad \mu_H(T, P_0) = \mu_Q(T, P_0) \tag{1}$$

where

$$\mu_H = \frac{\varepsilon_H + P_H - s_H T}{n_H} \,, \qquad \mu_Q = \frac{\varepsilon_Q + P_Q - s_Q T}{n_Q} \tag{2}$$

are the Gibbs' energies per baryon (average chemical potentials) for the hadron and quark phase respectively, ε_H (ε_Q), P_H (P_Q), s_H (s_Q) and n_H (n_Q) denote respectively the total (*i.e.*, including leptonic contributions) energy density, total pressure, total entropy density, and baryon number density for the hadron (quark) phase.

Above the "transition point" P_0 the hadronic phase is metastable, and the stable quark phase will appear as a results of a nucleation process. Virtual drops of the stable quark phase will arise from localized fluctuations in the state variables of the metastable hadronic phase. These fluctuations are characterized by a time scale $\nu_0^{-1} \sim 10^{-23}$ s. This time scale is set by the strong interactions (which are responsible for the deconfinement phase transition), and it is many orders of magnitude shorter than the typical time scale for the weak interactions. Therefore quark flavor must be conserved forming a vitual drop of QM. We will refer to this form of deconfined matter, in which the flavor content is equal to that of the β-stable hadronic system at the same pressure and temperature, as the Q*-phase. Soon afterward a critical size drop of Q*-matter is formed, the weak interactions will have enough time to act, changing the quark flavor fraction of the deconfined droplet to lower its energy, and a droplet of β-stable QM is formed (hereafter the Q-phase). This first seed of β-stable QM will trigger the conversion [29] of the pure hadronic star to a quark star (QS) (*i.e.* to a hybrid neutron star or to a strange star depending on the details of the EOS for quark matter used to model the phase transition).

The direct formation by fluctuations of a drop of β-stable QM is also possible in principle. However, it is strongly suppressed with respect to the formation of the Q*-phase drop by a factor

$\sim G_{\rm Fermi}^{2N/3}$ being N the number of particles in the critical size quark drop. This is so because the formation of a β-stable drop will imply the almost simultaneous conversion of $\sim N/3$ up and down quarks into strange quarks. For a critical size β-stable drop at the center of a neutron star it is found $N \sim 100 - 1000$, and therefore the suppresion factor is actually very tiny.

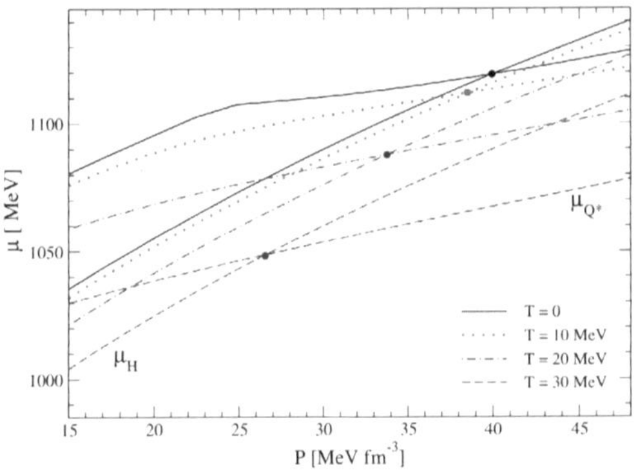

Figure 1. Gibbs energy per baryon of β-stable hadronic phase and Q*-phase, in neutrino-free matter, as a function of pressure, at different temperatures. Lines with the steeper slope refer to the hadronic phase. Full dots indicate the transition pressure P_0 for each temperature. GM1 EOS with $x_\sigma = 0.6$ for the hadronic phase and MIT bag model EOS with B = 85 MeV/fm^3 for the Q*-phase (GM1$_{0.6}$–B85 EOS).

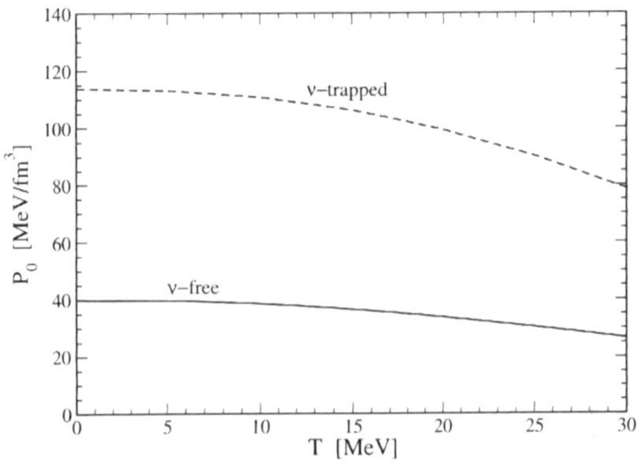

Figure 2. Phase equilibrium curve between the β-stable hadronic phase and the Q* phase. The continuous curve is relative to neutrino-free matter, the dashed curve to matter with trapped neutrinos. EOS: GM1 with $x_\sigma = 0.6$ plus MIT Bag model with B = 85 MeV/fm^3.

In Fig. 1 we plot the Gibbs' energies per baryon for the hadron-phase and for the Q*-phase in neutrino-free matter, at different temperatures (T = 0, 10, 20, 30 MeV). Results in Fig. 1 are obtained using the GM1 model with $x_\sigma = 0.6$ for the hadronic phase and the MIT Bag model with B = 85 MeV/fm^3 for the quark phase (hereafter the GM1$_{0.6}$–B85 EOS). Lines with the steeper slope refer to the hadron phase. As we see, the transition pressure P_0 (indicated by a full dot) decreases when the hadronic matter temperature is increased.

The phase equilibrium curve $P_0(T)$ between the β-stable hadronic phase and the Q^*-phase is shown in Fig. 2 for neutrino-free matter and matter with trapped neutrinos, making use of the GM1$_{0.6}$–B85 EOS. The region of the P_0–T plane above each curve represents the deconfined Q*-phase. As expected [30, 31, 11, 32] neutrino trapping in β-stable hadronic matter inhibits the quark deconfinement phase transition, thus the global effect of neutrino-trapping is to produce a shift of the phase equilibrium curve toward higher values of the pressure in the P_0–T plane.

In Fig. 3, we show the phase equilibrium curve, for neutrino-free matter, in the case of the NJL model (left panel) or the Chromo Dielectric model (right panel) to describe the deconfined phase. For the hadron phase we take the GM1 model with $x_\sigma = 0.7$ (GM1$_{0.7}$) in both cases. Notice that in the case of the NJL model the transition pressure $P_0(T)$ is substantially higher than the one in the case of the MIT Bag or Chromo Dielectric models to describe the deconfined phase. As discussed in detail in Ref. [17], this behaviour can be traced back the large value of the stange quark effective mass in the NJL model.

As is well known, for a first-order phase transition the derivative dP_0/dT is related to the

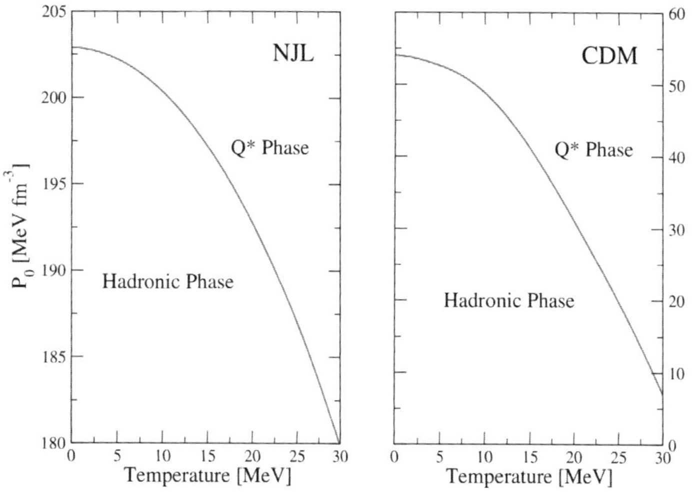

Figure 3. Phase equilibrium curve between the β-stable hadronic phase and the Q*-phase. The EOS for the hadronic phase is obtained using the GM1 model with $x_\sigma = 0.7$, the EOS for the Q^*-phase using the NJL model (left panel) or the Chromo Dielectric model (right panel).

specific latent heat $\mathcal{Q}$ of the phase transition by the Clapeyron-Clausius equation

$$\frac{dP_0}{dT} = -\frac{n_H n_{Q^*}}{n_{Q^*} - n_H}\frac{\mathcal{Q}}{T} \tag{3}$$

$$\mathcal{Q} = \tilde{W}_{Q^*} - \tilde{W}_H = T(\tilde{S}_{Q^*} - \tilde{S}_H) \tag{4}$$

where $\tilde{W}_H$ ($\tilde{W}_{Q^*}$) and $\tilde{S}_H$ ($\tilde{S}_{Q^*}$) denote the enthalpy per baryon and entropy per baryon for the hadron (quark) phase, respectively. The specific latent heat $\mathcal{Q}$ and the phase numbers densities n_H and n_{Q^*} at phase equilibrium are reported in Tables 1 and 2 in the case of the $\mathrm{GM1}_{0.6}$–B85 equation of state. Results in Table 1 refer to neutrino-free matter, whereas those in Table 2 refer to matter with trapped neutrinos. As expected for a first-order phase transition, one has a discontinuity jump in the phase number densities: in our particular case $n_{Q^*}(T, P_0) > n_H(T, P_0)$. This result, together with the positive value of $\mathcal{Q}$ (*i.e.* the deconfinement phase transition absorbs heat), tells us (see Eq. (3)) that the phase transition temperature decreases with pressure (as in the melting of ice).

The effect of neutrino trapping on the phase equilibrium properties of the system can be seen comparing the results reported in Tables 1 and 2. As we see, the phase number densities n_H and n_{Q^*} at phase equilibrium are shifted to higher values, and the specific latent heat $\mathcal{Q}$ is increased with respect to the neutrino-free matter case.

4. Quark matter nucleation in cold hadronic stars

In a cold ($T = 0$) and neutrino-free pure hadronic star the formation of the first drop of QM could take place solely via a quantum nucleation process. The basic quantity needed to calculate the nucleation time is the energy barrier separating the Q*-phase from the metastable hadronic phase. This energy barrier, which represents the difference in the free energy of the system with and without a Q*-matter droplet, can be written as [33, 34]

$$U(\mathcal{R}) = \frac{4}{3}\pi n_{Q^*}(\mu_{Q^*} - \mu_H)\mathcal{R}^3 + 4\pi\sigma\mathcal{R}^2 \tag{5}$$

where $\mathcal{R}$ is the radius of the droplet (supposed to be spherical), and σ is the surface tension for the surface separating the hadron from the Q*-phase. The energy barrier has a maximum at the critical radius $\mathcal{R}_c = 2\sigma/[n_{Q^*}(\mu_H - \mu_{Q^*})]$. We neglected the term associated with the

Table 1. The specific (*i.e.* per baryon) latent heat $\mathcal{Q}$ and the phase number densities n_H and n_{Q^*} at phase equilibrium. GM1 EOS with $x_\sigma = 0.6$ for the hadronic phase, MIT bag model with $B = 85$ MeV/fm^3 for the quark phase. Results for neutrino-free matter.

T MeV	$\mathcal{Q}$ MeV	n_{Q^*} fm^{-3}	n_H fm^{-3}	P_0 MeV/fm^3
0	0.00	0.453	0.366	39.95
5	0.56	0.451	0.364	39.74
10	2.40	0.447	0.358	38.58
15	5.71	0.439	0.348	36.55
20	10.60	0.428	0.334	33.77
25	17.17	0.414	0.316	30.36
30	25.44	0.398	0.294	26.53

Table 2. Same as Table 1, but with trapped neutrinos.

T MeV	$\mathcal{Q}$ MeV	n_{Q^*} fm^{-3}	n_H fm^{-3}	P_0 MeV/fm^3
0	0.00	0.603	0.516	113.77
5	0.65	0.601	0.514	113.11
10	2.87	0.594	0.509	110.69
15	6.78	0.580	0.499	106.17
20	12.65	0.560	0.483	99.18
25	20.21	0.534	0.462	90.12
30	29.88	0.502	0.434	78.65

curvature energy and also the terms connected with the electrostatic energy, since they are known to only introduce small corrections [34, 10]. The value of the surface tension σ for the interface separating the quark and hadron phase is poorly known, and typically values used in the literature range within $10-50$ MeV fm^{-2} [35]). We assume σ to be temperature independent and we take $\sigma = 30$ MeV fm^{-2}.

The quantum nucleation time τ_q can be straightforwardly evaluated within a semi-classical approach [33, 34, 9] and it can be expressed as

$$\tau_q = (\nu_0 p_0 N_c)^{-1} , \qquad (6)$$

where p_0 is the probability of tunneling the energy barrier $U(\mathcal{R})$ in its ground state, ν_0 is the oscillation frequency of a virtual drop of the Q*-phase in the potential well, and $N_c \sim 10^{48}$ is the number of nucleation centers expected in the innermost part ($r \le R_{nuc} \sim 100$ m) of the HS, where the pressure and temperature (when we will consider the finite T case) can be considered constant and equal to their central values.

Thus a pure hadronic star, having a central pressure P_c larger than the static transition pressure P_0 for the formation of the Q*-phase, is metastable to the "decay" (conversion) to a quark star (QS) *i.e.* to a more compact stellar configuration in which deconfined quark matter is present. These metastable HSs have a *mean-life time* which is related to the nucleation time to

Journal of Physics: Conference Series **336** (2011) 012021 doi:10.1088/1742-6596/336/1/012021

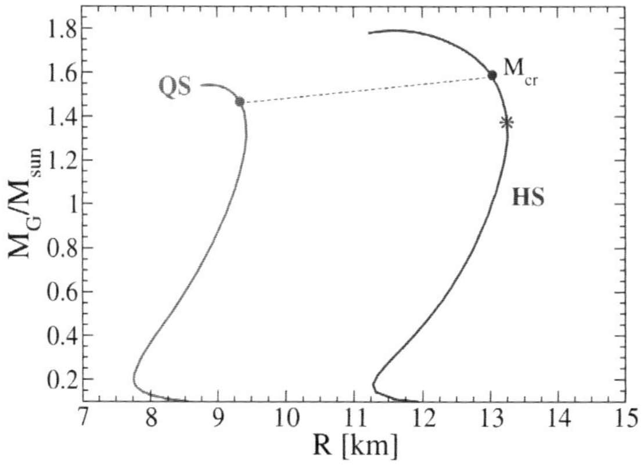

Figure 4. Mass-radius relation for a pure hadronic stars (HS) and hybrid star configurations (QS). The configuration marked with an asterisk represents the HS for which $\tau_q = \infty$ (*i.e.* $P_c = P_0$). The conversion process of the HS, with a gravitational mass equal to M_{cr}, into the final QS is denoted by the full circles connected by a dashed line. Results are relative to the GM1 model with $x_\sigma = 0.6$ for the hadronic phase and the MIT Bag model EOS with B = 85 MeV/fm^3 for the quark phase. The surface tension is $\sigma = 30$ MeV/fm^2. Stellar masses are in unit of the mass of the sun, $M_{sun} = 1.989 \times 10^{33}$ g.

form the first critical-size drop of deconfined matter in their interior (the actual *mean-life time* of the HS will depend on the mass accretion or on the spin-down rate which modifies the nucleation time via an explicit time dependence of the stellar central pressure). Following Refs. [8, 9, 10] we define as *critical mass* M_{cr} of the metastable HSs, the value of the gravitational mass for which the nucleation time is equal to one year: $M_{cr} \equiv M_{HS}(\tau_q = 1\text{yr})$. Pure hadronic stars with $M_{HS} > M_{cr}$ are very unlikely to be observed. Thus M_{cr} plays the role of an *effective maximum mass* for the hadronic branch of compact stars. Notice that the Oppenheimer–Volkov maximum mass M_{HS}^{max} is determined by the overall stiffness of the EOS for hadronic matter, whereas the value of M_{cr} will depend in addition on the bulk properties of the EOS for quark matter and on the properties at the interface between the confined and deconfined phases of matter (*e.g.*, the surface tension σ).

These findings are exemplified in Fig. 4, where we show the mass-radius (MR) curve for hadronic stars (HS) and that for quark stars (QS). The configuration marked with an asterisk on the hadronic MR curve represents the HS for which the central pressure is equal to P_0 and thus $\tau_q = \infty$. The full circle on the HS sequence represents the critical mass configuration M_{cr}, in the case $\sigma = 30$ MeV/fm^2. The full circle on the QS mass-radius curve represents the hybrid star which is formed from the conversion of the hadronic star with $M_{HS} = M_{cr}$. We assume [29] that during the stellar conversion process the total number of baryons in the star is conserved. The stellar conversion process, described so far, will start to populate the new branch of quark stars (the part of the QS sequence above the full circle). Long term accretion on the QS can next produce stars with masses up to the limiting mass M_{QS}^{max} for the quark star configurations.

5. Quark matter nucleation in proto-hadronic stars

A neutron star at birth (proto-neutron star) is very hot (T = 10 – 30 MeV) with neutrinos being still trapped in the stellar interior [36, 37, 30, 38, 39, 40]. Subsequent neutrino diffusion causes deleptonization and heats the stellar matter to an approximately uniform entropy per baryon $\tilde{S} = 1 - 2$ (in units of the Boltzmann's constant k_B). Depending on the stellar composition,

Journal of Physics: Conference Series **336** (2011) 012021 doi:10.1088/1742-6596/336/1/012021

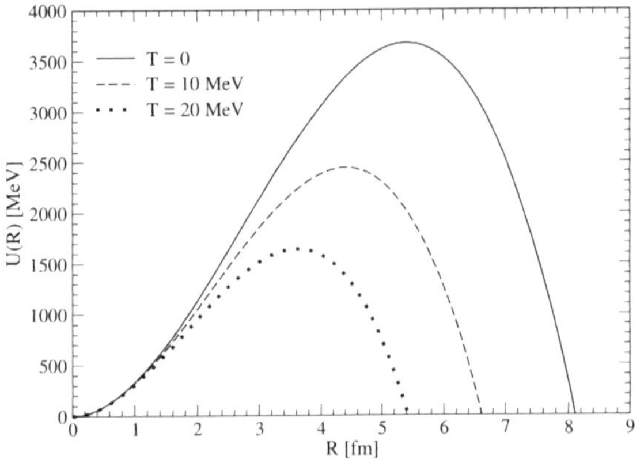

Figure 5. Energy barrier for a virtual drop of the Q*-phase in β-stable neutrino-free hadronic matter as a function of the droplet radius and for different temperatures for a fixed pressure $P = 57$ MeV/fm^3. Results are relative to the GM1$_{0.6}$–B85 equation of state and surface tension $\sigma = 30$ MeV/fm^2.

during this stage neutrino escape can lead the more "massive" stellar configurations to the formation of a black hole [41, 30]. However, if the mass of the star is sufficiently small, the star will remain stable and it will cool to temperatures well below 1 MeV within a cooling time $t_{cool} \sim$ a few 10^2 s, as the neutrinos continue to carry energy away from the stellar material [36, 30, 38]. Thus in a proto-neutron star, the quark deconfinement phase transition will be likely triggered by a thermal nucleation process [42, 43, 44, 45]. In fact, for sufficiently high temperatures, thermal nucleation is a much more efficient process with respect to the quantum nucleation mechanism. In the following we will establish the physical conditions under which a newborn hadronic star (proto-hadronic star, PHS) could survive the early stages of its evolution without "decaying" to a quark star.

According to the Langer theory [46, 47] of homogeneous nucleation the thermal nucleation rate can be written [47] as

$$I = \frac{\kappa}{2\pi}\Omega_0 \exp(-U(\mathcal{R}_c, T)/T) \tag{7}$$

where κ is the so-called dynamical prefactor, which is related to the growth rate of the drop radius $\mathcal{R}$ near the critical radius ($\mathcal{R}_c$), and Ω_0 is the so-called statistical prefactor, which measures the phase-space volume of the saddle-point region around $\mathcal{R}_c$. We have used [15, 16] for κ and Ω_0 the expressions derived in Refs. [48, 49], where the Langer nucleation theory has been extended to the case of first order phase transitions occurring in relativistic systems, as in the case of the quark deconfinement transition. The dominant factor in the nucleation rate (7) is the exponential, in which $U(\mathcal{R}_c, T)$ is the activation energy, *i.e.* the change in the free energy of the system required to activate the formation of a critical size droplet. This quantity has been calculated [15, 16] using the generalization of Eq. (5) to the case of $T \neq 0$ (thus using finite temperature EOS for the hadronic and the quark phases), and assuming a temperature independent surface tension.

The thermal nucleation time τ_{th}, relative to the innermost stellar region ($V_{nuc} = (4\pi/3)R_{nuc}^3$, with $R_{nuc} \sim 100$ m) where almost constant pressure and temperature occur, can thus be written as

$$\tau_{th} = (V_{nuc}\, I)^{-1} \,. \tag{8}$$

In Fig. 5, we represent the energy barrier for a virtual drop of the Q*-phase in the neutrino-free hadronic phase as a function of the droplet radius and for different temperatures at a fixed

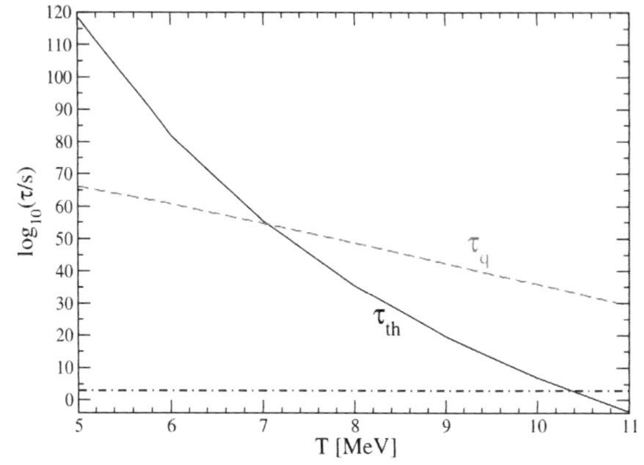

Figure 6. Thermal (τ_{th}) and quantum (τ_q) nucleation time of quark matter (Q*-phase) in β-stable neutrino-free hadronic matter as a function of temperature at fixed pressure P = 57 MeV/fm^3. The crossover temperature is T_{co} = 7.05 MeV. The limiting conversion temperature for the proto-hadronic star is, in this case, Θ = 10.3 MeV, obtained from the intersection of the thermal nucleation time curve (continuous line) and the dot-dashed line representing $\log_{10}(\tau/\mathrm{s}) = 3$. The surface tension is σ = 30 MeV/fm^2. Results are relative to $\mathrm{GM1}_{0.6}$–B85 EOS.

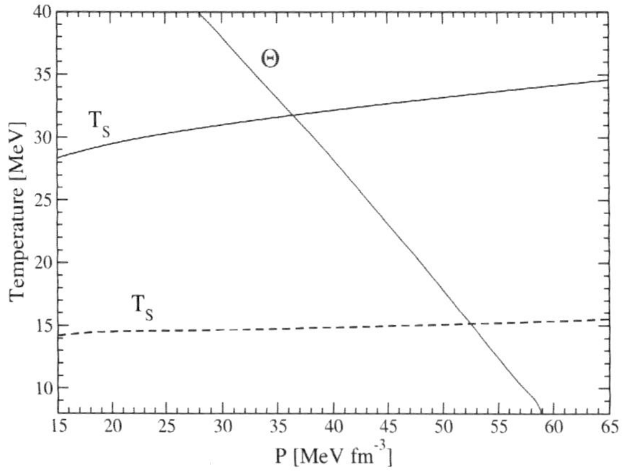

Figure 7. The limiting conversion temperature Θ for a newborn hadronic star as a function of the central stellar pressure. Newborn hadronic stars with a central temperature and pressure located on the right side of the curve $\Theta(P)$ will nucleate a Q*-matter drop during the early stages of their evolution, and will finally evolve to cold and deleptonized quark stars, or will collapse to black holes. The lines labeled T_S represent the stellar matter temperature as a function of pressure at fixed entropies per baryon $\tilde{S}/k_B = 1$ (dashed line) and 2 (solid line). Results are relative to $\mathrm{GM1}_{0.6}$–B85 EOS.

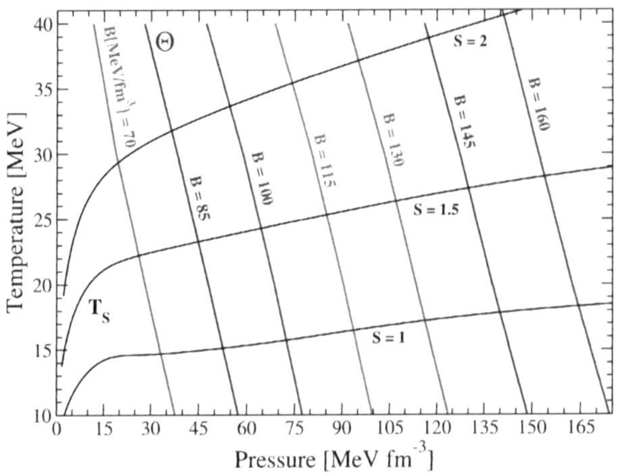

Figure 8. The limiting conversion temperature Θ for a newborn hadronic star as a function of the central stellar pressure for different values of the bag constant B. The lines labeled T_S represent the stellar matter temperature as a function of pressure at fixed entropies per baryon $\tilde{S}/k_B = 1$, 1.5, 2. Results for neutrino-free matter.

pressure $P = 57$ MeV/fm^3. Results in Fig. 5 are obtained using the GM1$_{0.6}$–B85 equation of state. As expected, from the results plotted in Fig. 1, the energy barrier $U(\mathcal{R}, T)$ and the droplet critical radius $\mathcal{R}_c$ decrease as the matter temperature is increased. This effect favors the Q*-phase formation and, in particular, increases (decreases) the quantum nucleation rate (nucleation time τ_q) with respect to the corresponding quantities calculated at $T = 0$.

In Fig. 6, we plot the quantum and thermal nucleation times of the Q*-phase in β-stable neutrino-free hadronic matter as a function of temperature and at a fixed pressure $P = 57$ MeV/fm^3. As expected, we find a crossover temperature T_{co} above which thermal nucleation is dominant with respect to the quantum nucleation mechanism. For the case reported in Fig. 6, we have $T_{co} = 7.05$ MeV and the corresponding nucleation time is $\log_{10}(\tau/\mathrm{s}) = 54.4$.

Having in mind the physical conditions in the interior of a PHS [36, 30], to establish if this star will survive the early stages of its evolution without decaying to a quark star, one has to compare the quark matter nucleation time $\tau = \min(\tau_q, \tau_{th})$ with the cooling time $t_{cool} \sim$ a few 10^2 s. If $\tau >> t_{cool}$ then quark matter nucleation will not likely occur in the newly formed star, and this star will evolve to a cold deleptonized configuration. We thus introduce the concept of *limiting conversion temperature* Θ for the proto-hadronic star and define it as the value of the stellar central temperature T_c for which the Q*-matter nucleation time is equal to 10^3 s. The limiting conversion temperature Θ will clearly depend on the value of the stellar central pressure (and thus on the value of the stellar mass).

The limiting conversion temperature Θ is plotted in Fig. 7 as a function of the stellar central pressure. A proto-hadronic star with a central temperature $T_c > \Theta$ will likely nucleate a Q*-matter drop during the early stages of its evolution, and will finally evolve to a cold and deleptonized quark star, or will collapse to a black hole (depending on the value of the stellar baryonic mass M_B and on the EOS).

In Fig. 8 we plot the limiting conversion temperature Θ for a newborn HS for different values of the bag constant. The increase in B produces a growth of the region of the P-T plane where the proto-hadronic star could survive Q* nucleation and thus evolve to a cold hadronic star.

For an isoentropic stellar core [36, 30], the central temperature of the proto-hadronic star is given, for the GM1$_{0.6}$ EOS model, by the lines labeled by T_S in Fig. 7, relative to the case $\tilde{S} = 1\ k_B$ (dashed curve) and $\tilde{S} = 2\ k_B$ (continuous curve). The intersection point (P_S, Θ_S) between the two curves $\Theta(P)$ and $T_S(P)$ thus gives the central pressure and temperature of the configuration that we denote as the *critical mass* configuration of the proto-hadronic stellar sequence. Taking $\sigma = 30$ MeV/fm^2 and $\tilde{S} = 2\ k_B$ we get $M_{cr} = 1.390\ M_{sun}$ for the gravitational critical mass and $M_{B,cr} = 1.492\ M_{sun}$ for the baryonic critical mass (being $M_{sun} = 1.989 \times 10^{33}$ g

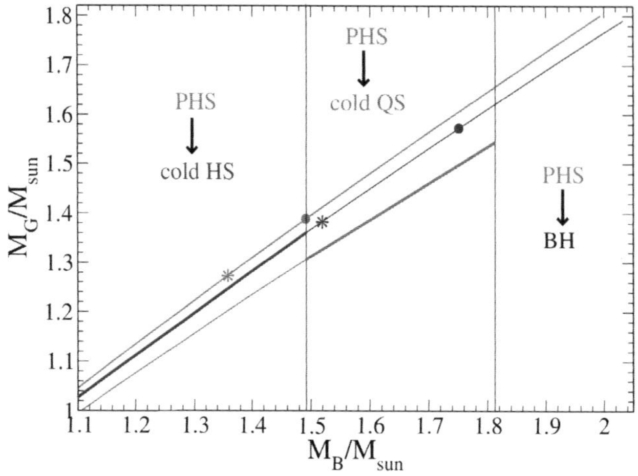

Figure 9. Stellar equilibrium sequences in the gravitational–baryonic mass plane for the $\mathrm{GM1}_{0.6}$–B85 EOS and $\sigma = 30$ MeV/fm^2. The upper (red) line represents the PHS sequence ($\tilde{S} = 2\ k_B$). The middle (blue) line represents the cold HS sequence. The asterisk and the full circle on these lines represent respectively the stellar configuration with $\tau = \infty$ and the critical mass configuration M_{cr}. The lower (green) line represent the cold QS sequence. Assuming $M_B =$ const, the evolution of a PHS in this plane occurs along a vertical line.

the mass of the sun).

The evolution of a PHS within our scenario is delineated in Fig. 9, where we plot the appropriate stellar equilibrium sequences in the gravitational–baryonic mass plane for the $\mathrm{GM1}_{0.6}$–B85 EOS and $\sigma = 30$ MeV/fm^2. The upper (red) line represents the PHS sequence, *i.e.* isoentropic HSs ($\tilde{S} = 2\ k_B$) and neutrino-free matter. The middle (blue) line represents the cold HS sequence. The asterisk and the full circle on these lines represent respectively the stellar configuration with $\tau = \infty$ and the critical mass configuration. Finally, the lower (green) line represent the cold QS sequence. We assume $M_B =$ const during these stages of the stellar evolution. Thus according to the results in Fig. 9, proto-hadronic stars with a baryonic mass $M_B < 1.492\ M_{sun}$ will survive Q*-matter *early nucleation* (*i.e.* nucleation within the cooling time $t_{cool} \sim$ a few 10^2 s) and in the end they will form stable ($\tau = \infty$) cold hadronic stars. Proto-hadronic stars with $1.492\ M_{sun} \le M_B < 1.813\ M_{sun}$ (the maximum baryonic mass $M^{max}_{B,QS}$ of the cold QS sequence for the present EOS) will experience early nucleation of a Q*-matter drop and will ultimately form a cold deleptonized quark star. The last possibility is for PHSs having $M_B > 1.813\ M_{sun}$. In this case the early nucleation of a Q*-matter drop will trigger a stellar conversion process to a cold QS configuation with $M_B > M^{max}_{B,QS}$, thus these PHSs will finally forms black holes.

The outcomes of this scenario are not altered by neutrino trapping effects in hot β-stable hadronic matter. In fact, for proto-hadronic stars with trapped neutrinos (νPHSs) (with a lepton fraction $Y_L = 0.4$ and $\tilde{S} = 2\ k_B$) we find [16] (for the same EOS and surface tension σ used in Fig. 9) a critical baryonic mass $M_{B,cr} = 1.96\ M_{sun}$. Thus νPHSs with $M_B \ge M_{B,cr}$ after neutrino escape and cooling will finally evolve to black holes. The fate of a νPHS with $M_B < M_{B,cr}$ is the same as the corresponding neutrino-free PHS with equal baryonic mass.

In Fig. 10 we plot the PHS, cold HS, and cold QS sequences in the gravitational–baryonic mass plane for the case of the NJL model for the quark phase and the GM1 model in the case of pure nucleonic matter (right panel) or hyperonic matter with $x_\sigma = 0.7$ (left panel). It is clearly

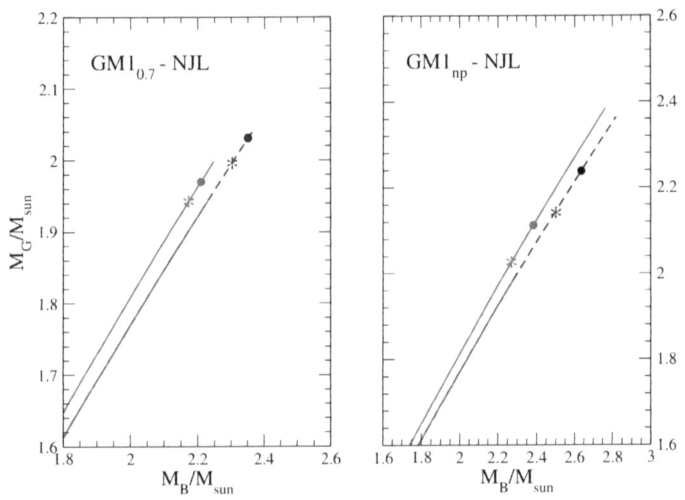

Figure 10. Same as in the previous figure but in the case of the NJL model for the quark phase and the Glendenning–Moszkowski model for hyperonic matter with $x_\sigma = 0.7$ (GM1$_{0.7}$) (left panel) and pure nucleonic matter GM1$_{\rm np}$ (right panel).

seen that in the case of the NJL model it is almost impossible to populate the QS branch. Cold quark stars can be formed in the case of $x_\sigma = 0.7$ (left panel) for a very narrow range of baryonic stellar masses $2.20 < M_B/M_\odot < 2.23M_\odot$.

6. Conclusions

In summary, in this work we have studied the quark deconfinement phase transition in hot β-stable hadronic matter and explored some of its consequences for the physics of neutron stars at birth. We calculated and compared the nucleation time due to thermal and quantum nucleation mechanisms, and computed the crossover temperature above which thermal nucleation dominates the finite temperature quantum nucleation mechanism. In addition, we introduced the new concept of *limiting conversion temperature* Θ for proto-hadronic stars and extended the concept of critical mass (introduced in Refs. [8, 9, 10]) to the case of finite temperature hadronic stars.

Our main finding is that proto-hadronic stars with a gravitational mass lower than the critical mass M_{cr} could survive the early stages of their evolution without decaying to a quark star. This outcome contrasts with the predictions of earlier studies [42, 43, 44, 45] where it was inferred that all the pure hadronic compact stars, with a central temperature above 2 – 3 MeV, are converted to quark stars within the first seconds after their birth. However, the prompt formation of a critical size drop of quark matter could take place when $M > M_{cr}$. These proto-hadronic stars evolve to cold and deleptonized quark stars or collapse to a black holes.

We found that for the NJL model the presence of hyperons disfavour the phase transition pushing the transition point to very high densities. In addition, in the case of the NJL model it is almost impossible to populate the quark star branch and quark matter nucleation will lead to the formation of a black hole. Thus within the NJL model for the quark phase, all compact stars are pure hadronic stars.

Accurate measurements of both the mass and radius of a few individual "neutron stars" [50, 51] or measurements of temperature profiles of accretion discs around rapidly spinning compact stars [52] could shed light on the validity of the scenario discussed in the present work.

References

[1] Lattimer J M and Prakash M 2001 *Astrophys. Jour.* **550** 426
[2] Bombaci I 2007 *Eur. Phys. J.* A **31** 810
[3] Peng G X, Li A and Lombardo U 2008 *Phys. Rev.* C **77** 065807
[4] Li Z H and and Schulze H-J 2008 *Phys. Rev.* C **78** 028801
[5] Vidaña I, Logoteta D, Providência C, Polls A and Bombaci I 2011*EPL (Europhysics letter)* 94 11002
[6] Hsu S D H and Schwetz M 1998 *Phys. Lett.* **B 432** 2003
[7] Fodor Z and Katz S D 2004 *JHEP* **04** 050
[8] Berezhiani Z, Bombaci I, Drago A, Frontera F and Lavagno A 2002 *Nucl. Phys. B - Proceedings Supplement* **113** 268
[9] Berezhiani Z, Bombaci I, Drago A, Frontera F and Lavagno A 2003 *Astrophys. Jour.* **586** 1250
[10] Bombaci I, Parenti I and Vidaña I 2004 *Astrophys. Jour.* **614**, 314
[11] Vidaña I, Bombaci I and Parenti I 2005 *J. Phys. G: Nucl. Part. Phys.* **31** S1165
[12] Lugones G and Bombaci I 2005 *Phys. Rev.* D **72** 065021
[13] Bombaci I, Lugones G and Vidaña I 2007 *Astron. and Astrophys.* **462** 1017
[14] Bombaci I, Panda P K, Providência C and Vidaña I 2008 *Phys. Rev.* D **77** 083002
[15] Bombaci I, Logoteta D, Panda P K, Providência C and Vidaña I 2009 *Phys. Lett.* B **680** 448
[16] Bombaci I, Logoteta D, Providência C and Vidaña I 2011 *Astron. and Astrophys.* **528** A71
[17] Logoteta D, Bombaci I, Providência C and Vidaña I 2011 *Phys. Rev.* D **XX** ppp (submitted)
[18] Drago A, Lavagno A and Pagliara G 2004 *Phys. Rev.* D **69** 057505
[19] Drago A, Pagliara G and Schaffner-Bielich J 2008 *J. Phys. G* **35** 014052
[20] Bambi C and Drago A 2008 *Astropart. Phys.* **29** 223
[21] Drago A, Lavagno A and Parenti I 2007 *Astrophys. Jour.* **659** 1519
[22] Glendenning N K and Moszkowski S A 1991 *Phys. Rev. Lett.* **67** 2414
[23] Glendenning N K 2000, Compact Stars: Nuclear Physics, Particle Physics, and General Relativity, Springer, New York.
[24] Farhi E and Jaffe R L 1984 *Phys. Rev.* D **30** 2379
[25] Nambu Y and Jona-Lasinio G 1961 *Phys Rev.* **122** 345
[26] Rehberg P, Klevansky S P and Hufner J 1996 *Phys. Rev.* C **53** 410
[27] Pirner H J, Chanfray G and Nachtmann O 1984 *Phys. Lett.* B **147** 249
[28] Drago A and Tambini E 1999 *J. Phys.* G **25** 971
[29] Bombaci I and Datta B 2000 *Astrophys. Jour.* **530** L69
[30] Prakash M, Bombaci I, Prakash M, Ellis P J, Lattimer J M and Knorren R 1997 Phys. Rep. **280** 1
[31] Lugones G and Benvenuto O G 1998 *Phys. Rev.* D **58** 083001
[32] Lugones G, Grunferld A G, Scoccola N N and Villavicencio C 2009 *Phys. Rev.* D **80** 045017
[33] Lifshitz I M and Kagan Y 1972 *Sov. Phys. JETP* **35** 206
[34] Iida K and Sato K 1998 *Phys. Rev.* C **58** 2538
[35] Heiselberg H, Pethick C J and Staubo E F 1993 *Phys. Rev. Lett.* **70** 1355
[36] Burrow A and Lattimer J M 1986 *Astrophys. Jour.* **307** 178
[37] Bombaci I, Prakash M, Prakash M, Ellis P J, Lattimer J M and Brown G E 1995 *Nucl. Phys.* A **583** C623
[38] Pons J A, Reddy S, Prakash M, Lattimer J M and Miralles J A 1999 *Astrophys. Jour.* **513** 780
[39] Vidaña I, Bombaci I, Polls A and Ramos A 2003 *Astron. and Astrophys.* **399** 687
[40] Margueron J, Vidaña I and Bombaci I 2003 *Phys. Rev.* C **68** 055806
[41] Bombaci I 1996 *Astron. and Astrophys.* **305** 871
[42] Horvath J E, Benvenuto O G and Vucetich H 1992 *Phys. Rev.* D **45** 3865
[43] Horvath J E 1994 *Phys. Rev.* D **49** 5590
[44] Olesen M L and Madsen J 1994 *Phys. Rev.* D **49** 2698
[45] Harko T, Cheng K S and Tang P S 2004 *Astrophys. Jour.* **608** 945
[46] Langer J S 1969 *Ann. Phys. (N.Y.)* **54** 258
[47] Langer J S and Turski L A 1973 *Phys. Rev.* A **8** 3230
[48] Csernai L and Kapusta J I 1992 *Phys. Rev.* D **46** 1379
[49] Venugopalan R and Vischer A P 1994 *Phys. Rev.* E **49** 5849
[50] Bhattacharyya S 2010 *Adv. Space Res.* **45** 949
[51] Steiner A W, Lattimer J M and Brown E F 2010 *Astrophys. Jour.* **722** 33
[52] Bhattacharyya S, Thampan A V and Bombaci I 2001 *Astr. and Astrophys.* **372** 925

Neutron star structure with hyperons and quarks

H-J Schulze

INFN Sezione di Catania, Dipartimento di Fisica, Via Santa Sofia 64, I-95123 Catania, Italy

Abstract. The high-density nuclear equation of state within the Brueckner-Hartree-Fock approach is discussed. Particular attention is paid to the effects of nucleonic three-body forces, the presence of hyperons, the joining with an eventual quark matter phase, and the extension to finite temperature. The resulting properties of neutron stars, in particular the mass-radius relation, are determined. It turns out that in this approach stars heavier than about 1.4 solar masses contain necessarily quark matter.

1. Introduction

During the last decades the increasing interest for the equation of state (EOS) of nuclear matter has stimulated a lot of theoretical activity. The properties of nuclear matter at high density and large isospin asymmetry are relevant for astrophysical objects, i.e., supernovae and neutron stars (NS). In particular, the structure of a NS is very sensitive to the compressibility and the symmetry energy. The NS mass, measured in binary systems, has been proposed as a constraint for the EOS of nuclear matter [1, 2, 3].

One of the most advanced microscopic approaches to the EOS of nuclear matter is the Brueckner theory. In recent years, it has made solid progress regarding the convergence of the underlying Brueckner-Bethe-Goldstone (BBG) expansion, the addition of microscopic three-body forces (TBF), the inclusion of hyperons, and the extension to finite temperature. These issues are reviewed in this contribution and results for NS structure based on the resulting EOS of dense hadronic matter are presented. Doing so, it turns out necessary to consider the transition to quark matter in the interior of sufficiently massive stars, which will also be discussed in detail.

2. Brueckner theory

The nonrelativistic BBG expansion for the energy of nuclear matter E/A can be cast as a power series in terms of the number of hole lines contained in the corresponding diagrams, which amounts to a density power expansion [4]. The two hole-line truncation is named the Brueckner-Hartree-Fock (BHF) approximation. At this order the energy E_2 is very much affected by the choice of the auxiliary single-particle (s.p.) potential. But, adding the three-hole line contributions E_3, the resulting EOS is almost insensitive to the choice of the auxiliary potential, and very close to the result E_2 with the continuous choice of the s.p. potential [5].

In spite of this satisfactory convergence, the saturation density misses the empirical value $\rho_0 \approx 0.17\ \mathrm{fm}^{-3}$ extracted from the central density of heavy nuclei. This confirms the belief that the concept of a many-nucleon system interacting via only two-body forces is not adequate to describe nuclear matter, especially at high density. To this purpose three-body forces have to be taken into account.

Published under licence by IOP Publishing Ltd

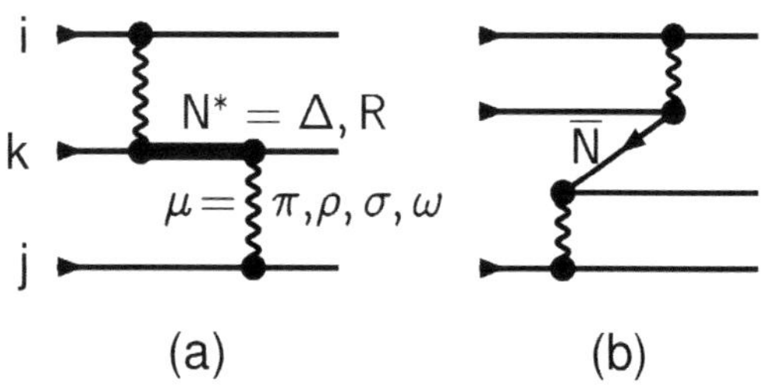

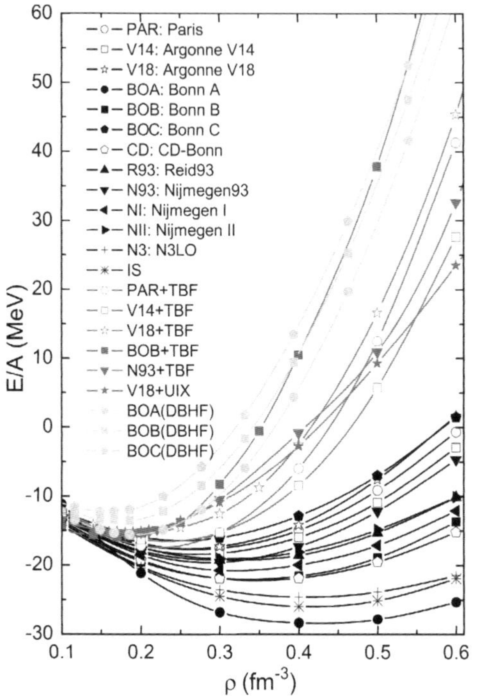

Figure 1. (Color online) Above: Diagrams contributing to the microscopic TBF. Right: Energy per nucleon of symmetric nuclear matter employing different potentials and TBF.

3. Three-body forces

The microscopic TBF of Refs. [6] are based on meson-exchange mechanisms accompanied by the excitation of nucleonic resonances, as represented by the diagram figure 1(a), namely the excitation of the isobar $\Delta(1232)$ resonance via the exchange of light (π,ρ) mesons, the lowest non-isobar nucleon resonance $N^*(1440)$ excited mainly by heavy meson (σ,ω) exchanges, as well as the generation of a virtual $N\bar{N}$ pair [diagram (b)]. The meson-exchange parameters of this model are chosen compatible with the underlying nucleon-nucleon (NN) potential. A second class of TBF widely used in the literature are the phenomenological Urbana TBF [7], which consider only $\Delta(1232)$ exchange and add a phenomenological repulsive scalar contribution.

The effect of these TBF is a stiffening of the EOS and a consequent remarkable improvement of the saturation properties of symmetric nuclear matter [8], as shown in figure 1 (right) for the EOS bases on different modern NN potentials with (red curves) and without (black curves) TBF.

4. Neutron star structure

In order to study the effects of different TBF on NS structure, the composition and the EOS of charge-neutral matter consisting of neutrons, protons, and leptons (e^-, μ^-) in beta equilibrium are calculated in the following standard way [9]: The Brueckner calculation yields the energy density of baryon/lepton matter as a function of the different partial densities,

$$\varepsilon(\rho_n, \rho_p, \rho_e, \rho_\mu) = (\rho_n m_n + \rho_p m_p) + (\rho_n + \rho_p)\frac{E}{A}(\rho_n, \rho_p) + \varepsilon_e + \varepsilon_\mu \,, \tag{1}$$

where $\varepsilon_{e,\mu}$ are the contributions of the non-interacting leptons. The various chemical potentials (of the species $i = n, p, e, \mu$) can then be computed straightforwardly,

$$\mu_i = \frac{\partial \varepsilon}{\partial \rho_i} \,, \tag{2}$$

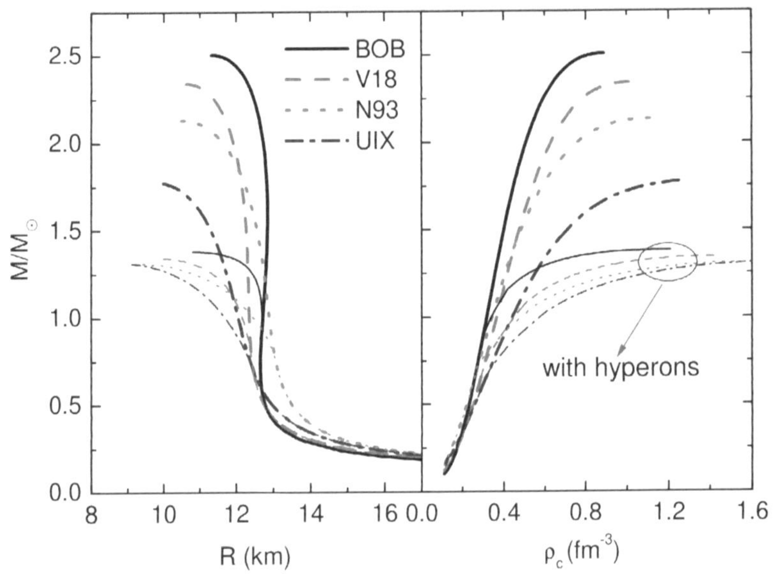

Figure 2. (Color online) The NS gravitational mass (in units of solar mass $M_\odot$) vs. the radius (left panel) and the central baryon density ρ_c (right panel) for different hadronic EOSs with (thin curves) or without (thick curves) hyperons.

and the equations for beta equilibrium and charge neutrality (b_i and q_i denoting baryon number and charge of species i),

$$\mu_i = b_i\mu_n - q_i\mu_e \quad , \quad \sum_i \rho_i q_i = 0 \, , \tag{3}$$

allow to determine the equilibrium composition $\{\rho_i(\rho)\}$ at given baryon density $\rho = \rho_n + \rho_p$ and finally the EOS,

$$p(\rho) = \rho^2 \frac{d}{d\rho} \frac{\varepsilon(\{\rho_i(\rho)\})}{\rho} = \rho\frac{d\varepsilon}{d\rho} - \varepsilon = \rho\mu_n - \varepsilon \, . \tag{4}$$

In order to calculate the mass-radius relation, one has then to solve the well-known Tolman-Oppenheimer-Volkov equations [9],

$$\frac{dp}{dr} = -\frac{Gm}{r^2}\frac{(\varepsilon + p)(1 + 4\pi r^3 p/m)}{1 - 2Gm/r} \quad , \quad \frac{dm}{dr} = 4\pi r^2 \varepsilon \, . \tag{5}$$

The solutions provide information on the interior structure of a star, $\rho_i(r)$, as well as the mass-radius relation, $M(R)$.

The results obtained with different potentials+compatible TBF are shown in figure 2 (thick curves). One notices maximum masses ranging from about 1.8 $M_\odot$ (UIX) up to about 2.5 $M_\odot$ (BOB), reflecting the current theoretical uncertainty regarding nuclear TBF. However, these results should be considered as only provisory, since it is well known that the inclusion of hyperons or quark matter may strongly affect the structure of the star, in particular reducing substantially the maximum mass. This is discussed now in more detail.

5. Hyperons in nuclear matter

While at moderate densities $\rho \approx \rho_0$ the matter inside a NS consists only of nucleons and leptons, at higher densities several other species of particles, like Λ, Σ, and Ξ hyperons, may appear due to the fast rise of the baryon chemical potentials with density. In the following it is assumed that a baryonic description of nuclear matter holds up to those densities encountered in the core of NS, and the hyperon degrees of freedom are included within the BHF approximation [10, 11].

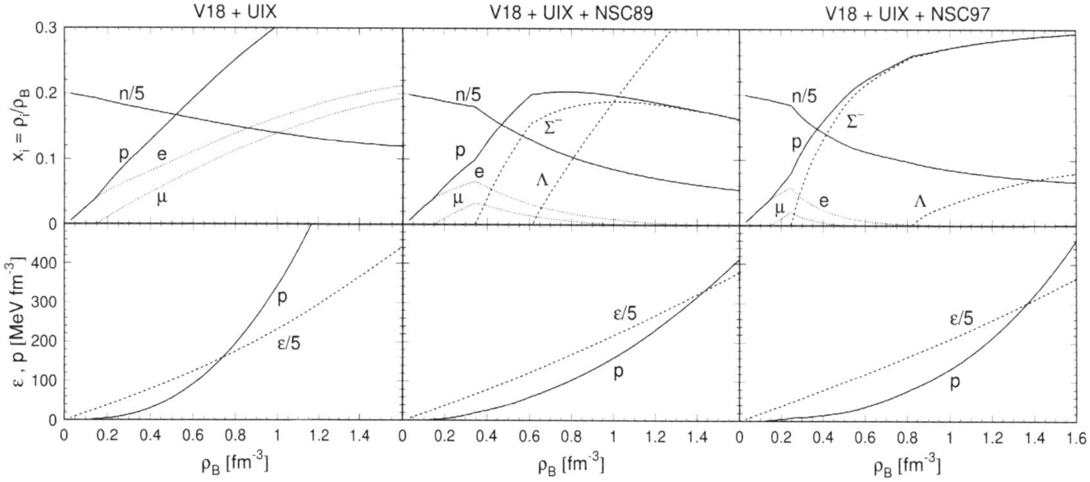

Figure 3. The composition and EOS of asymmetric and beta-stable hypernuclear matter.

To this purpose, one requires nucleon-hyperon (NY) and hyperon-hyperon (YY) potentials. In our work the Nijmegen soft-core NY potentials NSC89 [12] or NSC97f [13] are used, which are fitted to the available experimental NY scattering data, together with the nucleonic forces + TBF introduced before. With the NN and NY potentials, the various G matrices are evaluated by solving numerically the Bethe-Goldstone equation, written in operatorial form as

$$G_{ab}[W] = V_{ab} + \sum_c \sum_{p,p'} V_{ac} \Big| pp' \Big\rangle \frac{Q_c}{W - E_c + i\varepsilon} \Big\langle pp' \Big| G_{cb}[W] \,, \tag{6}$$

where the indices a, b, c indicate pairs of baryons and the Pauli operator Q_c and energy E_c determine the propagation of intermediate baryon pairs. In a given baryon-baryon channel $c = (12)$ one has

$$Q_{(12)} = [1 - n_1(k_1)][1 - n_2(k_2)] \,, \tag{7}$$
$$E_{(12)} = m_1 + m_2 + e_1(k_1) + e_2(k_2) \tag{8}$$

with the s.p. energy $e_i(k) = k^2/2m_i + U_i(k)$ and the distribution function $n_i(k) = \theta(k_F^{(i)} - k)$. The various s.p. potentials within the continuous choice are given by

$$U_1(k_1) = \text{Re} \sum_{2=n,p,\Lambda,\Sigma} \sum_{k_2} n_2(k_2) \Big\langle k_1 k_2 \Big| G_{(12)(12)}[E_{(12)}] \Big| k_1 k_2 \Big\rangle_A \,, \tag{9}$$

where k_i generally denote momentum and spin. Once the different s.p. potentials are known, the total nonrelativistic baryonic energy density, the EOS, and NS structure can be computed as described before, now making allowance for the species $i = n, p, \Lambda, \Sigma^-, e^-, \mu^-$.

Figure 3 shows the composition of the resulting beta-stable and charge-neutral asymmetric nuclear matter containing hyperons. One observes rather low Σ^- onset densities of about $(2-3)\rho_0$. Also, an almost equal percentage of nucleons and hyperons is present in the stellar core at high densities, and thus the inclusion of hyperons produces a much softer EOS.

The consequences for the structure of NS are illustrated by the thin curves in figure 2. One notes that while the purely nucleonic EOSs yielded quite different maximum masses, the presence of hyperons equalizes in a self-regulating way the results, leading now to maximum

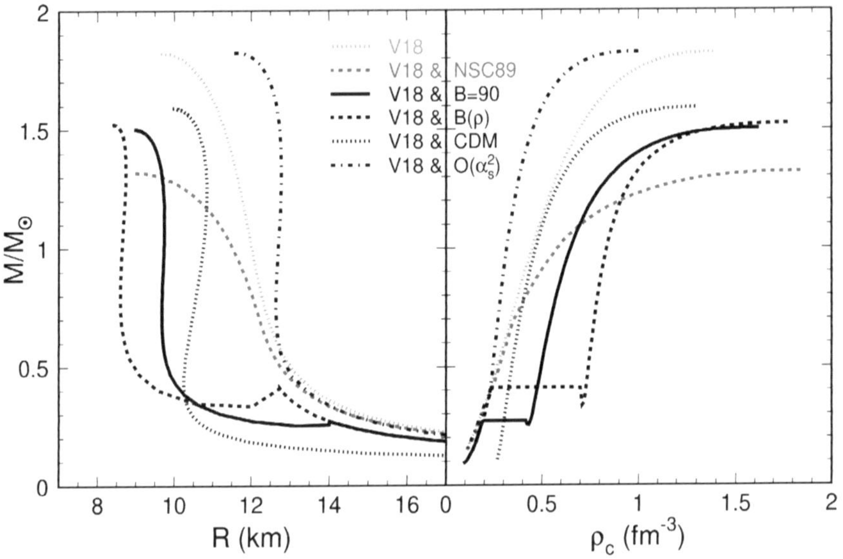

Figure 4. (Color online) The mass-radius and mass-central density relations for different hybrid neutron stars.

masses of 1.3–1.4 solar masses for all the nuclear TBF. This surprising result is due to the strong softening of the baryonic EOS when including hyperons as additional degrees of freedom, and one does not expect substantial changes when introducing refinements of the theoretical framework, such as hyperon-hyperon potentials, hyperonic TBF [14], relativistic corrections, etc.. The only remaining possibility in order to reach larger maximum masses appears the transition to another phase of dense (quark) matter inside the star [15, 16, 17, 18].

6. Quark matter

Unfortunately, the current theoretical description of quark matter is burdened with large uncertainties, seriously limiting the predictive power of any theoretical approach at high baryonic density. For the time being one can therefore only resort to phenomenological models for the quark matter EOS and try to constrain them as well as possible by the few experimental information on high-density baryonic matter. One important condition is due to the fact that certainly in symmetric nuclear matter no phase transition is observed below $\rho \approx 3\rho_0$.

The MIT bag model [19] provides a simple example for the description of the bulk properties of uniform quark matter, deconfined from the beta-stable hadronic matter mentioned in the previous section. The energy density of the $f = u, d, s$ quark gas can be expressed as a sum of the kinetic term and the one-gluon-exchange term [20, 21] proportional to the QCD coupling constant α_s,

$$\varepsilon_Q = B + \sum_f \varepsilon_f \,, \tag{10}$$

$$\varepsilon_f(\rho_f) = \frac{3m_f^4}{8\pi^2}\left[\left(2x_f^3 + x_f\right) y_f - \operatorname{arsinh} x_f\right] - \alpha_s \frac{m_f^4}{\pi^3}\left[x_f^4 - \frac{3}{2}\left(x_f y_f - \operatorname{arsinh} x_f\right)^2\right], \tag{11}$$

where m_f is the f current quark mass, $x_f = k_F^{(f)}/m_f$, and $y_f = \sqrt{1 + x_f^2} = E_F^{(f)}/m_f$. The baryon number density of f quarks is $\rho_f = {k_F^{(f)}}^3/3\pi^2$, and B is the energy density difference between the perturbative vacuum and the true vacuum, i.e., the bag "constant."

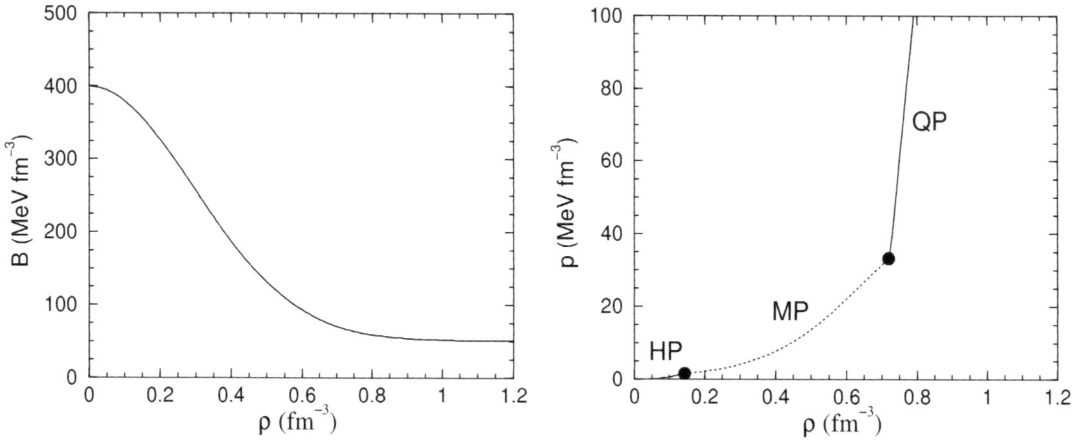

Figure 5. Left: Bag parameter B versus baryon number density. Right: EOS including both hadronic and quark components. The mixed phase region of quarks and hadrons (MP) is bordered by two dots, while HP and QP label the pure hadron and quark phases.

It has been found [15, 22] that within the MIT bag model (without color superconductivity) with a density-independent bag constant B, the maximum mass of a NS cannot exceed a value of about 1.6 solar masses. Indeed, the maximum mass increases as the value of B decreases, but too small values of B are incompatible with a hadron-quark transition density $\rho \gtrsim (2-3)\rho_0$ in nearly symmetric nuclear matter, as demanded by heavy-ion collision phenomenology.

In order to overcome these restrictions of the model, one can introduce an empirical density-dependent bag parameter $B(\rho)$, and this approach was followed in Refs. [15, 16, 17]. This allows one to lower the value of B at large density, providing a stiffer QM EOS and increasing the value of the maximum mass, while at the same time still fulfilling the condition of no phase transition below $\rho \approx 3\rho_0$ in symmetric nuclear matter. The following results are based on the MIT model using either a constant value of the bag parameter, $B = 100$ MeV/fm^{-3}, or a Gaussian parametrization for the density dependence,

$$B(\rho) = B_\infty + (B_0 - B_\infty)\exp\left[-\beta\Big(\frac{\rho}{\rho_0}\Big)^2\right] \tag{12}$$

with $B_\infty = 50$ MeV/fm^3, $B_0 = 400$ MeV/fm^3, and $\beta = 0.17$, displayed in figure 5 (left panel).

In the beta-stable pure quark phase, the individual quark chemical potentials are fixed by Eq. (3) with $b_q = 1/3$, which implies

$$\mu_s = \mu_d = \mu_u + \mu_l \,. \tag{13}$$

The charge neutrality condition and the total baryon number conservation read

$$0 = 2\rho_u - \rho_d - \rho_s - \rho_l \,, \tag{14}$$

$$\rho = \rho_u + \rho_d + \rho_s \,. \tag{15}$$

These equations determine the composition $\rho_f(\rho)$ and the pressure of the QM phase,

$$p_Q(\rho) = \rho^2 \frac{d(\varepsilon_Q/\rho)}{d\rho} \,, \tag{16}$$

which can be used together with the pressure of the hadronic phase to determine the phase transition using a Maxwell construction (MC), i.e., equating baryon chemical potentials and pressures of the two phases.

The corresponding results for the structure of hybrid NS are shown in figure 4 (black curves), displaying mass-radius and mass-central density relations. It is evident that the most striking effect of the inclusion of quark matter is the increase of the maximum mass relative to the hyperon star, now reaching about 1.5 $M_\odot$ with the MIT models with a bag constant $B = 100$ MeV/fm^{-3} or with $B(\rho)$, about 1.6 $M_\odot$ with the more sophisticated color dielectric model (CDM) [17, 23], and about 1.8 $M_\odot$ in a recent $O(\alpha_s^2)$ calculation [18].

7. Hadron-quark phase transition

A more sophisticated model for the phase transition between hadronic and quark phase inside the star is the Gibbs (Glendenning) construction [24], which determines a range of baryon density where both phases coexist as a mixed phase (MP). The essential point of this procedure is that both the hadron and the quark phase are allowed to be separately charged, still preserving the total charge neutrality. The chemical potentials of the different species are related to each other satisfying chemical and beta stability, Eq. (3), while the pressure is the same in the two phases to ensure mechanical stability,

$$p_H(\mu_e, \mu_n) = p_Q(\mu_e, \mu_n) = p_M(\mu_n) \,. \tag{17}$$

The equation for global charge neutrality involving the charge densities q_H and q_Q,

$$0 \;=\; \chi q_Q + (1-\chi) q_H \,, \tag{18}$$

determines the additional parameter χ, the volume fraction occupied by quark matter in the MP. From this, the baryon density ρ_M and the energy density ε_M of the MP can be calculated:

$$\rho_M \;=\; \chi \rho_Q + (1-\chi)\rho_H \,, \tag{19}$$
$$\varepsilon_M \;=\; \chi \varepsilon_Q + (1-\chi)\varepsilon_H \,. \tag{20}$$

The EOS resulting from this procedure is shown in figure 5 (right panel) for the MIT model, where the pure hadron, mixed, and pure quark matter portions are indicated. The MP begins actually at a quite low density around ρ_0 and persists throughout a wide range of baryon density.

As a further refinement, surface and Coulomb energies in the hadron-quark MP may be taken into account [25, 26], so that charge and baryon number density are no more uniform, but the MP can have exotic shapes called pasta structures (for a review, see Ref. [27]). As a result, the broad density region of the MP in the bulk Gibbs (BG) calculation described before is considerably reduced and the EOS approaches the one given by the MC [27].

Within a Wigner-Seitz approximation the whole space is divided into equivalent cells with a given geometrical symmetry, sphere for three dimensions (3D), cylinder for 2D, and slab for 1D. In each cell the quark and hadron phases are spatially separated by a sharp boundary with surface S and the surface energy is taken into account with a surface tension parameter σ (a typical value is $\sigma = 40$ MeV/fm^2). The energy density of the MP is then written as

$$\varepsilon = \frac{1}{V_W}\left[\int_{V_H} d\boldsymbol{r}\, \varepsilon_H(\boldsymbol{r}) + \int_{V_Q} d\boldsymbol{r}\, \varepsilon_Q(\boldsymbol{r}) + \int_{V_W} d\boldsymbol{r} \left(\varepsilon_e(\boldsymbol{r}) + \frac{[\nabla\phi(\boldsymbol{r})]^2}{8\pi e^2}\right) + \sigma S\right] , \tag{21}$$

where the volume of the Wigner-Seitz cell is $V_W = V_H + V_Q$. ϕ is the Coulomb potential and ε_H, ε_Q, and ε_e are energy densities of hadrons, quarks, and electrons, respectively, which are

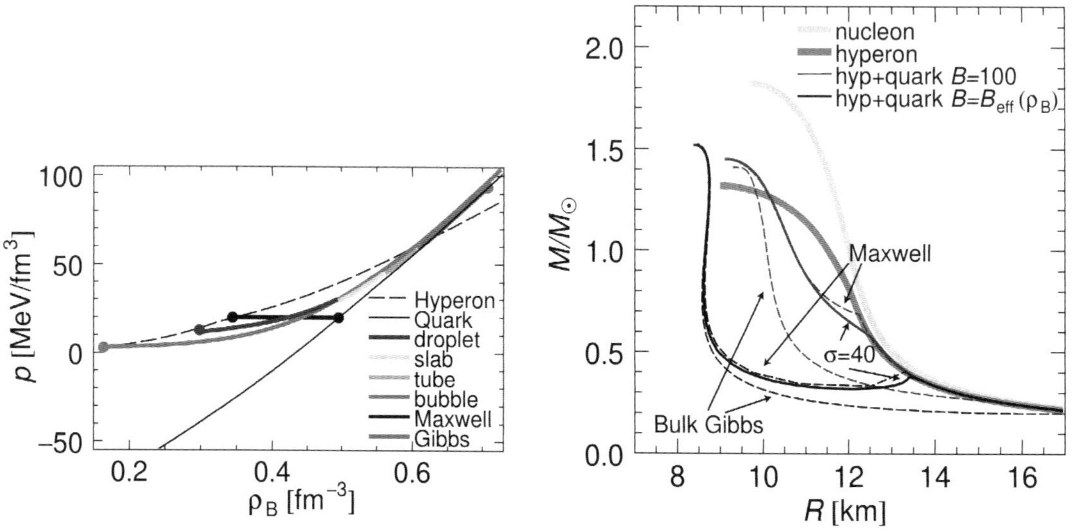

Figure 6. (Color online) Left: The pressure of the MP (thick curves) in comparison with the pure hadron and quark phases (thin curves), and the MC (black) and BG (gray) calculations. Right: NS mass-radius relations for different EOS and three different phase transition constructions. For the hybrid stars (blue and black curves), the dashed lines indicate the Maxwell (upper curves) or bulk Gibbs (lower curves) constructions and the solid lines the MP of the full calculation.

functions of local densities of $n, p, \Lambda, \Sigma^-, u, d, s, e$. For ε_H the BHF energy density is used and for the quark phase the MIT bag model introduced before. For a given baryon density ρ the optimum dimensionality of the cell, the cell size R_W, the lump size R, and the density profile of each component are searched for to minimize the energy density.

Figure 6 (left panel) compares the resulting EOS $p(\rho)$ with that of the pure hadron and quark phases. The thick black curve indicates the case of the MC, while the colored lines indicate the MP in its various geometric realizations, starting with quark droplets and terminating with quark bubbles. The charge screening effect tends to make matter locally charge neutral to reduce the Coulomb interaction energy. Combined with the surface tension, it renders the non-uniform structures mechanically less stable and limits the density region of the MP [27]. The structure and the composition of the MP are very different from those of the MC. In particular, although a relevant hyperon (Σ^-) fraction is present in the MC case, it is completely suppressed in the full calculation, because the hadron phase is allowed to be positively charged in the MP.

Figure 6 (right panel) compares the mass-radius relations obtained with the different models. The purely nucleonic EOS (thick green curve) yields a maximum NS mass of about $1.82\,M_\odot$, which is reduced to $1.32\,M_\odot$ when allowing the presence of hyperons (thick red curve). The inclusion of quark matter within the MIT model augments the maximum mass of hybrid stars to about $1.5\,M_\odot$, independently of the phase transition construction that is used: For the $B(\rho)$ model the maximum mass is $1.52\,M_\odot$, practically independent of the kind of phase transition, whereas for the $B = 100$ MeV/fm^{-3} model there is a slight variation of $M = 1.45, 1.45, 1.41\,M_\odot$ for the Maxwell, mixed, and bulk construction.

Whereas the maximum masses are practically independent of the phase transition construction, there are large differences for the internal composition of the star. This is illustrated in figure 7, which shows the total energy density, pressure, and particle fractions as a function of the radial coordinate for a $1.4\,M_\odot$ NS. One observes with the BG construction (left panel) a

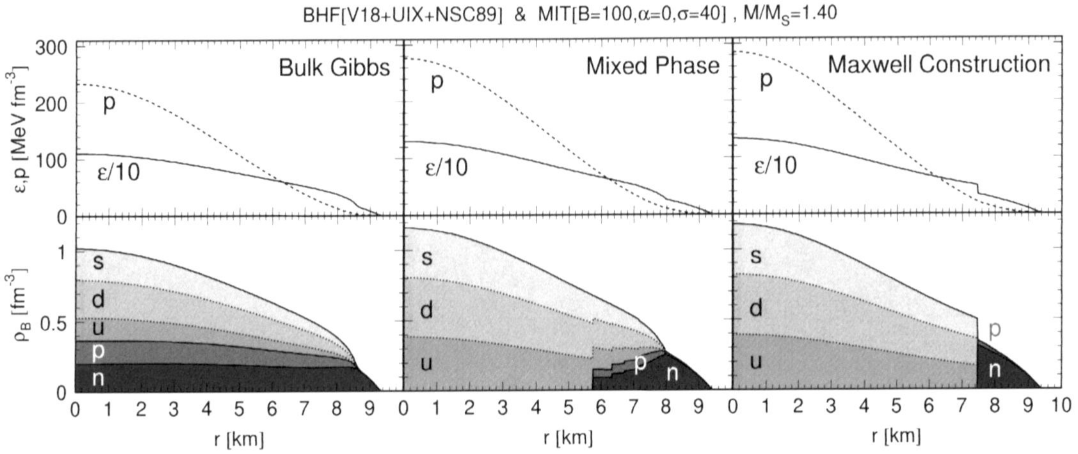

Figure 7. (Color online) Internal structure of a $1.4\,M_\odot$ NS obtained with the BG construction (left panel), the MP of the full calculation (central panel), and the MC (right panel). The upper panels show total energy density and pressure and the lower panels the overall particle fractions as functions of the radial coordinate of the star.

coexistence of hadrons and quarks in a significant range of the star, whereas with the MC (right panel) an abrupt transition involving a discontinuous jump of energy and baryon density occurs at a distance $r \approx 7.5$ km from the center of the star. The MP with the full calculation (central panel) lies between the two extreme cases, hadrons and quarks coexisting in a smaller range than in the BG case.

8. EOS at finite temperature: Protoneutron stars

A natural extension of the BHF formalism at zero temperature is the theoretical modeling of a protoneutron star (PNS), which represents the typical stellar state for some tens of seconds after supernova collapse, during which the system first deleptonizes and heats up the interior of the star in the process, before beginning to cool down by further neutrino diffusion [28].

We do not perform dynamical simulations, but focus on the consistent construction of the temperature-dependent nuclear EOS and the evaluation of its basic consequences during the prominent PNS stage [29, 30, 31, 32]. We therefore assume strongly idealized, static temperature and lepton-trapping profiles representing this environment, namely constant lepton fractions $Y_e \equiv x_e + x_{\nu_e} = 0.4$, $Y_\mu = 0$ and a constant entropy per baryon throughout the star, and investigate the sensitivity of the results to the chosen value of entropy S/A [33, 34].

The finite-temperature extension of the Bethe-Goldstone equation [4, 35, 36] consists in replacing the distribution function $n_i(k)$ appearing in the Pauli operator Q_c in Eq. (7) and the s.p. potential, Eq. (9), by the Fermi distribution $n_i(k) = (e^{[e_i(k)-\tilde{\mu}_i]/T} + 1)^{-1}$. For given partial densities ρ_i ($i = n, p, \Lambda, \Sigma$) and temperature T, Eqs. (6) and (9) have to be solved self-consistently along with the equations for the auxiliary chemical potentials $\tilde{\mu}_i$,

$$\rho_i = \sum_k n_i(k) \, . \tag{22}$$

Once the different s.p. potentials for the species $i = n, p, \Lambda, \Sigma^-$ are known, the free energy density of hypernuclear matter has the following simplified expression

$$f = \sum_i \left[\sum_k n_i(k) \left(\frac{k^2}{2m_i} + \frac{1}{2} U_i(k) \right) - T s_i \right] , \tag{23}$$

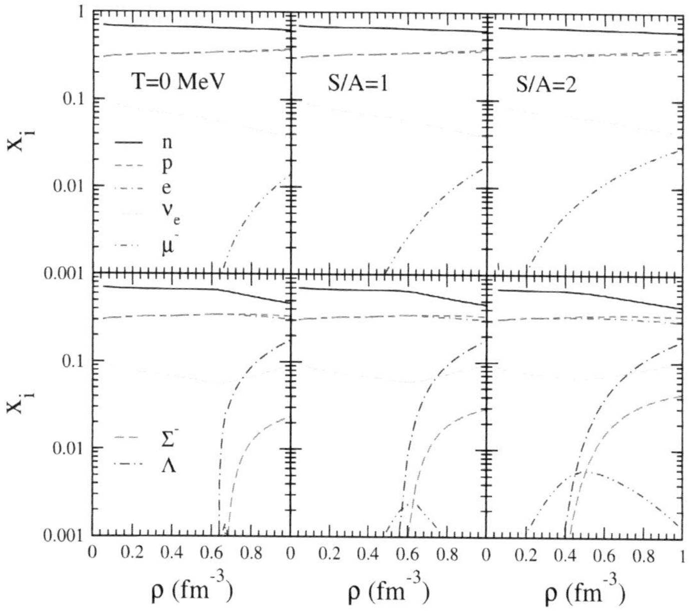

Figure 8. (Color online) Relative particle fractions as functions of baryon density in neutrino-trapped beta-equilibrated matter at entropies $S/A = 0, 1, 2$ without (upper panels) and with (lower panels) hyperons.

where

$$s_i = -\sum_k \left(n_i(k) \ln n_i(k) + [1 - n_i(k)] \ln[1 - n_i(k)] \right) \tag{24}$$

is the entropy density for component i treated as a free gas with s.p. spectrum $e_i(k)$ [4, 36]. A further simplification can be achieved by disregarding the effects of finite temperature on the s.p. potentials U_i, and using the $T = 0$ results in order to speed up the calculations (frozen correlations approximation) [29, 30, 32].

All thermodynamic quantities of interest can then be computed from the free energy density, Eq. (23); namely, the 'true' chemical potentials μ_i, pressure p, entropy density s, and internal energy density ε read as

$$\mu_i = \frac{\partial f}{\partial \rho_i}, \tag{25}$$

$$p = \rho^2 \frac{\partial (f/\rho)}{\partial \rho} = \sum_i \mu_i \rho_i - f, \tag{26}$$

$$s = -\frac{\partial f}{\partial T}, \tag{27}$$

$$\varepsilon = f + Ts. \tag{28}$$

As an illustration, figure 8 shows the particle fractions at entropies $S/A = 0, 1, 2$ in beta-stable trapped matter [32]. One observes the following qualitative features: (i) Finite temperature removes any particle thresholds, i.e., hyperons and leptons become more and more abundant at low densities with rising temperature/entropy. (ii) Hyperon fractions are lower in trapped matter than in untrapped matter, in particular the Σ^- is strongly suppressed, because due to the trapping condition it cannot easily replace the electron as is the case in untrapped matter. These features have direct consequences for the EOS $p(\varepsilon)$: For purely nucleonic matter the effects of trapping and temperature are not very large, but both soften the EOS. On the contrary, hyperons soften the EOS of untrapped matter much more than that of trapped matter, due to

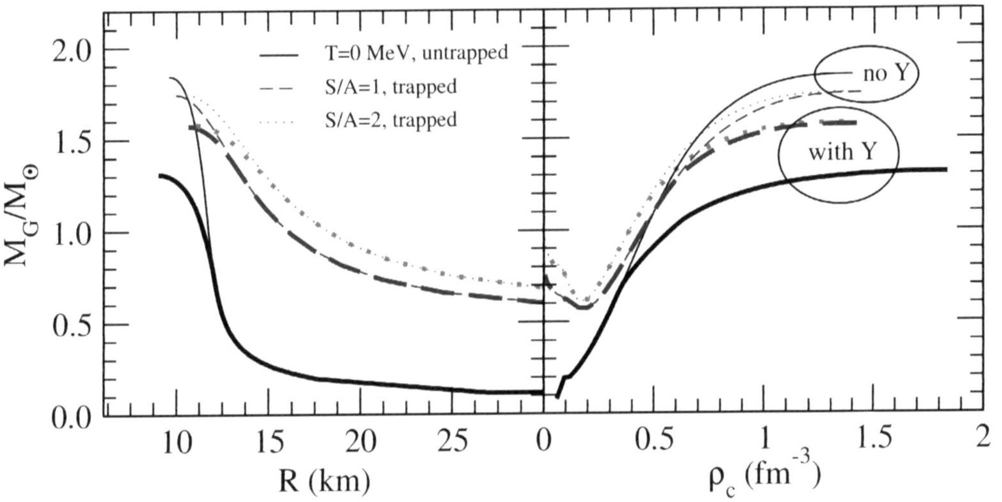

Figure 9. (Color online) Gravitational mass as a function of radius (left panel) and central baryon density (right panel) for cold NSs (solid curves) and neutrino-trapped PNSs at entropies $S/A = 1,2$ (broken curves). The thick (thin) curves describe configurations with (without) hyperons.

their higher concentration in the former. Altogether, finite entropy and in particular trapping affect hyperon-rich matter much more (and in an opposite sense) than nuclear matter.

The relation $p(\varepsilon)$ as input to Eq. (5) determines directly the mass – radius relations of (P)NSs shown in figure 9 (left panel). One observes for nucleonic stars a slight reduction of the maximum mass (from about 1.84 $M_\odot$ to 1.74 $M_\odot$) due to trapping and finite temperature, while for hyperon stars both trapping and also finite temperature increase notably the maximum mass (from about 1.31 $M_\odot$ to 1.58 $M_\odot$). The latter feature would permit a delayed collapse phenomenon, as is usually found for hyperon stars [28, 33, 34, 37]. However, this conclusion is rather academic, because the maximum mass of hyperon stars is 1.31 $M_\odot$ in this approach, so that most observed NSs [3] would actually be hybrid stars involving a transition to quark matter in the interior, as pointed out before. This is also pinpointed by the mass – central density relations, displayed in figure 9 (right panel), which shows for hyperon stars central densities up to about 10 ρ_0, where a realistic description of stellar matter should necessarily include quark matter degrees of freedom.

9. Conclusions

This contribution reported the theoretical description of hypernuclear matter in the BHF approach and the application to NS structure calculation. The important role of TBF at high density was pointed out, which is, however, strongly compensated by the inclusion of hyperons. The resulting hadronic NS have maximum masses of less than about 1.4 $M_\odot$, and the presence of quark matter inside the star is required in order to reach larger values.

It remains to be seen which values of the maximum mass can be reached with more sophisticated quark matter models in the future, in particular with respect to the recent claim of discovery of a two solar mass NS [38].

Acknowledgments

Many collaborators have contributed to the work described here. In particular, M Baldo, G F Burgio, J Cugnon, A Lejeune, A Li, Z H Li, U Lombardo, C Maieron, T Maruyama, A Polls, A Ramos, T Tatsumi, I Vidaña, X R Zhou, and W Zuo.

References

[1] Glendenning N K 2000 *Compact Stars, Nuclear Physics, Particle Physics, and General Relativity*, 2nd ed. (Springer, New York)
[2] Haensel P, Potekhin A Y, and Yakovlev D G 2007 *Neutron Stars 1* (Springer, New York)
[3] Lattimer J M and Prakash M 2007 *Phys. Rep.* **442** 109
[4] Baldo M 1999 *Nuclear Methods and the Nuclear Equation of State, International Review of Nuclear Physics*, Vol. 8 (World Scientific, Singapore)
[5] Song H Q, Baldo M, Giansiracusa G, and Lombardo U 1998 *Phys. Rev. Lett.* **81** 1584; Baldo M, Fiasconaro A, Song H Q, Giansiracusa G, and Lombardo U 2001 *Phys. Rev. C* **65** 017303
[6] Coon S A, Scadron M D, McNamee P C, Barrett B R, Blatt D W E, and McKellar B H J 1979 *Nucl. Phys. A* **317** 242; Grangé P, Lejeune A, Martzolff M, and Mathiot J-F 1989 *Phys. Rev. C* **40** 1040; Zuo W, Lejeune A, Lombardo U, and Mathiot J-F 2002 *Nucl. Phys. A* **706** 418; Li Z H, Lombardo U, Schulze H-J, and Zuo W 2008 *Phys. Rev. C* **77** 034316; Li Z H and Schulze H-J 2008 *Phys. Rev. C* **78** 028801
[7] Pudliner B S, Pandharipande V R, Carlson J, Pieper S C, and Wiringa R B, 1997 *Phys. Rev. C* **56** 1720; Zhou X R, Burgio G F, Lombardo U, Schulze H-J, and Zuo W 2004 *Phys. Rev. C* **69** 018801
[8] Li Z H, Lombardo U, Schulze H-J, Zuo W, Chen L W, and Ma H R 2006 *Phys. Rev. C* **74** 047304
[9] Shapiro S and Teukolsky S A 2000 *Black Holes, White Dwarfs, and Neutron Stars* (John Wiley & Sons, New York)
[10] Schulze H-J, Baldo M, Lombardo U, Cugnon J, and Lejeune A 1998 *Phys. Rev. C* **57** 704
[11] Baldo M, Burgio G F, and Schulze H-J 2000 *Phys. Rev. C* **61** 055801; Schulze H-J, Polls A, Ramos A, and Vidaña I 2006 *Phys. Rev. C* **73** 058801
[12] Maessen P M M, Rijken T A, and De Swart J J 1989 *Phys. Rev. C* **40** 2226
[13] Stoks V G J and Rijken T 1999 *Phys. Rev. C* **59** 3009
[14] Vidaña I, Logoteta D, Providência C, Polls A, and Bombaci I 2011 *Europhys. Lett.* **94** 11002
[15] Burgio G F, Baldo M, Sahu P K, and Schulze H-J 2002 *Phys. Rev. C* **66** 025802
[16] Baldo M, Buballa M, Burgio G F, Neumann F, Oertel M, and Schulze H-J 2003 *Phys. Lett. B* **562** 153; Baldo M, Burgio G F, Castorina P, Plumari S, and Zappalà D 2007 *Phys. Rev. C* **75** 035804
[17] Maieron C, Baldo M, Burgio G F, and Schulze H-J 2004 *Phys. Rev. D* **70** 043010
[18] Kurkela A, Romatschke P, and Vuorinen A 2010 *Phys. Rev. D* **81** 105021
[19] Chodos A, Jaffe R L, Johnson K, Thorn C B, and Weisskopf V F 1974 *Phys. Rev. D* **9** 3471
[20] Witten E 1984 *Phys. Rev. D* **30** 272; Baym G, Kolb E W, McLerran L, Walker T P, and Jaffe R L 1985 *Phys. Lett. B* **160** 181; Glendenning N K 1990 *Mod. Phys. Lett. A* **5** 2197
[21] Fahri E and Jaffe R L 1984 *Phys. Rev. D* **30** 2379
[22] Alford M and Reddy S 2003 *Phys. Rev. D* **67** 074024
[23] Pirner H J, Chanfray G, and Nachtmann O 1984 *Phys. Lett. B* **147** 249; Drago A, Tambini U, and Hjorth-Jensen M 1996 *Phys. Lett. B* **380** 13
[24] Glendenning N K 1992 *Phys. Rev. D* **46** 1274
[25] Heiselberg H, 1993 *Phys. Rev. D* **48** 1418; Glendenning N K 2001 *Phys. Rep.* **342** 393
[26] Maruyama T, Chiba S, Schulze H-J, and Tatsumi T 2007 *Phys. Rev. D* **76** 123015; 2008 *Phys. Lett. B* **659** 192
[27] Maruyama T, Tatsumi T, Endo T, and Chiba S 2006 *Recent Res. Devel. in Physics* **7** 1
[28] Prakash M, Bombaci I, Prakash M, Ellis P J, Lattimer J M, and R. Knorren, 1997 *Phys. Rep.* **280** 1
[29] Nicotra O E, Baldo M, Burgio G F, and Schulze H-J 2006 *Astron. Astrophys.* **451** 213; Nicotra O E, Baldo M, Burgio G F, and Schulze H-J 2006 *Phys. Rev. D* **74** 123001
[30] Burgio G F and Schulze H-J 2009 *Phys. Atom. Nuc.* **72** 1197
[31] Burgio G F and Schulze H-J 2010 *Astron. Astrophys.* **518** A17
[32] Burgio G F, Schulze H-J, and Li A 2011 *Phys. Rev. C* **83** 025804
[33] Pons J A, Reddy S, Prakash M, Lattimer J M, and Miralles J A 1999 *Astrophys. J.* **513** 780
[34] Strobel K and Weigel M K 2001 *Astron. Astrophys. Suppl.* **367** 582
[35] Bloch C and De Dominicis C 1958 *Nucl. Phys.* **7** 459; 1959 *ibidem* **10** 181,509
[36] Baldo M and Ferreira L S 1999 *Phys. Rev. C* **59** 682
[37] Vidaña I, Bombaci I, Polls A, and Ramos A 2003 *Astron. Astrophys.* **399** 687
[38] Demorest P B, Pennucci T, Ransom S M, Roberts M S E, and Hessels J W T 2010 *Nature* **467** 1081

Journal of Physics: Conference Series **336** (2011) 012023 doi:10.1088/1742-6596/336/1/012023

Hadron-quark phase transition in dense matter

G Y Shao[1], M Di Toro[1,2], B Liu [3,4], M Colonna[1], V Greco[1,2], Y X Liu [5] and S Plumari[1,2]

[1] INFN-Laboratori Nazionali del Sud, Via S. Sofia 62, I-95123 Catania, Italy
[2] Physics and Astronomy Dept, University of Catania,Via S. Sofia 64, I-95123 Catania, Italy
[3] IHEP, Chinese Academy of Sciences, Beijing 100049, China
[4] Theoretical Physics Center for Scientific Facilities, Chinese Academy of Sciences, Beijing 100049, China
[5] Department of Physics and State Key Laboratory of Nuclear Physics and Technology, Peking University, Beijing 100871, China

E-mail: shaogy@pku.edu.cn

Abstract. The two-Equation of State (EoS) model is used to describe the hadron-quark phase transition in asymmetric matter formed at high density in heavy-ion collisions. For the quark phase, the three-flavor Nambu–Jona-Lasinio (NJL) effective theory is used to investigate the influence of dynamical quark mass effects on the phase transition. At variance to the MIT-Bag results, with fixed current quark masses, the main important effect of the chiral dynamics is the appearance of an end point for the coexistence zone. We show that a first order hadron-quark phase transition may take place in the region T= (50 − 80) MeV and $\rho_B = (2-4)\,\rho_0$, which is possible to be probed in the new planned facilities, such as FAIR at GSI-Darmstadt and NICA at JINR-Dubna. From isospin properties of the mixed phase some possible signals are suggested. The importance of chiral symmetry and dynamical quark mass on the hadron-quark phase transition is stressed.

1. Introduction

The exploration of the phase diagram of strongly interacting matter and the search for signals of the hadron-quark phase transition are challenges both in theory and experiment. Intensive studies on these fields have been performed in the last decades based on Lattice QCD at zero or small chemical potential and effective chiral models.

Most effective models describe the phase transition based on quark degrees of freedom. As a matter of fact, at low temperature and small chemical potential, QCD dynamics are governed by hadrons. Therefore, it is natural to describe the strongly interacting matter with hadronic degrees of freedom at low T and small μ and quarks at high T and large μ. This picture can be realized in the two equation of state (Two-EoS) model in which hadronic and quark phases are connected by Gibbs condition. The Two-EoS model is widely used in describing the phase transition in the interior of compact star in weak equilibrium (e.g.,[1, 2, 3, 4, 5, 6, 7, 8, 9]).

We remark that, even in the Two-EoS approach, only a few papers have studied the phase diagram of hadron-quark transitions at high baryon density in connection to the phenomenology of heavy-ion collision in ten A GeV range (intermediate energies) [10, 11, 12, 13, 14]. In these studies, current mass (or massless) u, d quarks are taken for the quark phase, where the MIT-Bag like models are used. In order to obtain more reliable results and predict possible observables

Published under licence by IOP Publishing Ltd

in the planned experiments, in [15] we take the Nambu-Jona Lasinio (NJL) model to describe the quark phase with the interaction between quarks, where the chiral symmetry breaking and restoration are well described. We pay more attention to isospin asymmetric matter, and suggest some observable signals, possibly reached through heavy-ion reactions at new planned facilities, such as FAIR at GSI-Darmstadt and NICA at JINR-Dubna, where heavy ion beams (even unstable, with large isospin asymmetry) will be available with good intensities in the 1-30 A GeV energy region.

2. The Two-EoS model

In our Two-EoS approach, the hadron matter and quark matter are described by the non-linear Walecka model and the NJL model, respectively. For the mixed phase between pure hadronic and quark phase, the two phases are connected each other with the Gibbs conditions deduced from thermal, chemical and mechanical equilibriums. If we define χ as the fraction of quark matter in the mixed phase, the Gibbs conditions for the mixed phase are

$$\begin{aligned} \mu_B^H(\rho_B, \rho_3, T) &= \mu_B^Q(\rho_B, \rho_3, T) \\ \mu_3^H(\rho_B, \rho_3, T) &= \mu_3^Q(\rho_B, \rho_3, T) \\ P^H(\rho_B, \rho_3, T) &= P^Q(\rho_B, \rho_3, T), \end{aligned} \tag{1}$$

where $\rho_B = (1-\chi)\rho_B^H + \chi\rho_B^Q$ is the total baryon density in the mixed phase, and the total isospin density is $\rho_3 = (1-\chi)\rho_3^H + \chi\rho_3^Q$. With the initial condition of asymmetric parameter α^H in heavy-ion collision, the global asymmetry parameter α for the mixed phase should be

$$\alpha \equiv -\frac{\rho_3}{\rho_B} = \frac{(1-\chi)\rho_3^H + \chi\rho_3^Q}{(1-\chi)\rho_B^H + \chi\rho_B^Q} = \alpha^H \mid_{\chi=0} = \alpha^Q \mid_{\chi=1}, \tag{2}$$

according to the charge conservation. For details, please refer to [15] and refs. therein.

3. Numerical results and discussion

We will pay more attention to the phase diagram of asymmetric matter and focus the attention on experiments with neutron-rich beams, where large intensities can be reached. Here the largest accessible isospin asymmetry parameter α is just above 0.2 (e.g., we have $\alpha = 0.227$ in $^{238}U+^{238}U$ collision, see also the Table II in Ref. [13]). Therefore we will consider $\alpha = 0.2$ in our calculation, as taken in Refs. [11, 12, 13, 14]. Of course the use of more asymmetric unstable beams will enhance the isospin effects described here.

3.1. Some results with the MIT-Bag quark EoS

We firstly display some results with the MIT-Bag model in which free fermions are considered with the current mass $m_u = m_d = 5.5\,\mathrm{MeV}$. In figure 1 we plot the $T - \rho_B$ and $T - \mu_B$ phase diagrams. The solid curves are the phase-transition lines from nuclear matter to quark matter for $(B_{\mathtt{MIT}})^{1/4} = 160\,\mathrm{MeV}$ (190 MeV), and the corresponding dash-dot curves are the transition lines to pure quark matter. *NLρ* hadron *EoS* [11] has been used in the calculation. From this figure, it easy to see that the decrease of bag constant reduces the onset density of quark matter, and in general that the phase transition curve highly depends on the value of bag constant, which is one of the motivations of this study to use the NJL model to investigate the phase transition.

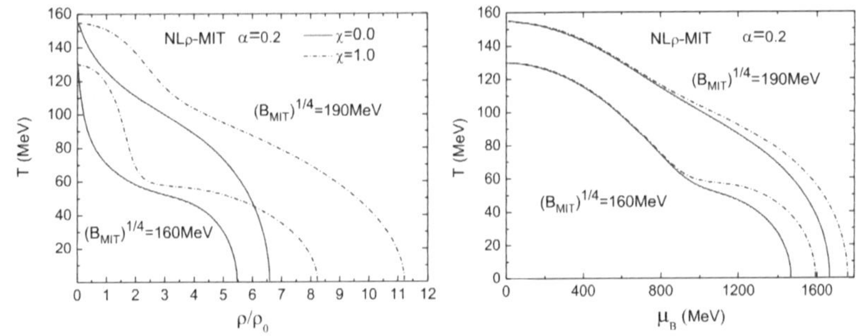

Figure 1. (Color online) Left(Right): the $T - \rho_B$ ($T - \mu_B$) plane of asymmetric matter with the isospin ratio $\alpha = 0.2$ for $(B_{\mathrm{MIT}})^{1/4} = 160\,\mathrm{MeV}$ and $190\,\mathrm{MeV}$. The region between the solid and dash-dot curve gives the binodal surface of the mixed phase, where the hadron and quark matter coexist. $NL\rho$ parametrization is used for the hadron phase.

3.2. Results with the NJL quark EoS

In figure 2 we present the $P - \rho_B$ phase diagram, within the NLρ-NJL two-EoS scheme [15], respectively for symmetric and asymmetric ($\alpha = 0.2$) matter. Clearly the pressure is a constant in the mixed phase of symmetric matter, just the same as with Maxwell construction. At variance in the asymmetric case, we have a monotonous increase. It is interesting to note that pressure rising is faster in the first part of the mixed phase (more evident for the lower temperatures T=30, 60 MeV). This is due to the isospin distillation effect, i.e., a large $\rho_d - \rho_u$ asymmetry for reduced quark fractions, see next subsection, which is increasing the quark pressure.

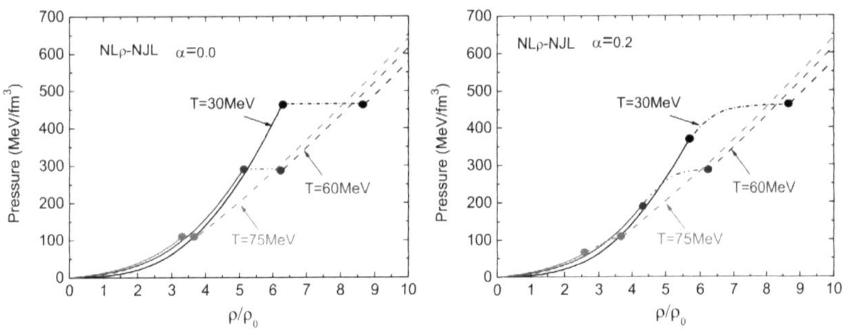

Figure 2. (Color online) Left: Pressure of symmetric matter as a function of baryon number density at different temperatures. Right: Pressure of asymmetric matter with $\alpha = 0.2$.

From figure 2, we can also see that the size of the mixed phase shrinks with increasing temperature. Meanwhile the onset density becomes smaller, opening the possibility of probing the coexistence phase in heavy-ion collision at intermediate energies. This effect is further enhanced by the isospin asymmetry as we will discuss in the next subsection.

3.2.1. Symmetry energy effects

In figure 3 we plot the (identical) $T - \rho_B$ phase diagrams of symmetric and asymmetric matter with different parameters in hadron phase. For different cases, in the mixed phase region, the binodal surface, is rather different. In general the onset density of the mixed phase in asymmetric matter is smaller than that in symmetric matter, similar to the results obtained by the MIT-Bag model [10, 11, 12, 13, 14].

If the $NL\rho\delta$ parameter set [11] is used, the onset density will be further reduced, as shown in figure 3. This is nicely due to the fact that at high baryon density the symmetry energy of hadron matter with the parameter set $NL\rho\delta$ is much larger than in the $NL\rho$ case, see the following. All that also indicates that isospin effects can be very important for hadron-quark phase transition in heavy-ion collision in order to shed light on nuclear interactions in an hot and dense medium.

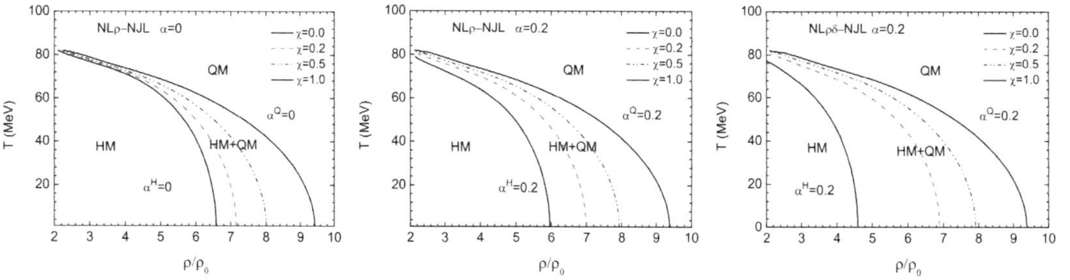

Figure 3. (Color online) Left: Phase diagram in the $T - \rho_B$ plane for symmetric matter with the parameter set $NL\rho$. Middle: for asymmetric matter with the parameter set $NL\rho$. Right: for asymmetric matter with the parameter set $NL\rho\delta$.

To understand the role of the symmetry term on the hadron-quark phase transition, we further investigate the symmetry energy E_{sym} in the two phases, defined in general by [11, 14]

$$(E/A)_{\alpha_i} = (E/A)_{\alpha_i=0} + E_{sym}{\alpha_i}^2 \tag{3}$$

where $\alpha_i = \alpha^H$ for hadron matter, α^Q for quark matter.

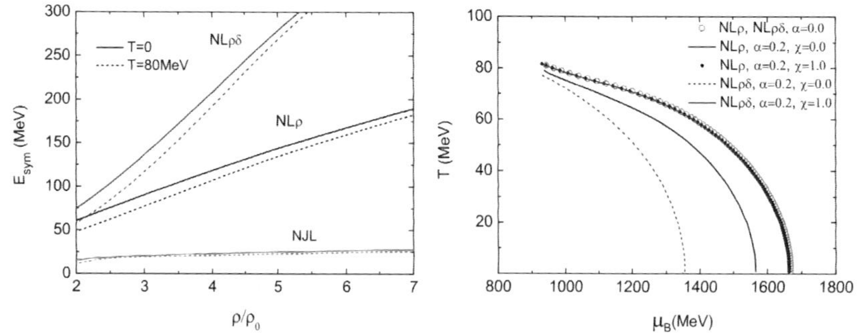

Figure 4. (Color online) Left: Symmetry energy of nuclear matter with the parameter sets $NL\rho$, $NL\rho\delta$ and that of quark matter in the NJL model. Right: Phase diagram in the $T - \mu_B$ plane for symmetric and asymmetric matter.

We display the numerical results in the left panel of figure 4. This figure shows that the symmetry energy of hadron matter with the parameter set $NL\rho\delta$ is larger than that of the $NL\rho$ case. This leads to a smaller onset density of the transition. It is interesting to note that this is a genuine relativistic effects since the scalar covariant nature of the isovector δ meson contributes to increase the symmetry energy at high densities directly with a larger repulsion and indirectly via a splitting of the neutron/proton effective masses, with $M_n^* < M_p^*$, see details in refs. [16, 17].

In the right panel figure 4 we plot the $T - \mu_B$ phase diagram for symmetric and asymmetric matter with both the parameter sets $NL\rho$ and $NL\rho\delta$. For the symmetric matter at a given temperature, μ_B is constant in the mixed phase, i.e., varying the quark matter fraction χ, then we have only one transition line (the empty circles). At variance, like in the MIT-Bag calculation of figure 1, in the asymmetric case we have a monotonous μ_B increase with increasing χ, so there are two curves in $T - \mu_B$ plane presenting the start and end, respectively, of the transition from nuclear to quark matter. What is important is that the chemical potentials of the onset of the transition ($\chi = 0$) in asymmetric matter are always smaller than the corresponding ones of symmetric matter. This is more evident with the parameter set $NL\rho\delta$ (dotted curve), which again shows the importance of symmetry energy effects.

3.2.2. Isospin Distillation inside the mixed phase

In figure 5, we show the variation of the isospin asymmetry parameters of hadron (α^H) and quark (α^Q) matter inside the mixed phase at various temperatures, with the global asymmetry $\alpha = 0.2$. We clearly see the much larger values of α^Q when the quark phase starts forming, it reduces to 0.2 when the pure quark phase is reached ($\chi = 1$)with $\chi = 1$. This is a nice Isospin Distillation effect ruled by the symmetry energy gap in the two phases, as confirmed by the enhancement in figure 5, where the more repulsive $NL\rho\delta$ EoS is used for the hadron part.

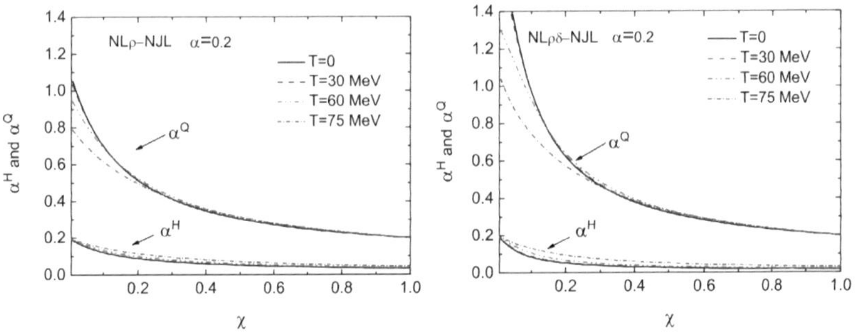

Figure 5. (Color online) Asymmetry of quark matter and nuclear matter inside the mixed phase at different temperatures. The parameter set $NL\rho$ ($NL\rho\delta$) is used for the hadron phase in the left (right) panel.

This behavior of the quark isospin asymmetry inside the mixed phase of the hadron-phase transition will affect the following hadronization in the expansion stage, finally producing some observable signals in heavy-ion experiments. As suggested before in Ref. [11, 12], an inversion in the trend of the emission of neutron rich clusters, π^-/π^+, K^0/K^+ yield ratios in high density regions, and an enhancement of the production of isospin-rich resonances and subsequent decays may be found. Besides, there is a controversial point of view about the enhancement of the yield ratio $\bar{\Lambda}/\bar{p}$ [18, 19, 20, 21, 22].

3.2.3. New effect of the dynamical quark masses: a critical end point?

If we compare MIT-Bag (with fixed current quark masses) and NJL (with chiral restoration mechanism) results for the mixed phase we remark only one main difference. As shown in figure 1, phase diagrams in $T - \rho_B$ and $T - \mu_B$ planes are derived, using the MIT-Bag model, with a critical temperature reached only at zero baryon density. At variance the binodal curves at high temperature and low density cannot be obtained with the NJL model and we see a narrowing of the coexistence region up to a kind of *critical end point*, figure 3. This important result derives from two qualitative new features of the NJL effective theory: i) the quark masses variation due to the chiral restoration, ii) the dependence of the effective bag-constant B_{eff}, Eq. (17) in Ref. [15], on temperature and baryon density.

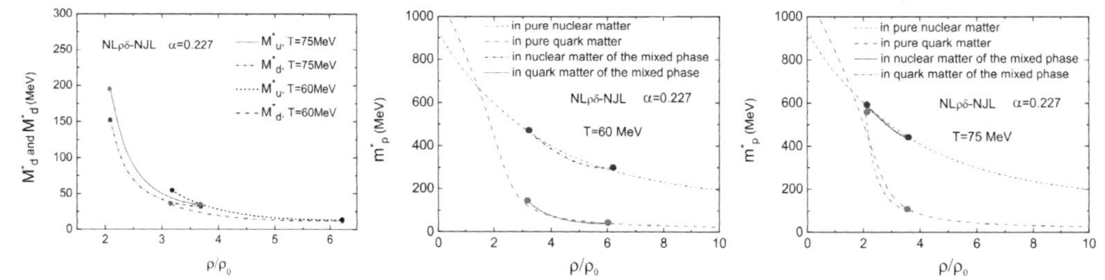

Figure 6. (Color online) Left: Dynamical masses of u, d quarks inside the mixed phase at different temperatures. Middle (Right): Effective proton masses in the hadron (upper) and quark (lower) matter inside the mixed phase at T =60 (75) MeV. The solid dots indicate the two density limits of the coexistence region.

- The chiral dynamics largely increases the quark masses at low densities and finite temperatures, see figure 6 where we plot the u, d effective masses inside the mixed phase at different temperatures, for asymmetric matter. If we reach the limit of a nucleon effective mass in the hadron phase smaller than the corresponding combination of quark effective masses, e.g., for protons $M_p^* < 2M_u^* + M_d^*$, we would have unphysical solutions for the Gibbs conditions with $\rho_B^Q < \rho_B^H$. This is confirmed by figure 6, where we plot the proton effective mass in the hadron phase (solid line) and in the quark phase (dot-dashed line) inside the coexistence region at T = 60 MeV and T = 75 MeV. We see that at the higher temperature we are close to a crossing, indication of the lack of physical solution at higher temperatures, as already seen in the corresponding $NL\rho\delta$ results of the previous figure 3.
- Actually we clearly see that in the same (T, ρ) region we do not have solutions of the Gibbs conditions. This is due to the second qualitative new feature of the NJL approach, the density and temperature dependence of the "effective bag constant", also related to the dynamical quark mass variation. As a consequence at low densities and high temperatures, for small values of the bag constant we cannot get mixed phase solutions since the hadron pressure (mostly thermal) cannot equilibrate the quark pressure in the coexistence zone. In figure 7 we show the density dependence of $(B_{eff})^{1/4}$ at T = 60 and T = 75 MeV (the solid circles give the mixed phase limits). For the T = 75 MeV case we see a sudden drop around the onset of the mixed phase, good indication of a lack of solution for larger temperatures.

No physical solution existing with the Gibbs conditions at high temperature and low density means that there exists a dynamical competition between hadronic and quark matter. Generally, for the quark phase, the dynamical effects are relevant for both chiral symmetry and (de)confinement. Only chiral dynamics is included in the NJL model in this study. This leads to the pressure of quark phase at low density and high temperature given by the NJL model being always higher than that of hadron phase at the same baryon and isospin chemical potential. So the absence of Gibbs equations solutions in the hadron-NJL model reflects that the consideration of only the chiral dynamics is not enough at low density and high temperature. The confinement

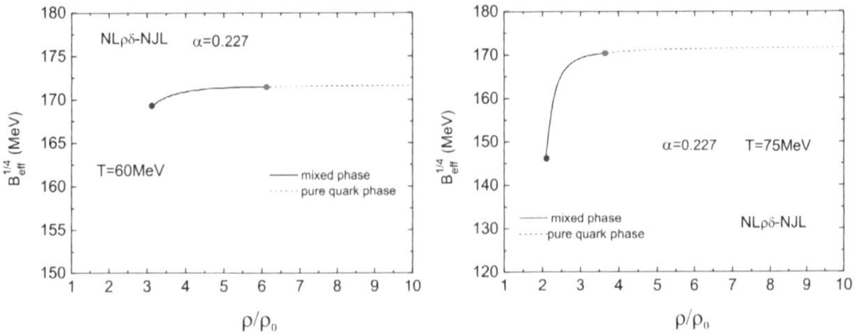

Figure 7. (Color online) Left(Right): Baryon density dependence of the NJL "effective bag constant", $(B_{eff})^{1/4}$, at $T = 60$ (75) MeV. The line ending with solid circles indicates coexistence region.

effect has to be included, which will affect the chiral dynamics and the equation of state of quark matter as shown in figure 8.

It is expected that the situation at high temperature and low density will be altered when the (de)confinement effect is included. This encourages us to perform a further study, following with the PNJL model for quark phase to get more accurate results about the "critical end point". One can refer to [23] for the PNJL parameter set as well as the updated result.

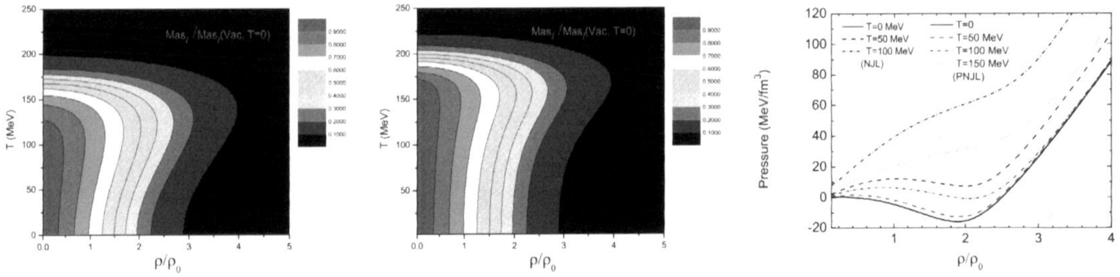

Figure 8. (Color online) Left (Middle): The contour of u, d quark mass as functions of temperature and baryon density in the NJL (PNJL) model. Right: Pressure of quark matter in the (P)NJL model. The logarithmic from of Polyakov loop effective potential with rescaled $T_0 = 210$ MeV [24] is used in the calculation of PNJL model.

3.3. Hadron-quark phase transition in neutron star matter

We also give a short discussion about the phase transition in neutron star matter under β equilibrium. As we know that the largest asymmetry parameter in heavy-ion experiment is $\alpha \simeq 0.227$. However, for neutron star matter the number density of neutron is much larger than that of proton, so the asymmetry parameter α can be larger than 0.227. Based on the conclusion obtained in last section, the onset density of hadron-quark phase transition should be much smaller in the interior of neutron star. This judgement is confirmed by the numerical calculation shown in figure 9 for symmetric matter and asymmetric matter. Figure 9 also shows that the onset density of quark matter with $NL\rho\delta$ is smaller than that with $NL\rho$. This is because the symmetry energy with $NL\rho\delta$ is larger than that of $NL\rho$, consistent with above discussion.

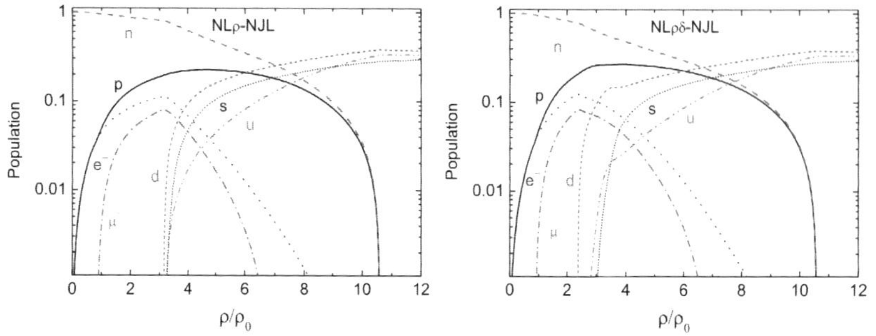

Figure 9. (Color online) Left (right): Population of neutron star matter under β equilibrium as functions of baryon density with Parameter set of $NL\rho$ ($NL\rho\delta$). The NJL model is used for quark phase

4. Summary

In this work, we investigate the hadron-quark phase transition in dense matter, particular in isospin asymmetric matter. The novelty of this study is to consider chiral dynamics of the quark masses using the 3-flavor NJL model for interacting quark instead of the MIT-Bag model. We obtain the binodal surface of a first order hadron-quark phase transition in the region of $\rho_B > 2\rho_0$ and temperature T less than about 80 MeV. The resulting effect is more relevant just in the region T$= (50-80)$ MeV and $\rho_B = (2-4)\,\rho_0$, available with the new planed facilities, where some important observables are suggested and may be found in the new planned facilities, for example, FAIR at GSI-Darmstadt and NICA at JINR-Dubna, with realistic asymmetries for stable/unstable beams.

References

[1] Glendenning N K 1992 *Phys. Rev. D* **46** 1274
[2] Glendenning N K and Schaffner-Bielich J 1998 *Phys. Rev. Lett.* **81** 4564
[3] Burgio G F, Baldo M, Sahu P K and Schulze H J 2002 *Phys. Rev. C* **66** 025802
[4] Maruyama T, Chiba S, Schulze H J and Tatsumi T 2007 *Phys. Rev. D* **76** 123015
[5] Yang F and Shen H 2008 *Phys. Rev. C* **77** 025801
[6] Shao G Y and Liu Y X 2009 *Phys. Lett. B* **682** 171
[7] Shao G Y and Liu Y X 2010 *Phys. Rev. C* **82** 055801
[8] Xu J, Chen L W, Ko C M and Li B A 2010 *Phys. Rev. C* **81** 055803
[9] Shao G Y 2011 *Phys. Lett. B* **704** 343
[10] Müller H 1997 *Nucl. Phys. A* **618** 349
[11] Di Toro M, Drago A, Gaitanos T, Greco V and Lavagno A 2006 *Nucl. Phys. A* **775** 102
[12] M. Di Toro *et al.* 2011 *Phys. Rev. C* **83** 014911
[13] Cavagnoli R, Providência C and Menezes D P 2011 *Phys. Rev. C* **83** 045201
[14] Pagliara G and Schaffner-Bielich J 2010 *Phys. Rev. D* **81** 094024
[15] Shao G Y, Di Toro M, Liu B, Colonna M, Greco V, Liu Y X and Plumari S 2011 *Phys. Rev. D* **83** 094033
[16] Baran V, Colonna M, Greco V and Di Toro M 2005 *Phys. Rep.* **410** 335205
[17] Liu B, Greco V, Baran V, Colonna M and Di Toro M 2002 *Phys. Rev. C* **65** 045201
[18] Stephans G S F and Wu Y 1997 *J. Phys. G* **23** 1895
[19] Armstrong T A *et al.* 1999 *Phys. Rev. C* **59** 2699
[20] Back B B *et al.* 2001 *Phys. Rev. Lett.* **87** 242301
[21] Rapp R and Shuryak E V 2001 *Phys. Rev. Lett.* **86** 2980
[22] Greiner C and Leupold S 2001 *J. Phys. G* **27** L95
[23] Shao G Y, Di Toro M, Greco V, Colonna M, Plumari S, Liu B and Liu Y X 2011 *Phys. Rev. D* **84** 034028
[24] Rßöner S, Ratti C and Weise W 2007 *Phys. Rev. D* **75** 034007

Journal of Physics: Conference Series **336** (2011) 012024 doi:10.1088/1742-6596/336/1/012024

Nuclear effects in charged-current quasielastic neutrino-nucleus scattering

Maria B Barbaro

Università di Torino, Dipartimento di Fisica Teorica and INFN, Via P Giuria 1, 10125 Torino, ITALY

E-mail: barbaro@to.infn.it

Abstract.
After a short review of the recent developments in studies of neutrino-nucleus interactions, the predictions for double-differential and integrated charged current-induced quasielastic cross sections are presented within two different relativistic approaches: one is the so-called SuSA method, based on the superscaling behavior exhibited by electron scattering data; the other is a microscopic model based on relativistic mean field theory, and incorporating final-state interactions. The role played by the meson-exchange currents in the two-particle two-hole sector is explored and the results are compared with the recent MiniBooNE data.

1. Introduction

The analysis and interpretation of ongoing and future neutrino oscillation experiments strongly rely on the nuclear modeling for describing the interaction of neutrinos and anti-neutrinos with the detector. Moreover, neutrino-nucleus scattering has recently become a matter of debate in connection with the possibility of extracting information on the nucleon axial mass. Specifically, the data on muon neutrino charged-current quasielastic (CCQE) cross sections obtained by the MiniBooNE collaboration [1] are substantially underestimated by the Relativistic Fermi Gas (RFG) prediction. This has been ascribed either to effects in the elementary neutrino-nucleon interaction, or to nuclear effects. The most poorly known ingredient of the single nucleon cross section is the cutoff parameter M_A employed in the dipole prescription for the axial form factor of the nucleon, which can be extracted from ν and $\overline{\nu}$ scattering off hydrogen and deuterium and from charged pion electroproduction. If M_A is kept as a free parameter in the RFG calculation, a best fit of the MiniBooNE data yields a value of the order of 1.35 GeV/c^2, much larger than the average value $M_A \simeq 1.026 \pm 0.021$ GeV/c^2 extracted from the (anti)neutrino world data [2]. This should be taken more as an indication of incompleteness of the theoretical description of the data based upon the RFG, rather than as a true indication for a larger axial mass. Indeed it is well-known from comparisons with electron scattering data that the RFG model is too crude to account for the nuclear dynamics. Hence it is crucial to explore more sophisticated nuclear models before drawing conclusions on the value of M_A.

Several calculations have been recently performed and applied to neutrino reactions. These include, besides the approach that will be presented here, models based on nuclear spectral functions [3, 4, 5, 6, 7, 8, 9, 10], relativistic independent particle models [11, 12, 13], relativistic Green function approaches [14, 15, 16, 17, 18], models including NN correlations [19, 20, 21],

Published under licence by IOP Publishing Ltd

coupled-channel transport models [22, 23, 24, 25], RPA calculations [26, 27, 28] and models including multinucleon knock-out [29, 30, 31, 32]. The difference between the predictions of the above models can be large due to the different treatment of both initial and final state interactions. As a general trend, the models based on impulse approximation, where the neutrino is supposed to scatter off a single nucleon inside the nucleus, tend to underestimate the MiniBooNE data, while a sizable increase of the cross section is obtained when two-particle-two-hole (2p-2h) mechanisms are included in the calculations. Furthermore, a recent calculation performed within the relativistic Green function (RGF) framework has shown that at this kinematics the results strongly depend on the phenomenological optical potential used to describe the final state interaction between the ejected nucleon and the residual nucleus [18]. With an appropriate choice of the optical potential the RGF model can reproduce the MiniBooNE data without the need of modifying the axial mass (see Giusti's contribution to this volume [33]).

The kinematics of the MiniBooNE experiment, where the neutrino flux spans a wide range of energies reaching values as high as 3 GeV, demands relativity as an essential ingredient. This is illustrated in figure 1, where the relativistic and non-relativistic Fermi gas results for the CCQE double differential cross section of 1 GeV muon neutrinos on ^{12}C are shown as a function of the outgoing muon momentum and for two values of the muon scattering angle. The relativistic effects, which affect both the kinematics and the dynamics of the problem, have been shown to be relevant even at moderate momentum and energy transfers [34, 35]. Hence in our approach

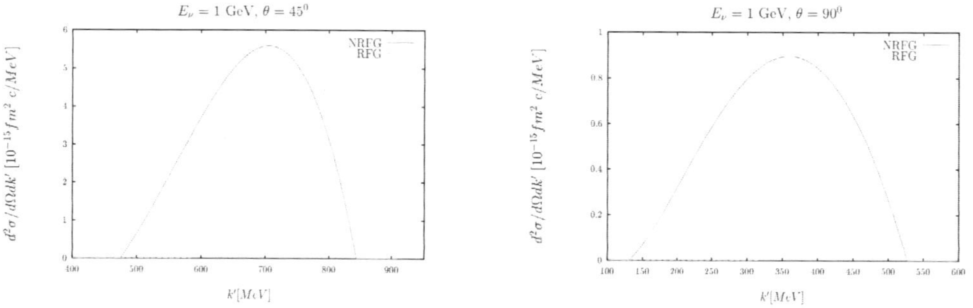

Figure 1. (Color online) ν_μCCQE double differential cross sections on ^{12}C displayed versus the outgoing muon momentum for non-relativistic (NRFG) and relativistic (RFG) Fermi gas.

we try to retain as much as possible the relativistic aspects of the problems.

In spite of its simplicity, the RFG has the merit of incorporating an exact relativistic treatment, fulfilling the fundamental properties of Lorentz covariance and gauge invariance. However, it badly fails to reproduce the electron scattering data, in particular when it is compared with the Rosenbluth-separated longitudinal and transverse responses. Comparison with electron scattering data must be a guiding principle in selecting reliable models for neutrino reactions. A strong constraint in this connection is represented by the "superscaling" analysis of the world inclusive (e, e') data: in Refs. [36, 37, 38, 39, 40] it has been proved that, for sufficiently large momentum transfers, the reduced cross section (namely the double differential cross section divided by the appropriate single nucleon factors), when represented versus the scaling variable ψ [41], is largely independent of the momentum transfer (first-kind scaling) and of the nuclear target (second-kind scaling). The simultaneous occurrence of the two kinds of scaling is called susperscaling. Moreover, from the experimental longitudinal response a phenomenological quasielastic scaling function has been extracted that shows a clear asymmetry with respect to the quasielastic peak (QEP) with a long tail extended to positive values of the scaling variable, i.e., larger energy transfers. On the contrary the RFG model, as well as most models based on impulse approximation, give a symmetric superscaling function with a maximum value 20-30% higher than the data [42].

In this contribution, after recalling the basic formalism for CCQE reactions and their connection with electron scattering, we shall illustrate two models which provide good agreement with the above properties of electron scattering data: one of them, the relativistic mean field (RMF) model, comes from microscopic many-body theory, the other, the superscaling approximation (SuSA) model, is extracted from (e, e') phenomenology. We shall then include the contribution of 2p-2h excitations in the SuSA model and finally compare our results with the MiniBooNE double differential, single differential and total cross sections. Most of the results which will be presented are contained in Refs. [43] and [44].

2. Formalism

Charged current quasielastic muonic neutrino scattering (ν_μ, μ^-) off nuclei is very closely related to quasielastic inclusive electron scattering (e, e'). However two major differences occur between the two processes:

(i) in the former case the probe interacts with the nucleus via the weak force, in the latter the interaction is (dominantly) electromagnetic. While the weak vector current is related to the electromagnetic one via the CVC theorem, the axial current gives rise to a more complex structure of the cross sections, with new response functions which cannot be related to the electromagnetic ones. As a consequence, while in electron scattering the double-differential cross section can be expressed in terms of two response functions, longitudinal and transverse with respect to three-momentum carried by the virtual photon, for the CCQE process it can be written according to a Rosenbluth-like decomposition as [45]

$$\left[\frac{d^2\sigma}{dT_\mu d\cos\theta}\right]_{E_\nu} = \sigma_0 \left[\hat{V}_L R_L + \hat{V}_T R_T + \hat{V}_{T'} R_{T'}\right], \tag{1}$$

where T_μ and θ are the muon kinetic energy and scattering angle, E_ν is the incident neutrino energy, σ_0 is the elementary cross section, $\hat{V}_i$ are kinematic factors and R_i are the nuclear response functions, the indices L, T, T' referring to longitudinal, transverse, transverse-axial, components of the nuclear current, respectively. The expression (1) is formally analogous to the inclusive electron scattering case, but: *i)* the "longitudinal" response R_L takes contributions from the charge (0) and longitudinal (3) components of the nuclear weak current, which, at variance with the electromagnetic case, are not related to each other by current conservation, *ii)* R_L and R_T have both "VV" and "AA" components (stemming from the product of two vector or axial currents, respectively), *iii)* a new response, $R_{T'}$, arises from the interference between the axial and vector parts of the weak nuclear current. In figure 2 we show the separate contributions of the three responses in (1), evaluated in the RFG model for the ^{12}C target nucleus, for two different values of the scattering angle. It can be seen that in the forward direction the transverse response dominates over the longitudinal and transverse-axial ones, whereas at higher angles the L-component becomes negligible and the T' and T responses are almost equal. This cancellation has important consequences on antineutrino-nucleus scattering, where the response $R_{T'}$ has opposite sign.

(ii) In (e, e') experiments the energy of the electron is well-known, and therefore the detection of the outgoing electron univoquely determines the energy and momentum transferred to the nucleus. In neutrino experiments the neutrino energy is not known, but distributed over a range of values (for MiniBooNE from 0 to 3 GeV with an average value of about 0.8 GeV). The cross section must then be evaluated as an average over the experimental flux

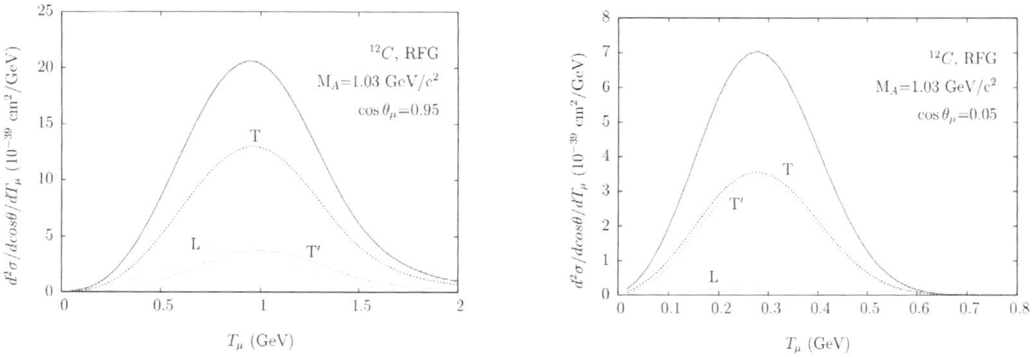

Figure 2. (Color online) Separate contributions of the RFG longitudinal (L), transverse (T) and axial-vector interference (T') responses to the double differential ν_μCCQE cross sections displayed versus the muon kinetic energy at two different angles. The neutrino energy is averaged over the MiniBooNE flux and the axial mass parameter has the standard value.

$\Phi(E_\nu)$

$$\frac{d^2\sigma}{dT_\mu d\cos\theta} = \frac{1}{\Phi_{tot}} \int \left[\frac{d^2\sigma}{dT_\mu d\cos\theta}\right]_{E_\nu} \Phi(E_\nu) dE_\nu \ , \tag{2}$$

which may require to account for effects not included in models devised for quasi-free scattering. This is, for instance, the situation at the most forward scattering angles, where a significant contribution in the cross section comes from very low-lying excitations in nuclei [43], as illustrated in figure 3: here the double differential cross section is evaluated in the SuSA model (see later) at the MiniBooNE kinematics and the lowest angular bin and compared with the result obtained by excluding the energy transfers lower than 50 MeV from the flux-integral (2). It appears that at these angles 30-40% of the cross section

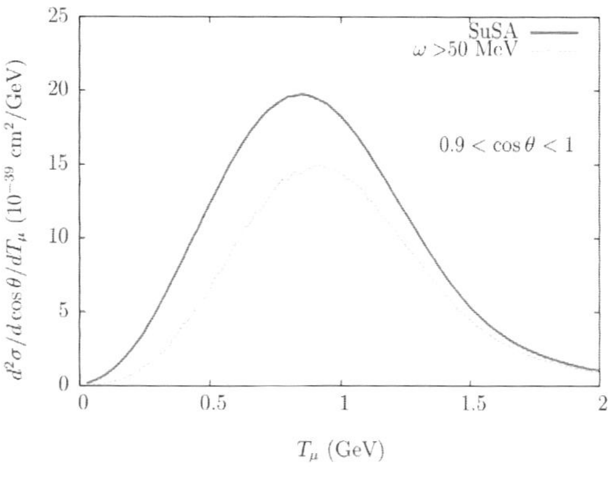

Figure 3. (Color online) Solid lines (red online): flux-integrated ν_μCCQE cross sections on ^{12}C calculated in the SuSA model for a specific bin of scattering angle. Dashed lines (green online): a lower cut $\omega = 50$ MeV is set in the integral over the neutrino flux.

corresponds to very low energy transfers, where collective effects dominate. Moreover, processes involving meson exchange currents (MEC), which can excite both one-particle-one-hole (1p1h) and two-particle-two-hole (2p-2h) states via the exchange of a virtual meson, should also be taken into account, since they lead to final states where no pions are present, classified as "quasielastic" in the MiniBooNE experiment.

3. Models

In this Section we briefly outline the main ingredients of the RMF and SuSA model and we illustrate our calculation of the contribution of 2p-2h meson exchange currents.

3.1. RMF

In the RMF model a fully relativistic description of both the kinematics and the dynamics of the process is incorporated.

Details on the RMF model applied to inclusive QE electron and CCQE neutrino reactions can be found in Refs. [46, 47, 48, 49, 50, 51]. Here we simply recall that the weak response functions are given by taking the appropriate components of the weak hadronic tensor, constructed from the single-nucleon current matrix elements

$$\langle J^\mu_W \rangle = \int d\mathbf{r} \overline{\phi}_F(\mathbf{r}) \hat{J}^\mu_W(\mathbf{r}) \phi_B(\mathbf{r}) , \tag{3}$$

where ϕ_B and ϕ_F are relativistic bound-state and scattering wave functions, respectively, and $\hat{J}^\mu_W$ is the relativistic one-body current operator modeling the coupling between the virtual W-boson and a nucleon. The bound nucleon states are described as self-consistent Dirac-Hartree solutions, derived by using a Lagrangian containing σ, ω and ρ mesons [52, 53, 54]. The outgoing nucleon wave function is computed by using the same relativistic mean field (scalar and vector energy-independent potentials) employed in the initial state and incorporates the final state interactions (FSI) between the ejected proton and the residual nucleus.

The RMF model successfully reproduces the scaling behaviour of inclusive QE (e, e') processes and, more importantly, it gives rise to a superscaling function with a significant asymmetry, namely, in complete accord with data [46, 47]. This is a peculiar property associated to the consistent treatment of initial and final state interactions. It has been shown in Refs. [46, 47] that other versions of the RMF model, which deal with the FSI through a real relativistic optical potential, are not capable of reproducing the asymmetry of the scaling function.

Moreover, contrary to most other models based on impulse approximation, where scaling of the zeroth kind - namely the equality of the longitudinal and transverse scaling functions - occurs, the RMF model provides L and T scaling functions which differ by typically 20%, the T one being larger. This agrees with the analysis [39] of the existing L/T separated data, which has shown that, after removing inelastic contributions and two-particle-emission effects, the purely nucleonic transverse scaling function is significantly larger than the longitudinal one.

3.2. SuSA

The SuSA model is based on the phenomenological superscaling function extracted from the world data on quasielastic electron scattering [37]. The model has been extended to the Δ-resonance region in Ref. [45] and to neutral current scattering in Ref. [55], but here we restrict our attention to the quasielastic charged current case.

Assuming the scaling function f extracted from (e, e') data to be universal, i.e., valid for electromagnetic and weak interactions, in [45, 56] CCQE neutrino-nucleus cross sections have been evaluated by multiplying f by the corresponding elementary weak cross section. Thus in the SuSA approach all the nuclear responses in (1) are expressed as follows

$$R_K = N \frac{2E_F}{k_F q} U_K(q, \omega) f(\psi) \quad , \quad (K = L, T, T') \tag{4}$$

where U_K are the elementary lepton-nucleon responses, E_F and k_F are the Fermi energy and momentum, N is the number of nucleons (neutrons in the ν_μCCQE case) and $f(\psi)$ in the universal superscaling function, depending only on the scaling variable ψ [41].

The SuSA approach provides nuclear-model-independent neutrino-nucleus cross sections and reproduces the longitudinal electron data by construction. However, its reliability rests on some basic assumptions.

First, it assumes that the scaling function - extracted from *longitudinal* (e, e') data - is appropriate for all of the weak responses involved in neutrino scattering (charge-charge, charge-longitudinal, longitudinal-longitudinal, transverse and axial), and is independent of the vector or axial nature of the nuclear current entering the hadronic tensor. In particular it assumes the equality of the longitudinal and transverse scaling functions (scaling of the zeroth kind), which, as mentioned before, has been shown to be violated both by experiment and by some microscopic models, for example relativistic mean field theory.

Second, the charged-current neutrino responses are purely isovector, whereas the electromagnetic ones contain both isoscalar and isovector components and the former involve axial-vector as well as vector responses. One then has to invoke a further kind of scaling, namely the independence of the scaling function of the choice of isospin channel — so-called scaling of the third kind.

Finally, the SuSA approach neglects violations of scaling of first and second kinds. These are known to be important at energies above the QE peak and to reside mainly in the transverse channel, being associated to effects which go beyond the impulse approximation: inelastic scattering, meson-exchange currents and the associated correlations needed to conserve the vector current. The inclusion of these contributions in the SuSA model is discussed in the next paragraph.

3.3. 2p-2h MEC

Meson exchange currents are two-body currents carried by a virtual meson which is exchanged between two nucleons in the nucleus. They are represented by the diagrams in figure 4, where the external lines correspond to the virtual boson (γ or W) and the dashed lines to the exchanged meson: in our approach we only consider the pion, which is believed to give the dominant contribution in the quasielastic regime. The thick lines in diagrams (d)-(g) represent the propagation of a Δ-resonance. The explicit relativistic expressions for the current matrix elements can be found, e.g., in Ref. [57].

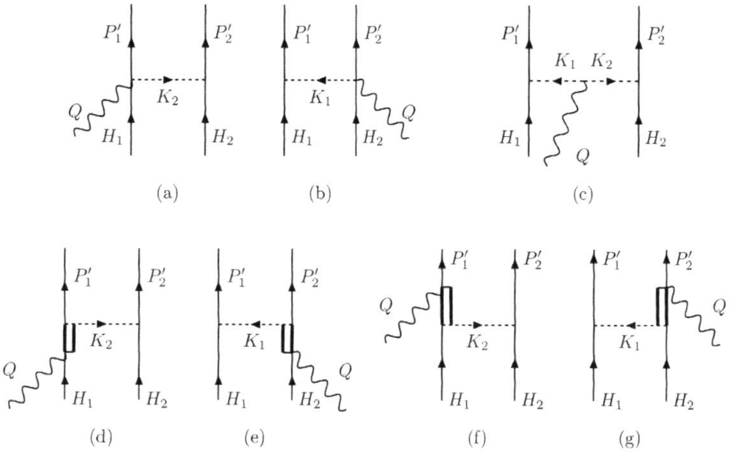

Figure 4. Two-body meson-exchange currents. (a) and (b): "contact", or "seagull" diagram; (c): "pion-in-flight" diagram; (d)-(g): "Δ-MEC" diagram.

Being two-body currents, the MEC can excite both one-particle one-hole (1p-1h) and two-particle two-hole (2p-2h) states. In the 1p-1h sector, MEC studies of electromagnetic (e, e') process have been performed for low-to-intermediate momentum transfers (see, *e.g.*, [58, 35, 59, 60]), showing a small reduction of the total response at the quasielastic peak, mainly due to diagrams involving the electroexcitation of the Δ resonance.

However in a perturbative scheme where all the diagrams containing one and only one pionic line are retained, the MEC are not the only diagrams arising, but pionic correlation contributions, where the virtual boson is attached to one of the two interacting nucleons, should also be considered. These are represented by the same diagrams as in figure 4(d)-(g), where now the thick lines are nucleon propagators. Only when all the diagrams are taken into account gauge invariance is fulfilled and the full two-body current is conserved. Correlation diagrams have been shown to roughly compensate the pure MEC contribution [58, 35, 59, 60], so that in first approximation we can neglect the 1p-1h sector and restrict our attention to 2p-2h final states.

The contribution to the inclusive electron scattering cross section arising from two-nucleon emission via meson exchange current interactions was first calculated in the Fermi gas model in Refs. [61, 62], where sizable effects were found at large energy transfers. In these references a non-relativistic reduction of the currents was performed, while fully relativistic calculations have been developed more recently in Refs. [63, 64, 57]. It has been found that the MEC give a significant positive contribution which leads to a partial filling of the "dip" between the quasielastic peak and the analogous peak associated with the excitation of the Δ resonance. Moreover, the MEC have been shown to break scaling of both first and second kinds [65].

Here we use the fully relativistic model of [64], where all the MEC many-body diagrams containing two pionic lines that contribute to the electromagnetic 2p-2h transverse response were taken into account. Similar results for the 2p-2h MEC were obtained in Ref. [57], where the correlation diagrams were also included.

In order to apply the model to neutrino scattering, we observe that in lowest order the 2p2h sector is not directly reachable for the axial-vector matrix elements. Hence the MEC affect only the transverse polar vector response, R_T^{VV}. Note that, at variance with the 1p-1h sector, where the contribution of the MEC diagrams originates from the interference between 1-body and 2-body amplitudes and has therefore no definite sign (in fact it turns out to be negative due to the dominance of the diagrams involving the Δ), the 2p-2h contribution of MEC to the nuclear responses is the square of an amplitude, hence it is positive by definition. Therefore the net effect of 2p-2h MEC to neutrino scattering is to increase the transverse vector response function, as will be illustrated in the next section.

4. Results

In this Section we present the predictions of the above models and their comparison with the MiniBooNE data. More results can be found in Refs. [43] and [44].

In figures 5 and 6 the flux-integrated double-differential cross section per target nucleon for the ν_μCCQE process on ^{12}C is evaluated for the three nuclear models above described: the RMF model and the SuSA approach with and without the contribution of 2p-2h MEC. In figure 5 the cross sections are displayed versus the muon kinetic energy T_μ at fixed scattering energy θ, in figure 6 they are displayed versus $\cos\theta$ at fixed T_μ.

It appears that the SuSA predictions systematically underestimate the experimental cross section, the discrepancy being larger at high scattering angle and low muon kinetic energy. The inclusion of 2p-2h MEC tends to improve the agreement with the data at low angles, but it is not sufficient to account for the discrepancy at higher angles. The RMF calculation, which, as already mentioned, incorporates violations of scaling of the zeroth kind with a substantial enhancement of the vector transverse response, yields cross sections which are in general larger than the SuSA ones. In particular, in the region close to the peak in the cross section, the RMF result becomes larger than the one obtained with SuSA+MEC. Furthermore, the RMF does better than SuSA in fitting the shape of the experimental curves versus both the scattering angle and the muon energy: this is partly due to the fact that the RMF is better describing the low-energy excitation region whereas the SuSA model has no predictive power at very low angles, where the cross section is dominated by low excitation energies and the superscaling ideas are not supposed to apply.

Concerning the SuSA+MEC results, a possible explanation of the theory/data disagreement is the fact that, as already mentioned, a fully consistent treatment of two-body currents should take into account not only the genuine MEC contributions, but also the correlation diagrams that are necessary in order to preserve the gauge invariance of the theory. This, however, is not an easy task because in an infinite system like the RFG the correlation diagrams give rise to divergences which need to be regularized [57]. The divergences arise from a double pole in some of the diagrams, associated to the presence of on-shell nucleon propagators. Different prescriptions have been used in the literature in order to overcome this problem [66, 67, 68, 57], leading to a substantial model-dependence of the results. In particular in Ref. [57] the divergence has been cured by introducing a parameter ϵ which accounts for the finite size of the nucleus (and therefore the finite time of propagation of a nucleon inside the nucleus) and the ϵ-dependence of the contribution of correlation diagrams has been explored. The study has shown that for reasonable values of the parameter the correlations add to the pure MEC in the high-energy tail and are roughly of the same order of magnitude, but now contributing to both the longitudinal and the transverse channels. The inclusion of these terms in neutrino reactions is in progress [69] and is expected to give a further enhancement of the cross sections.

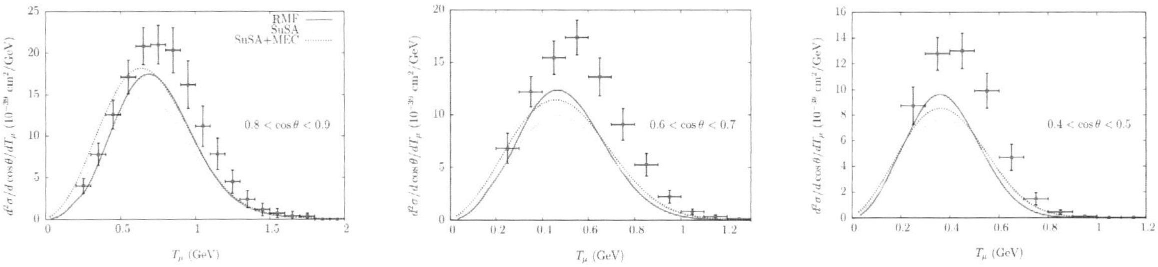

Figure 5. (Color online) Flux-integrated double-differential cross section per target nucleon for the ν_μ CCQE process on ^{12}C evaluated in the RMF (red) model and in the SuSA approach with (blue line) and without (green line) the contribution of MEC and displayed versus the muon kinetic energy T_μ for three specific bins of the scattering angle. The data are from MiniBooNE [1].

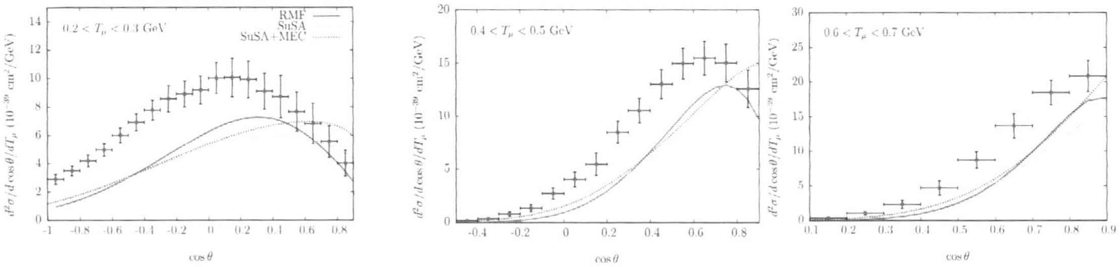

Figure 6. (Color online) Flux-integrated double-differential cross section per target nucleon for the ν_μ CCQE process on ^{12}C evaluated in the RMF (red) model and in the SuSA approach with (blue line) and without (green line) the contribution of MEC and displayed versus the muon scattering angle for three bins of the muon kinetic energy T_μ. The data are from MiniBooNE [1].

The single differential cross sections with respect to the muon kinetic energy and scattering angle, respectively, are presented in figures 7 and 8, where the relativistic Fermi gas result is also shown for comparison: again it appears that the RMF gives slightly higher cross sections than SuSA, due to the L/T unbalance, but both models still underestimate the data for most kinematics. The inclusion of 2p-2h excitations leads to a good agreement with the data at high T_μ, but strength is still missing at the lower muon kinetic energies (namely higher energy transfers) and higher angles.

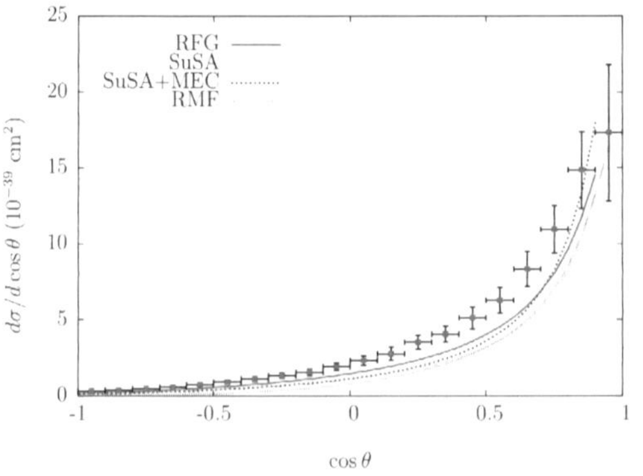

Figure 7. (Color online) Flux-averaged ν_μCCQE cross section on ^{12}C integrated over the scattering angle and displayed versus the muon kinetic energy. The data are from MiniBooNE [1].

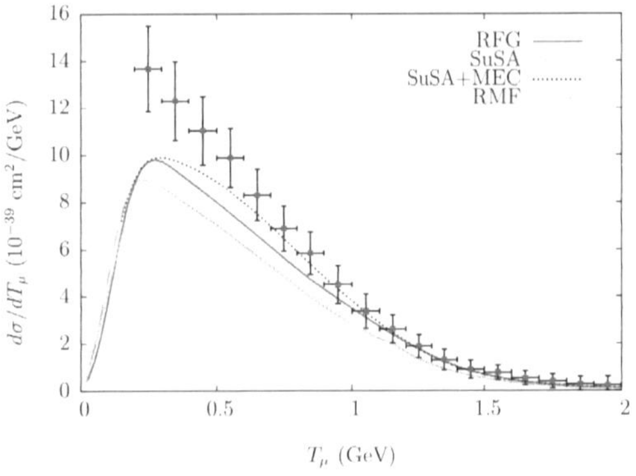

Figure 8. (Color online) Flux-averaged ν_μCCQE cross section on ^{12}C integrated over the muon kinetic energy and displayed versus the scattering angle. The data are from MiniBooNE [1].

Finally, in figure 9 the total (namely integrated over over all muon scattering angles and energies) CCQE cross section per neutron is displayed versus the neutrino energy and compared with the experimental flux-unfolded data. Besides the models above discussed, we show for comparison also the results of the relativistic mean field model when the final state interactions are ignored (denoted as RPWIA - relativistic plane wave impulse approximation) or described through a real optical potential (denoted as rROP). Note that the discrepancies between the various models, observed in figures 5 and 6, tend to be washed out by the integration, yielding very similar results for the models that include FSI (SuSA, RMF and rROP), all of them giving a lower total cross section than the models without FSI (RFG and RPWIA). On the other hand the SuSA+MEC curve, while being closer to the data at high neutrino energies, has a somewhat different shape with respect to the other models, in qualitative agreement with the relativistic calculation of [31]. It should be noted, however, that the result is affected by an uncertainty of about 5% associated with the treatment of the 2p-2h contribution at low momentum transfers and that pionic correlations are not included.

5. Conclusions

Two different relativistic models, one (SuSA) phenomenological and the other (RMF) microscopic, have been applied to the study of charged-current quasielastic neutrino scattering and the impact of 2p-2h meson exchange currents on the cross sections has been investigated. The results can be summarized as follows:

(i) Both the SuSA and the RMF models, in contrast with the relativistic Fermi gas, are fitting with good accuracy the longitudinal quasielastic electron scattering response at intermediate to high energy and momentum transfer.
The SuSA and RMF models give very similar results for the integrated neutrino cross section and both substantially under-predict the MiniBooNE experimental data. However the

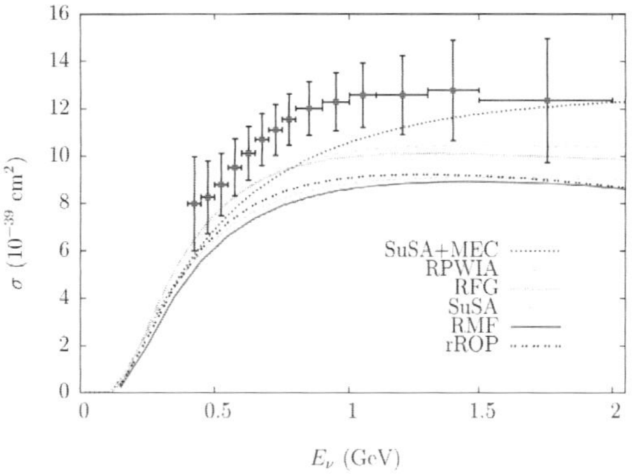

Figure 9. (Color online) Total CCQE cross section per neutron versus the neutrino energy. The curves corresponding to different nuclear models are compared with the flux unfolded MiniBooNE data [1].

comparison with the double differential experimental cross section reveals some differences between the two models, which are washed out by the integration. Indeed the RMF, although being lower than the data, reproduces better the slopes of the cross section versus the muon energy and scattering angle. This is essentially due to the enhancement of the transverse response, which arises from the self-consistent mean field approach of RMF (in particular from the consistent treatment of initial and final state interactions) and is absent in the superscaling approach.

(ii) In relativistic or semi-relativistic models final state interactions have been shown to play an essential role for reproducing the shape and size of the electromagnetic response [50, 46, 51] and cannot be neglected, in our scheme, in the study of neutrino interactions. The effect of final state interactions in the SuSA and RMF models is to lower the cross section, giving a discrepancy with the data larger than the RFG.

(iii) In the transverse channel, the analysis of (e, e') data points to the importance of meson-exchange currents which, through the excitation of two-particle-two-hole states, are partially responsible of filling the "dip" region between the QE and Δ peaks. The 2p-2h MEC can be even more relevant in the CCQE process, where "quasielastic" implies simply that no pions are present in the final state but, due to the large energy region spanned by the neutrino flux, processes involving the exchange of virtual pions can give a sizable contribution. In fact the inclusion of 2p-2h MEC contributions yields larger cross sections and accordingly better agreement with the data, although the theoretical curves still lie below the data at high angles and low muon energy. It should be stressed, however, that the present calculation, though exact and fully relativistic, is incomplete. In order to preserve gauge invariance the full two-body current, including not only the MEC but also the corresponding correlation diagrams, must be included. These have recently been shown to yield a sizable contribution at high energies in (e, e') scattering [57] and are likely to improve the agreement of our models with the MiniBooNE data.

(iv) In all our calculations the standard value M_A =1.03 GeV/c^2 has been used. It has been suggested that a larger value of the axial mass (1.35 GeV/c^2) would eliminate the disagreement with the data. However the fit was done using a RFG analysis, and more sophisticated nuclear models must be explored before drawing conclusions on the actual value of the axial mass. For instance in Ref. [70] it is shown that the MiniBooNE data can as well be fitted by effectively incorporating some nuclear effects in the magnetic form factor of the bound nucleon, without changing the axial mass.
Although our scope here is not to extract a value for the axial mass of the nucleon, but rather to understand which nuclear effects are effectively accounted for by a large axial cutoff parameter, let us mention that a best fit of the RMF and SuSA results to the MiniBooNE

Journal of Physics: Conference Series **336** (2011) 012024 doi:10.1088/1742-6596/336/1/012024

experimental cross section gives an effective axial mass $M_A^{\text{eff}} \simeq 1.5$ GeV/c^2 and values in the range $1.35 < M_A^{\text{eff}} < 1.65$ GeV/c^2 yield results compatible with the MiniBooNE data within the experimental errors. A similar analysis in the model including the 2p-2h contribution will be possible only when the above mentioned correlation diagrams will be consistently evaluated [69].

Acknowledgments
I would like to thank J E Amaro, J A Caballero, T W Donnelly, J M Udias and C W Williamson for the fruitful collaboration which lead to the results reported in this contribution.

References

[1] Aguilar-Arevalo A A *et al.* [MiniBooNE Collaboration] 2010 *Phys. Rev. D* **81** 092005
[2] Bernard V, Elouadrhiri L and Meissner U G 2002 *J. Phys. G* **28** R1
[3] Barbaro M B De Pace, Donnelly T W, Molinari A and Musolf M J 1996 *Phys. Rev. C* **54** 1954
[4] Benhar O, Farina N, Nakamura H, Sakuda M and Seki R 2005 *Phys. Rev. D* **72** 053005
[5] Benhar O and Meloni D 2007 *Nucl. Phys. A* **789** 379
[6] Benhar O and Meloni D 2009 *Phys. Rev. D* **80** 073003
[7] Ankowski A M, Benhar O and Farina N 2010 *Phys. Rev. D* **82** 013002
[8] Benhar O, Coletti P and Meloni D 2010 *Phys. Rev. Lett.* **105** 132301
[9] Juszczak C, Sobczyk J T and Zmuda J 2010 *Phys. Rev. C* **82** 045502
[10] Ankowski A M and Benhar O 2011 *Phys. Rev. C* **83** 054616
[11] Alberico W M *et al.* 1997 *Nucl. Phys. A* **623** 471
[12] Alberico W M *et al.* 1998 *Phys. Lett. B* **438** 9
[13] Alberico W M *et al.* 1999 *Nucl. Phys. A* **651** 277
[14] Meucci A, Giusti C and Pacati F D 2004 *Nucl. Phys. A* **739** 277
[15] Meucci A, Giusti C and Pacati F D 2004 *Nucl. Phys. A* **744** 307
[16] Meucci A, Giusti C and Pacati F D 2006 *Nucl. Phys. A* **773** 250
[17] Meucci A, Caballero J A, Giusti C and Udias J M 2011 *Phys. Rev. C* **83** 064614
[18] Meucci A, Barbaro M B, Caballero J A, Giusti C and Udias J M 2011 *arXiv:1107.5145 [nucl-th]*
[19] Antonov A N *et al.* 2006 *Phys. Rev. C* **74** 054603
[20] Antonov A N, Ivanov M V, Barbaro M B, Caballero J A, Moya de Guerra E and Gaidarov M K 2007 *Phys. Rev. C* **75** 064617
[21] Ivanov M V, Barbaro M B, Caballero J A, Antonov A N, Moya de Guerra E and Gaidarov M K 2008 *Phys. Rev. C* **77** 034612
[22] Leitner T, Alvarez-Ruso L and Mosel U 2006 *Phys. Rev. C* **73** 065502
[23] Leitner T, Alvarez-Ruso L and Mosel U 2006 *Phys. Rev. C* **74** 065502
[24] Buss O, Leitner T, Mosel U and Alvarez-Ruso L 2007 *Phys. Rev. C* **76** 035502
[25] Leitner T and Mosel U 2010 *Phys. Rev. C* **81** 064614
[26] Nieves J, Amaro J E and Valverde M 2004 *Phys. Rev. C* **70** 055503 [2005 *Erratum-ibid.* C **72** 019902]
[27] Nieves J, Valverde M and Vicente Vacas M J 2006 *Phys. Rev. C* **73** 025504
[28] Valverde M, Amaro J E and Nieves J 2006 *Phys. Lett. B* **638** 325
[29] Martini M, Ericson M, Chanfray G and Marteau J 2009 *Phys. Rev. C* **80** 065501
[30] Martini M, Ericson M, Chanfray G and Marteau J 2010 *Phys. Rev. C* **81** 045502
[31] Nieves J, Ruiz Simo I and Vicente Vacas M J 2011 *Phys. Rev. C* **83** 045501
[32] Nieves J, Ruiz Simo I and Vicente Vacas M J 2011 *arXiv:1106.5374 [hep-ph]*
[33] Giusti C and Meucci A 2011 *arXiv:1107.5425 [nucl-th]*
[34] Amaro J E, Barbaro M B, Caballero J A, Donnelly T W and Molinari A 1998 *Nucl. Phys. A* **643** 349
[35] Amaro J E, Barbaro M B, Caballero J A, Donnelly T W and Molinari A 2002 *Phys. Rep.* **368** 317
[36] Day D B, McCarthy J S, Donnelly T W and Sick I 1990 *Ann. Rev. Nucl. Part. Sci.* **40** 357
[37] Jourdan J 1996 *Nucl. Phys. A* **603** 117
[38] Donnelly T W and Sick I 1999 *Phys. Rev. Lett.* **82** 3212
[39] Donnelly T W and Sick I 1999 *Phys. Rev. C* **60** 065502
[40] Maieron C, Donnelly T W and Sick I 2002 *Phys. Rev. C* **65** 025502
[41] Alberico W M, Molinari A, Donnelly T W, Kronenberg E L and Van Orden J W 1988 *Phys. Rev. C* **38** 1801
[42] Barbaro M B, Amaro J E, Caballero J A and Donnelly T E 2006 *Proceedings of the XXV International Worshop on Nuclear Theory "Nuclear Theory 25" (Rila)*, ed S. Dimitrova (Sofia: Heron Press) *[arXiv:nucl-th/0609057]*

[43] Amaro J E, Barbaro M B, Caballero J A, Donnelly T W and Williamson C F 2011 *Phys. Lett. B* **696** 151
[44] Amaro J E, Barbaro M B, Caballero J A, Donnelly T W and Udias J M 2011 *Phys. Rev. D* **84** 033004
[45] Amaro J E, Barbaro M B, Caballero J A, Donnelly T W, Molinari A and Sick I 2005 *Phys. Rev. C* **71** 015501
[46] Caballero J A 2006 *Phys. Rev. C* **74** 015502
[47] Caballero J A, Amaro J E, Barbaro M B, Donnelly T W and Udias J M 2007 *Phys. Lett. B* **653** 366
[48] Caballero J A, Amaro J E, Barbaro M B, Donnelly T W, Maieron C and Udias J M 2005 *Phys. Rev. Lett.* **95** 252502
[49] Amaro J E, Barbaro M B, Caballero J A and Donnelly T W 2007 *Phys. Rev. Lett.* **98** 242501
[50] Maieron C, Martinez M C, Caballero J A and Udias J M 2003 *Phys. Rev. C* **68** 048501
[51] Amaro J E, Barbaro M B, Caballero J A, Donnelly T W and Udias J M 2007 *Phys. Rev. C* **75** 034613
[52] Horowitz C J and Serot B D 1981 *Nucl. Phys.* A **368** 503; 1979 *Phys. Lett. B* **86** 146
[53] Serot B D and Walecka J D 1986 *Adv. Nucl. Phys.* **16** eds J. W. Negele and E. W. Vogt. (New York: Plenum Press) p 1
[54] Sharma M M, Nagarajan M A and Ring P 1993 *Phys. Lett. B* **312** 377
[55] Amaro J E, Barbaro M B, Caballero J A and Donnelly T W 2006 *Phys. Rev. C* **73** 035503
[56] Amaro J E, Barbaro M B, Caballero J A, Donnelly T W and Maieron C 2005 *Phys. Rev. C* **71** 065501
[57] Amaro J E, Maieron C, Barbaro M B, Caballero J A and Donnelly T W 2010 *Phys. Rev. C* **82** 044601
[58] Alberico W M, Donnelly T W and Molinari A 1990 *Nucl. Phys. A* **512** 541
[59] Amaro J E, Barbaro M B, Caballero J A, Donnelly T W and Molinari A 2003 *Nucl. Phys. A* **723** 181
[60] Amaro J E, Barbaro M B, Caballero J A, Donnelly T W, Maieron C and Udias J M 2010 *Phys. Rev. C* **81** 014606
[61] Donnelly T W, Van Orden J W, De Forest T J and Hermans W C 1978 *Phys. Lett. B* **76** 393
[62] Van Orden J M and Donnelly T W 1981 *Ann. Phys.* **131** 451
[63] Dekker M J, Brussaard P J and Tjon J A 1994 *Phys. Rev. C* **49** 2650
[64] De Pace A, Nardi M, Alberico W M, Donnelly T W and Molinari A 2003 *Nucl. Phys. A* **726** 303
[65] De Pace A, Nardi M, Alberico W M, Donnelly T W and Molinari A 2004 *Nucl. Phys. A* **741** 249
[66] Alberico W M, Ericson M and Molinari A 1984 *Annals Phys.* **154** 356
[67] Alberico W M, De Pace A, Drago A and Molinari A 1991 *Riv. Nuovo Cim.* **14N5** 1
[68] Gil A, Nieves J and Oset E 1997 *Nucl. Phys. A* **627** 543
[69] Amaro J E *et al.* work in progress
[70] Bodek A, Budd H and Christy E 2011 *arXiv:1106.0340 [hep-ph]*

Journal of Physics: Conference Series **336** (2011) 012025 doi:10.1088/1742-6596/336/1/012025

Models for quasielastic electron and neutrino-nucleus scattering

Carlotta Giusti and Andrea Meucci

Dipartimento di Fisica Nucleare e Teorica, Università degli Studi di Pavia and INFN, Sezione di Pavia, via Bassi 6 I-27100 Pavia, Italy

E-mail: Carlotta.Giusti@pv.infn.it

Abstract. Models developed for the exclusive and inclusive quasielastic electron-nucleus scattering have been extended to quasielastic neutrino-nucleus scattering. Different descriptions of final-state interactions are compared. For the inclusive electron scattering the relativistic Green's function model is compared with a model based on the use of relativistic purely real mean field potentials in the final state. Both approaches lead to a redistribution of the strength but conserving the total flux. Results for electron and neutrino scattering are presented and discussed in different conditions and kinematics. The results of the relativistic Green's function and of the relativistic mean field models are compared with the double-differential charged-current quasielastic neutrino cross sections recently measured by the MiniBooNE collaboration using a carbon target.

1. Introduction

Several decades of experimental and theoretical work on electron scattering have provided a wealth of information on nuclear structure and dynamics [1, 2]. Additional information on nuclear properties is in principle available from neutrino-nucleus scattering. The main aim of neutrino experiments, however, is not to investigate nuclear properties, but to determine neutrino properties. In neutrino oscillation experiments neutrino-nucleus scattering is used to detect neutrinos and a proper analysis of data requires that the nuclear response to neutrino interactions is well under control and that the unavoidable theoretical uncertainties on nuclear effects are reduced as much as possible.

Different models developed and successfully tested in comparison with electron-scattering data have been extended to describe nuclear effects in neutrino-nucleus scattering. Although the two situations are different, electron scattering is the best available guide to determine the predictive power of a nuclear model. Nonrelativistic and relativistic models have been developed to describe nuclear effects with different approximations. In general, they can be considered as different and alternative models. Relativity is however important at all energies, in particular at high energies, and in the energy regime of many neutrino experiments a fully relativistic approach is required, where not only relativistic kinematics is considered, but also nuclear dynamics and current operators should be described within a relativistic framework.

Models for the exclusive and inclusive electron and neutrino scattering in the quasielastic region are presented in this contribution. In the quasielastic region the nuclear response is dominated by the mechanism of one-nucleon knockout, where the probe interacts with a quasifree nucleon that is emitted from the nucleus with a direct one-step mechanism and the remaining

Published under licence by IOP Publishing Ltd

nucleons are spectators. In electron-scattering experiments the outgoing nucleon can be detected in coincidence with the scattered electron. In the exclusive $(e,e'p)$ reaction the residual nucleus is left in a specific discrete eigenstate and the final state is completely specified. In the inclusive (e,e') scattering only the scattered electron is detected, the final nuclear state is not determined, and the cross section includes all the available final nuclear states.

For an incident neutrino (antineutrino) neutral-current (NC) and charged-current (CC) scattering can be considered

$$\begin{array}{rcll} \nu(\bar{\nu}) + A & \rightarrow & \nu'(\bar{\nu}') + N + (A-1) & \qquad \text{NC} \\ \nu(\bar{\nu}) + A & \rightarrow & l^-(l^+) + p(n) + (A-1) & \qquad \text{CC} \end{array}$$

In NC scattering only the emitted nucleon can be detected and the cross section is integrated over the energy and angle of the final lepton. Also the state of the residual $(A-1)$-nucleus is not determined and the cross section is summed over all the available final states. The same situation occurs for the CC reaction if only the outgoing nucleon is detected. The cross sections are therefore inclusive in the leptonic sector and semi-inclusive in the hadronic sector. The exclusive CC process where the charged final lepton is detected in coincidence with the emitted nucleon can be considered as well. The inclusive CC scattering where only the charged lepton is detected can be treated with the same models used for the inclusive (e,e') reaction.

For all these processes the cross section is obtained in the one-boson exchange approximation from the contraction between the lepton tensor, that under the assumption of the plane-wave approximation for the initial and final lepton wave functions depends only on the lepton kinematics, and the hadron tensor $W^{\mu\nu}$, that contains the nuclear response and whose components are given by bilinear products of the matrix elements of the nuclear current J^μ between the initial and final nuclear states, *i.e.*,

$$W^{\mu\nu} = \sum_f \langle \Psi_f \mid J^\mu(\boldsymbol{q}) \mid \Psi_i \rangle \langle \Psi_i \mid J^{\nu\dagger}(\boldsymbol{q}) \mid \Psi_f \rangle \, \delta(E_i + \omega - E_f), \tag{1}$$

where ω and $\boldsymbol{q}$ are the energy and momentum transfer, respectively. Different but consistent models to calculate the components of the hadron tensor in quasielastic electron- and neutrino-nucleus scattering are outlined in the next sections.

2. The exclusive $(e,e'p)$ reaction

For the exclusive $(e,e'p)$ reaction nonrelativistic and relativistic models based on the distorted-wave impulse approximation (DWIA) have been developed to calculate the matrix elements in Eq. (1). The DWIA expression of the matrix elements is the results of the following assumptions [1, 2, 3]:

i) An exclusive process is considered, where the residual nucleus is left in a discrete eigenstate n of its Hamiltonian.

ii) The final nuclear state is projected onto the channel subspace spanned by the vectors corresponding to a nucleon, at a given position, and the residual nucleus in the state n. This assumption neglects effects of coupled channels and is justified by the considered asymptotic configuration of the final state.

iii) The (one-body) nuclear-current operator does not connect different channel subspaces. Thus, also the initial state is projected onto the selected channel subspace. This assumption is the basis of the direct-knockout mechanism and is related to the impulse approximation.

The amplitudes of Eq. (1) are then obtained in a one-body representation as

$$\lambda_n^{1/2} \langle \chi^{(-)} \mid j^\mu(\boldsymbol{q}) \mid \varphi_n \rangle \ , \tag{2}$$

where j^μ the one-body nuclear current, $\chi^{(-)}$ is the single-particle (s.p.) scattering state of the emitted nucleon, φ_n the overlap between the ground state of the target and the final state n, i.e., a s.p. bound state, and the spectroscopic factor λ_n is the norm of the overlap function, that gives the probability of removing from the target a nucleon leaving the residual nucleus in the state n. In the model the s.p. bound and scattering states are consistently derived as eigenfunctions of a non Hermitian energy dependent Feshbach-type optical potential and of its Hermitian conjugate at different energies. In standard DWIA calculations phenomenological ingredients are usually employed: the scattering states are eigenfunctions of a phenomenological optical potential determined through a fit to elastic nucleon-nucleus scattering data and the s.p. bound states are obtained from mean-field potentials, or can be calculated in a phenomenological Woods-Saxon well.

The model can be formulated in a similar way within nonrelativistic DWIA and relativistic RDWIA frameworks [4]. In RDWIA, calculations are performed with a relativistic nuclear-current operator and four-vector relativistic wave functions for the s.p. bound and scattering states. Both DWIA and RDWIA models have been quite successful in describing a large amount of $(e, e'p)$ data in a wide range of nuclei and in different kinematics [2, 4, 5, 6]. The shape of the experimental recoil-momentum distributions corresponding to a particular state of the residual nucleus are well described by model calculations. Then, the spectroscopic factors are usually extracted from the comparison between experimental and theoretical results as the reduction factors that must be applied to the calculated cross sections to reproduce the magnitude of the experimental cross sections.

Recently, the nonrelativistic DWIA and relativistic RDWIA models widely and successfully adopted to describe data in a wide range of stable nuclei have been applied to nuclei with neutron excess, with the aim to study the evolution of the $(e, e'p)$ cross sections as a function of the asymmetry between the number of neutrons and protons [7].

3. Semi-inclusive neutrino-nucleus scattering

For the semi-inclusive neutral-current and charged-current processes where only the outgoing nucleon is detected, the transition amplitudes can be calculated in the same RDWIA model of Eq. (2). Since the outgoing lepton is not detected, the cross section must be integrated over the energy and angle of the outgoing lepton. The state of the residual nucleus is not determined and the cross section includes all the states n. In the calculations [8, 9, 10] a pure shell-model description is assumed, i.e., n is a one-hole state and the sum over n is over all the occupied shell-model states. In [8, 9, 10] final-state interactions are described by a complex optical potential whose imaginary part gives an absorption that reduces the calculated cross section. A similar reduction is found in $(e, e'p)$ calculations. The imaginary part of the optical potential accounts for the fact that in the elastic scattering, if other channels are open besides the elastic one, part of the incident flux is lost in the elastically scattered beam and goes to the inelastic channels that are open. In the exclusive scattering, where only one channel is considered, it is correct to include the absorptive imaginary part of the optical potential and account for the flux lost in the considered channel, but in the inclusive scattering, where all the channels are included, the flux lost in a channel must be found in the other channels, and in the sum over all the channels the flux can be redistributed but must conserved. Thus, the use of the absorptive imaginary part of the optical potential in the inclusive scattering seems inconsistent with the requirement of flux conservation. For the semi-inclusive process where the emitted nucleon is detected, some of the reaction channels which are responsible for the imaginary part of the potential like multi-step processes, fragmentation of the nucleus, etc., are not included in the experimental cross section. There are, however, contributions that are not considered in this model and that can be contained in the experimental cross section. In [8, 9, 10] the semi-inclusive neutrino scattering is treated as a process where the cross section is obtained from the sum of all the integrated exclusive one-nucleon knockout channels. We can expect that the description of final-state interactions by means of a complex optical potential that gives absorption can produce cross sections that are smaller than the experimental data [11]. However, since measurements of cross sections are a hard experimental task, ratios of cross sections have been proposed as alternative quantities that can provide useful informations. In [8, 9, 10] it is shown that ratios of cross sections are less sensitive to distortion effects and several ratios are considered in view of the possible determination of the strange quark content of the nucleon.

4. Inclusive lepton-nucleus scattering

For the inclusive scattering where only the outgoing lepton is detected the Green's function model [12, 13, 14, 15, 16, 17, 18] has been developed to describe final-state interactions consistently with the exclusive scattering and using the same complex optical potential.

In the Green's function model, under suitable approximations, that are basically related to the impulse approximation, the components of the hadron tensor are written in terms of the s.p. optical model Green's function, whose self-energy the Feshbach optical potential. The explicit calculation of the s.p. Green's function can be avoided by its spectral representation, that is based on a biorthogonal expansion in terms of a non Hermitian optical potential $\mathcal{H}$ and of its Hermitian conjugate $\mathcal{H}^\dagger$. Calculations require matrix elements of the same type as the DWIA ones in Eq. (2), but involve eigenfunctions of both $\mathcal{H}$ and $\mathcal{H}^\dagger$, where the imaginary part gives in one case an absorption and in the other case a gain of flux, and in the sum over n the total flux is redistributed and conserved. The Green's function approach guarantees a consistent treatment of final-state interactions in the exclusive and in the inclusive scattering and gives a good description of (e, e') data [12, 13].

For the inclusive electron scattering both nonrelativistic Green's function [12, 16] and relativistic Green's function (RGF) [13] approaches have been considered. For charged-current neutrino scattering the RGF model has been adopted. A numerical example is shown in figure 1 for the $^{12}\mathrm{C}(\nu_\mu, \mu^-)$ cross sections calculated for an incident neutrino energy of $E_\nu = 1000$ MeV and two values of the muon scattering angle, *i.e.*, $\vartheta = 45^\mathrm{o}$ (upper panel) and 135° (lower panel), as a function of the muon kinetic energy T_μ. In the case of the RGF model two different parameterizations of the relativistic optical potential have been used in the calculations, *i.e.*, the energy-dependent and A-dependent EDAD1 (RGF1) and the energy-dependent but A-independent EDAI (RGF-EDAI), which are fitted to proton-nucleus elastic scattering data in an energy range up to 1040 MeV [19]. In figure 1 the RGF1 and RGF-EDAI results are compared with the results of the relativistic plane-wave impulse approximation (RPWIA), where final-state interactions are neglected, and with the cross sections obtained when only the real part of the relativistic optical potential (rROP) is retained and the imaginary part is neglected. The rROP approximation conserves the flux but, independently of its numerical results, it is conceptually wrong because the optical potential has to be complex owing to the presence of inelastic channels.

The differences between the rROP, RGF1, and RGF-EDAI results are essentially due to the imaginary part of the optical potential. Different parameterizations of the optical potential are available, that are able to give equivalently good descriptions of elastic proton-nucleus scattering data, but correspond to different optical potentials that mainly differ for their imaginary part. The imaginary part accounts for the overall effects of inelastic channels and is not univocally determined from the elastic phenomenology. The real terms are very similar and the cross sections calculated in rROP are practically insensitive to the choice of the optical potential. In contrast, the imaginary part is sensitive to the parameterization of the relativistic optical potential and gives the differences shown in the figure. The RGF1 and RGF-EDAI results at $\vartheta = 45^\mathrm{o}$ have somewhat different shapes for high values of T_μ, *i.e.*, low values of the energy transferred. At $\vartheta = 135^\mathrm{o}$ the differences are reduced but the magnitude of the cross sections is

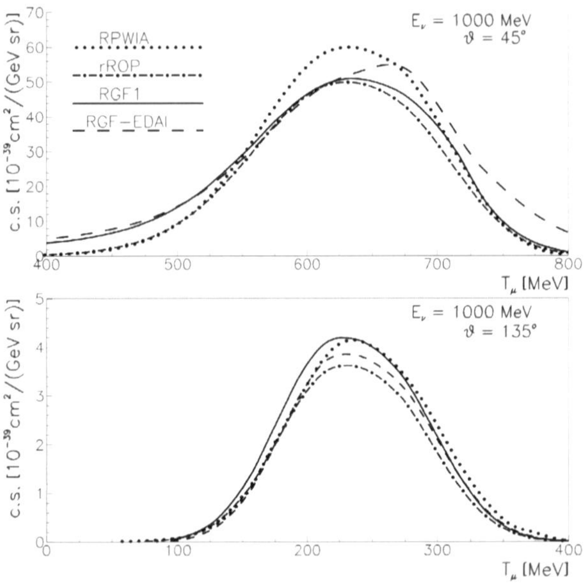

Figure 1. The cross sections of the ^{12}C(ν_μ, μ^-) reaction for an incident neutrino energy of E_ν = 1000 MeV and a muon scattering angle $\vartheta = 45^o$ (upper panel) and 135° (lower panel) as a function of the muon kinetic energy T_μ. Results for RPWIA (dotted), rROP (dot-dashed), and the relativistic RGF approach with two different optical potentials [19], i.e., EDAD1 (RGF1) and EDAI (RGF-EDAI), are compared.

significantly reduced.

5. Comparison of relativistic models

The analysis of data for neutrino experiments requires a precise knowledge of lepton-nucleus cross sections, where uncertainties on nuclear effects are reduced as much as possible. To this aim, it is important to check the consistency of different models and the validity of the adopted approximations. The results of the relativistic models developed by the Pavia and the Madrid-Sevilla groups for the inclusive scattering have been compared in [20] for the inclusive electron scattering and in [21] for the inclusive charged-current neutrino scattering. As a first step, the consistency of the RPWIA and rROP calculations performed by the two groups with independent numerical programs and with the same theoretical ingredients has been checked. Almost identical results are obtained in RPWIA and very similar results in rROP. This result gives us confidence on the reliability of our calculations and allows us to extend the comparison to the different descriptions of final-state interactions.

An example is shown in figure 2, where the ^{12}C(e, e') cross sections obtained by the two groups in RPWIA and rROP are compared with the RGF results obtained with two different energy-dependent and A-dependent parameterizations for the relativistic optical potential of [19], *i.e.*, the EDAD1 (RGF1), already considered for the calculations shown in figure 1, and the EDAD2 (RGF2), and with the results of the relativistic mean field (RMF) model [22] of the Madrid-Sevilla group, where the scattering wave functions are calculated with the same strong real mean-field potential used for the initial bound states. The RMF model fulfills the dispersion relation and maintains the continuity equation.

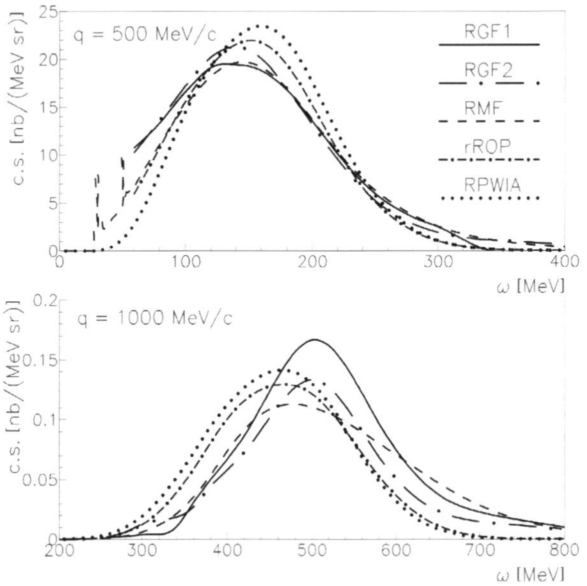

Figure 2. Differential cross section of the ^{12}C(e, e') reaction for an incident electron energy $\varepsilon = 1$ GeV and $q = 500$ and 1000 MeV/c. Results for RPWIA (dotted), rROP (dot-dashed), RGF1 (solid), RGF2 (long dot-dashed), and RMF (dashed) are compared.

The differences between the RMF and RGF results increase with the momentum transfer. Also the discrepancies between the RGF1 and RGF2 cross sections depend on the momentum transfer. At $q = 500$ MeV/c the three results are similar, both in magnitude and shape, larger differences are obtained at $q = 1000$ MeV/c. The shape of the RMF cross section shows an asymmetry, with a long tail extending towards higher values of ω, that is essentially due to the strong energy-independent scalar and vector potentials present in the RMF model. The asymmetry towards higher ω is less significant but still visible for RGF1 and RGF2, whose cross sections show a similar shape but with a significant difference in the magnitude. At $q =$ 1000 MeV/c both RGF1 and RGF2 cross sections are higher than the RMF one in the region where the maximum occurs. A stronger enhancement is obtained with RGF1, which at the peak overshoots the RMF cross section up to 40% and it is even higher than the RPWIA result.

The behavior of the RMF and RGF results as a function of q and ω is linked to the structure of the relativistic potentials involved in the RMF and RGF models. Whereas RMF is based on the use of a strong energy-independent real potential, RGF makes use of a complex energy-dependent optical potential. In RGF calculations the behavior of the optical potential changes with the momentum and energy transferred in the process, and higher values of q and ω correspond to higher energies for the optical potential. The RGF results are consistent with the general behavior of the optical potentials and are basically due to their imaginary part. The real terms of the relativistic optical potentials are very similar for the different parameterizations and the rROP cross sections do not show sensitivity to the particular parameterization considered. On the other hand, the energy-dependent scalar and vector components of the real part of the relativistic optical potential get smaller with increasing energies. Thus the rROP result approaches the RPWIA one for large values of ω. In contrast, the imaginary (scalar and vector) part presents its maximum strength around 500 MeV, being also sensitive to the particular parameterization of the optical potential. This explains the differences observed between the

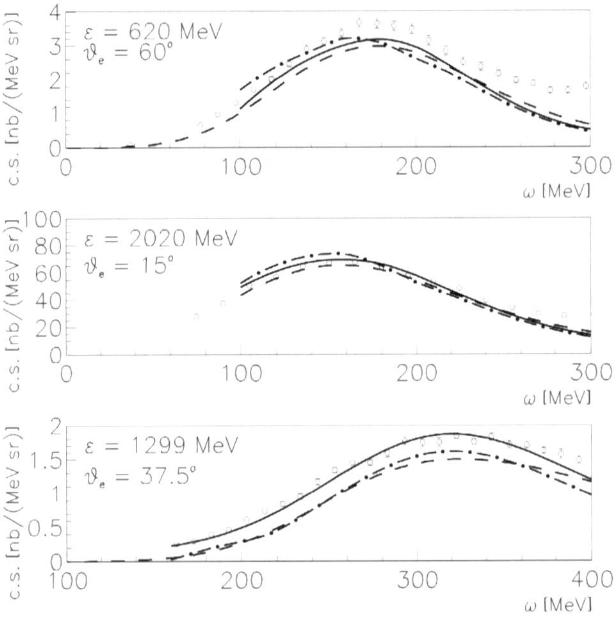

Figure 3. Differential cross section of the ^{12}C(e, e') reaction for different beam energies and electron-scattering angles in RGF1 (solid), RGF2 (long dot-dashed), RMF (dashed). Experimental data from [23, 24, 25].

rROP and the two RGF results as a function of ω and q.

In figure 3 the RGF1, RGF2, and RMF results are compared with the experimental cross sections for three different kinematics [23, 24, 25]. The three models lead to similar cross sections. The main differences are obtained for higher values of the momentum transfer, about 800 MeV/c (bottom panel), where the RGF1 cross section (solid line) is larger than the RGF2 (dot-dashed) and RMF (dashed) ones. The experimental cross section is slightly underpredicted in the top panel and well described in the middle panel by all calculations. Finally, results in the bottom panel show a fair agreement with data in the case of RGF1, whereas RGF2 and RMF underpredict the experiment. Although satisfactory on general grounds, the comparison with data gives only an indication and cannot be conclusive until contributions beyond the quasielastic peak, like meson-exchange currents and Δ effects, which may play a significant role in the analysis of data even at the maximum of the quasielastic peak, are carefully evaluated [26, 27, 28].

In figure 4 the ^{12}C (ν_μ, μ^-) cross section obtained by the two groups in RPWIA and rROP in a kinematics with a fixed value of the incident neutrino energy, $E_\nu = 1000$ MeV, and two values of the momentum transfer, *i.e.*, $q = 500$ and 1000 MeV/c, are compared with the corresponding RMF, RGF1 and RGF-EDAI results.

The cross sections shown in figure 4 are calculated with same incident lepton energy and momentum transfer as in the (e, e') calculations of figure 2. This is a kinematics more typical of electron-scattering experiments than of neutrino experiments, but the calculations have been carried out to perform a more direct comparison between the results of the models in electron and neutrino scattering. Actually, in the case of charged-current neutrino scattering the muon mass gives a different kinematics, with different values of the energy transfer and, as a consequence,

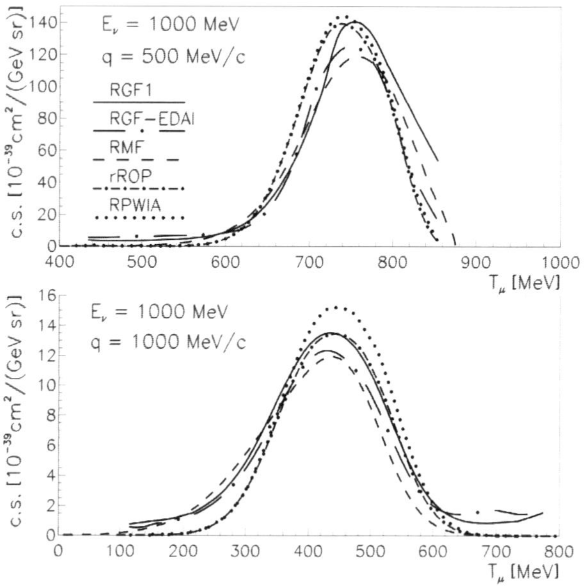

Figure 4. Differential cross section of the $^{12}\mathrm{C}(\nu_\mu, \mu^-)$ reaction for $\mathrm{E}_\nu = 1000$ MeV and $q =$ 500 MeV/c and 1000 MeV/c. Results for RPWIA (dotted), rROP (dot-dashed) RGF1 (solid), RGF- EDAI (long dot-dashed), and RMF (dashed) are compared.

of the energies of the outgoing nucleon. This means that in the RGF model the optical potential is calculated for electron and neutrino scattering at different energies. We have checked that if we reproduce the kinematics of electron scattering and perform calculations for the (ν_μ, μ^-) reaction with vanishing muon mass, the main difference with respect to the calculations shown in figure 4 is a shift of the cross section by about 100 MeV towards higher values of T_μ, without any significant change in the shape or in the strength.

Also for the results in figure 4 the shape of the RMF cross section shows an asymmetry with a tail extending towards higher values of ω (corresponding to lower values of the kinetic energy of the outgoing muon T_μ). An asymmetric shape towards higher ω is shown also by the RGF cross sections, while no visible asymmetry is given by the RPWIA and rROP results. Also in this case the significant differences obtained between the RGF and rROP cross sections are consistent with the general behavior of the phenomenological energy-dependent relativistic optical potentials and are basically due to their different imaginary part. As already shown for (e, e') reactions, the RGF yields in general a larger cross section than the RMF. The RGF and the RMF yield similar predictions, within a few percent for low-q, while as q goes up the RGF yields increasingly larger cross sections than RMF. This may reflect the influence of the pionic degrees of freedom, that can be included in a phenomenological way in the imaginary part of the optical potential [20, 21].

The results in figure 4 present some differences with respect to the corresponding cross sections of the inclusive electron scattering shown in figure 2. In both cases the differences between the results of the different models are generally larger for increasing value of the momentum transfer. For neutrino scattering, however, such a behavior is less evident and clear. In particular, the RGF1 cross section at $q = 1000$ Mev/c does not show the strong enhancement in the region of the maximum that is found for the (e, e') calculations of figure 2, where the RGF1 result is even larger than the RPWIA one. In the case of neutrino scattering the RGF results in the region

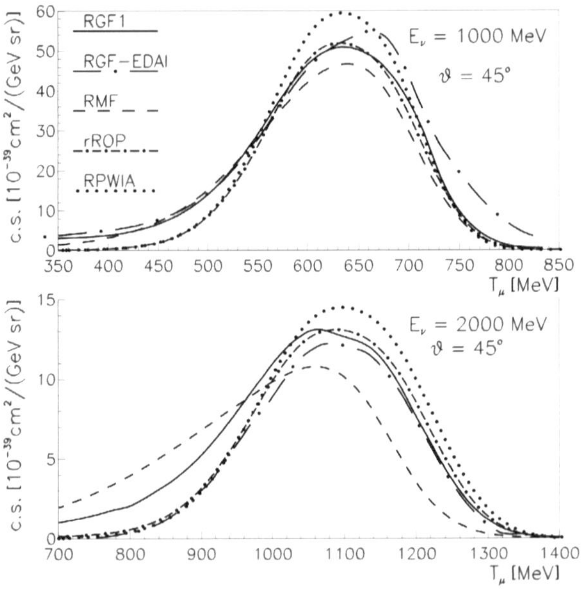

Figure 5. Differential cross section of the ^{12}C(ν_μ, μ^-) reaction for E$_\nu$ = 1000 MeV and 2000 MeV, and scattering angle $\vartheta = 45^o$. Results for RPWIA (dotted), rROP (dot-dashed) RGF1 (solid), RGF-EDAI (long dot-dashed), and RMF (dashed) are compared.

of the maximum are generally larger than the RMF ones, but smaller than the RPWIA cross sections.

In spite of many similarities, inclusive electron scattering and charged-current neutrino-nucleus scattering are two different processes and caution should be drawn on their comparison. The different currents and their possible interplay with the other ingredients of the models do not allow an easy comparison. The numerical differences between the RGF results for electron and neutrino scattering can mainly be ascribed to the combined effects of the weak current, in particular its axial term, and the imaginary part of the relativistic optical potential. We have checked that these effects can explain the fact that in neutrino scattering the RGF1 result does not give the strong enhancement in the region of the maximum that is obtained for the (e, e') cross section in figure 2.

The ^{12}C (ν_μ, μ^-) cross section for the kinematics with $E_\nu = 1000$ and 2000 MeV and $\vartheta = 45^o$ are presented in figure 5. In these kinematics, that are more similar to those actually explored at present neutrino experiment facilities, the momentum transfer is not fixed and its value around the peak of the cross section is usually large, *i.e.*, $q \approx 700$ MeV$/c$ for $E_\nu = 1000$ MeV and $q \approx 1400$ MeV$/c$ for $E_\nu = 2000$ MeV. The shape of the RMF cross section shows also in this case an asymmetry, with a long tail extending towards lower values of T_μ, which is due to the strong energy-independent scalar and vector potentials present in the RMF approach. The asymmetry increases with larger incident neutrino energies. For the RGF cross sections the asymmetry is less significant but still visible. while almost no asymmetry is found for the RPWIA and rROP cross sections. The RGF1 and RGF-EDAI cross sections have somewhat different shapes, that are particularly visible for low ω at $E_\nu = 1000$ MeV and for higher ω at $E_\nu = 2000$ MeV. These differences are essentially due to the imaginary part of the relativistic optical potential, that is sensitive to the particular parametrization considered, while the real terms are very similar for different parameterizations and give very similar results.

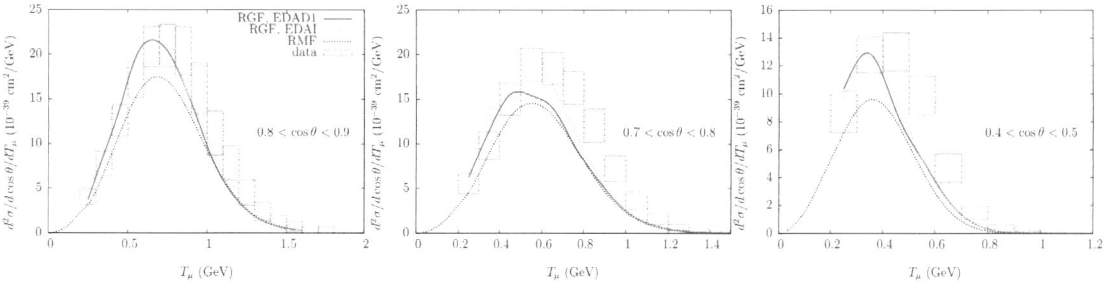

Figure 6. (Color online) Flux-averaged double-differential cross section per target nucleon for the CCQE ^{12}C(ν_μ, μ^-) reaction calculated in the RMF (blue line), the RGF1 (red), and RGF-EDAI (green) and displayed versus T_μ for various bins of $\cos\theta$. In all the calculations the standard value of the nucleon axial mass, *i.e.*, $M_A = 1.03$ GeV$/c^2$ has been used. The data are from MiniBooNE [29].

6. Comparison with Charged-Current Quasielastic MiniBooNE data

The double-differential cross sections for muon neutrino CC quasielastic (CCQE) scattering have recently been measured by the MiniBooNE collaboration [29]. These data have raised a strong debate over the role of the theoretical ingredients entering the description of the reaction. The experimental cross section is underestimated by the relativistic Fermi Gas (RFG) model, and also by the results of other more sophisticated models, unless the value of nucleon axial mass M_A is significantly enlarged (1.35 GeV$/c^2$ in the RFG) with respect to the accepted world average value (1.03 GeV$/c^2$ [30, 31]). Before drawing conclusions about the need to increase the axial mass it is however important to evaluate carefully the contributions of all nuclear effects.

Within the quasielastic kinematic domain the treatment of final-state interactions is essential for the comparison with data. The comparison between the results of the RMF and RGF models, where final-state interactions are described by very different ingredients, can be helpful for a deeper understanding of the role played by final-state interactions in the analysis of CCQE data and its influence in studies of neutrino oscillations at intermediate to high energies. The predictions of the two models have been compared with the recent CCQE MiniBooNE data [32].

The CCQE double-differential ^{12}C (ν_μ, μ^-) cross sections averaged over the neutrino flux as a function of T_μ for various bins of $\cos\theta$, where θ is the muon scattering angle, are shown in figure 6. The RMF results yield reasonable agreement with data for small angles and low muon energies, the discrepancy becoming larger as θ and T_μ increase. The shape followed by the RMF cross sections fits well the slope shown by the data. A good agreement with the experimental shape is shown also by the RGF cross sections. The RMF and RGF models yield close predictions at larger values of T_μ, for all the bins of $\cos\theta$ shown in the figure. The RGF cross sections are generally larger than the RMF ones. The differences increase approaching the peak region, where the additional strength shown by the RGF produces cross sections in reasonable agreement with the data.

Also the differences between the RGF results with the two optical potentials are enhanced in the peak region, but remain in general within the uncertainties of the experimental errors. The EDAD1 and EDAI potentials yield close predictions for the bin $0.4 < \cos\theta < 0.5$, the differences are visible but anyhow small for the bin $0.8 < \cos\theta < 0.9$, being the RGF-EDAI cross section larger than the RGF1 one, while the difference is sizeable for the bin $0.7 < \cos\theta < 0.8$, with the RGF1 result closer to the RMF than to the RGF-EDAI one.

In figure 7 the flux-averaged double-differential cross sections are plotted versus $\cos\theta$ for two bins of T_μ, *i.e.*, $0.2 < T_\mu < 0.3$ GeV and $0.6 < T_\mu < 0.7$ GeV. The shape of the experimental cross section is well described by the models. The RMF results generally underestimate the

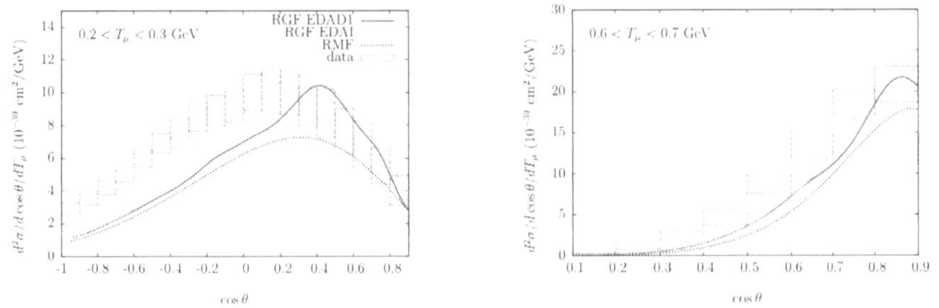

Figure 7. (Color online) Flux-averaged double-differential cross section per target nucleon for the CCQE $^{12}C(\nu_\mu, \mu^-)$ reaction displayed versus $\cos\theta$ for two bins of T_μ. The results obtained with RMF (blue line), RGF1 (red), and RGF-EDAI (green) are compared. In all the calculations the standard value of the nucleon axial mass, *i.e.*, $M_A = 1.03$ GeV/c^2 has been used. The data are from MiniBooNE [29].

data, especially for the lower muon energy values, the agreement improves as T_μ increases. The additional strength produced by the complex optical potential in the RGF gives a reasonable agreement with the size of the experimental cross section. The agreement is better for smaller angles while the data are slightly underpredicted as θ increases. The discrepancy is larger with RGF1, that gives in general a lower cross section than RGF-EDAI, and in the bin $0.2 < T_\mu < 0.3$ GeV.

These results give further and clear indication that before drawing conclusions about the comparison with data and about the need to increase the value of the nucleon axial mass a careful evaluation of all nuclear effects is required. Important contributions can be produced by final-state interactions. Both RGF and RMF models, where final-state interactions are described with very different theoretical ingredients, give a good description of the shape of the experimental cross sections. The RMF generally underpredicts the data, but for lower values of the muon energy and scattering angle. In contrast, the RGF model can give cross sections of the same magnitude as the experimental ones without the need to increase the standard value of the axial mass. The larger cross sections in the RGF model are produced by the imaginary part of the relativistic optical potential, that includes the overall effect of inelastic (nucleonic and non-nucleonic) channels and that is not univocally determined from the elastic phenomenology. The choice of the optical potential and a more refined determination of its imaginary part can deserve further investigation. Other contributions, not included in the present calculations, might play an important role in the analysis of CCQE data. Results of different models indicate that significant effects can be expected from correlations and meson-exchange currents [33, 34] or multiple knockout excitations [35, 36, 37, 38]. More refined calculations including all these contributions should be performed, but any new contribution should be included consistently in a model.

Acknowledgments

The authors wish to acknowledge the collaborations that have produced the results reported in this contribution. In particular, we are grateful to Franco Pacati, Franco Capuzzi, Juan Antonio Caballero, José Manuel Udías, and Maria Barbaro for fruitful collaborations.

References

[1] Boffi S, Giusti C and Pacati F D 1993 *Phys. Rep.* **226** 1
[2] Boffi S, Giusti C, Pacati F D and Radici M 1996 *Electromagnetic Response of Atomic Nuclei* Oxford Studies in Nuclear Physics, Vol. 20 (Oxford: Clarendon Press)

[3] Boffi S, Cannata F, Capuzzi F, Giusti C and Pacati F D 1982 *Nucl. Phys. A* **379** 509
[4] Meucci A, Giusti C and Pacati F D 2001 *Phys. Rev. C* **64** 014604
[5] Udías J M, Sarriguren P, Moya de Guerra E, Garrido E and Caballero J A 1993 *Phys. Rev. C* **48** 2731
[6] Radici M, Meucci A and Dickhoff W H 2003 *Eur. Phys. J. A* **17** 65
[7] Giusti C, Meucci A, Pacati F D, Cò G and De Donno V 2011 *Phys. Rev. C* **84** 024615
[8] Meucci A, Giusti C and Pacati F D 2004 *Nucl. Phys. A* **744** 307
[9] Meucci A, Giusti C and Pacati F D 2006 *Nucl. Phys. A* **773** 250
[10] Meucci A, Giusti C and Pacati F D 2008 *Phys. Rev. C* **77** 034606
[11] Meucci A, Giusti C and Pacati F D 2011 *Preprint* arXiv:1110.3928 [nucl-th]
[12] Capuzzi F, Giusti C and Pacati F D 1991 *Nucl. Phys. A* **524** 681
[13] Meucci A, Capuzzi F, Giusti C and Pacati F D 2003 *Phys. Rev. C* **67** 054601
[14] Meucci A, Giusti C and Pacati F D 2004 *Nucl. Phys. A* **739** 277
[15] Meucci A, Giusti C and Pacati F D 2005 *Nucl. Phys. A* **756** 359
[16] Capuzzi F, Giusti C, Pacati F D and Kadrev D N 2005 *Ann. Phys.* **317** 492
[17] Meucci A, Giusti C and Pacati F D 2006 *Acta Phys. Polon. B* **37** 2279
[18] Meucci A, Giusti C and Pacati F D 2009 *Acta Phys. Polon. B* **40** 2579
[19] Cooper E D, Hama S, Clark B C and Mercer R L 1993 *Phys. Rev. C* **47** 297
[20] Meucci A, Caballero J A, Giusti C, Pacati F D and Udías J M 2009 *Phys. Rev. C* **80** 024605
[21] Meucci A, Caballero J A, Giusti C and Udías J M 2011 *Phys. Rev. C* **83** 064614
[22] Caballero J A, Amaro J E, Barbaro M B, Donnelly T W, Maieron C and Udías J M 2005 *Phys. Rev. Lett.* **95** 252502
[23] Barreau P *et al.* 1983 *Nucl. Phys. A* **402** 277
[24] Day D B *et al.* 1993 *Phys. Rev. C* **48** 1849
[25] Sealock R *et al.* 1989 *Phys. Rev. Lett.* **62** 1350
[26] Barbaro M B, Caballero J A, Donnelly T W and Maieron C 2004 *Phys. Rev. C* **69** 035502
[27] Amaro J E, Barbaro M B , Caballero J A, Donnelly T W, Molinari A and Sick I 2005 *Phys. Rev. C* **71** 015501
[28] Ivanov M V *et al.* 2008 *Phys. Rev. C* **77** 034612
[29] Aguilar-Arevalo A A *et al.* [MiniBooNE Collaboration] 2010 *Phys. Rev. D* **81** 092005
[30] Bernard V, Elouadrhiri L and Meissner U G 2002 *J. Phys. G* **28** R1
[31] Bodek A, Avvakumov S, Bradford R and Budd H 2008 *Eur. Phys. J. C* **53** 349
[32] Meucci A, Barbaro M B, Caballero J A, Giusti C and Udías J M 2011 *Phys. Rev. Lett.* **107** 172501
[33] Amaro J E, Barbaro M B, Caballero J A and Donnelly 2011 *Phys. Lett. B* **696** 151
[34] Amaro J E, Barbaro M B, Caballero J A, Donnelly and Udías J M 2011 *Phys. Rev. D* **84** 033004
[35] Martini M, Ericson M, Chanfray G and Marteau J 2009 *Phys. Rev. C* **80** 065501
[36] Martini M, Ericson M, Chanfray G and Marteau J 2010 *Phys. Rev. C* **81** 045502
[37] Nieves J, Ruiz Simo I and Vicente Vacas M J 2011 *Phys. Rev. C* **83** 045501
[38] Nieves J, Ruiz Simo I and Vicente Vacas M J 2011 *Preprint* arXiv:1106.5374 [nucl-th]

Muon capture on light nuclei

Laura Elisa Marcucci

Department of Physics "E. Fermi", University of Pisa, and INFN-Pisa, 56127 Pisa, Italy

E-mail: `laura.marcucci@df.unipi.it`

Abstract. We review the latest theoretical predictions for the muon capture reactions $^2\mathrm{H}(\mu^-,\nu_\mu)nn$ and $^3\mathrm{He}(\mu^-,\nu_\mu)^3\mathrm{H}$, obtained using a realistic model for the nuclear Hamiltonian and weak transition operator, derived either in a meson-exchange or in a chiral effective field theory (χEFT) framework. The total rates are found to be 392.0 ± 2.3 s^{-1} for $A = 2$, and 1484 ± 13 s^{-1} for $A = 3$, where the spread accounts for the model dependence and cutoff sensitivity in the χEFT currents. The contributions of the two-body axial currents, obtained imposing the partially-conserved-axial-current hypothesis at the two-body level, ivestigated in the χEFT framework, are found to be negligible. The first steps toward a fully consistent χEFT calculation are also discussed.

1. Introduction and theoretical formalism

There is a significant body of experimental and theoretical work on muon captures on light nuclei. This is motivated by the fact that these processes, experimentally accessible, are related to astrophysically relevant reactions, such as the weak captures on proton and ^{3}He (the *pp* and *hep* reactions), whose rates instead cannot be measured experimentally. Thus, it becomes crucial to study all these processes within the same theoretical framework, and test the nuclear wave functions and, indirectly, the interactions from which these are obtained, and the models of the nuclear weak current, by comparing with the available experimental data [1].

In the present work, we focus our attention on muon capture on deuteron and ^{3}He, i.e., on the reactions

$$\mu^- + d \rightarrow n + n + \nu_\mu \,, \tag{1}$$
$$\mu^- + {}^3\mathrm{He} \rightarrow {}^3\mathrm{H} + \nu_\mu \,. \tag{2}$$

These reactions have been studied extensively through the years, experimentally and theoretically. The observables of interest are the doublet capture rate Γ^D for reaction (1), i.e., the rate obtained when the stopped muons are captured from the doublet hyperfine state, and the total capture rate Γ_0 for reaction (2). The experimental situation for reaction (2) is quite clear: a very precise determination yielded Γ_0=1496(4) s^{-1} [2], a value consistent with those of the earlier measurements, although these were affected by considerably larger uncertainties. On the other hand, Γ^D is poorly known: the available experimental data are 365(96) s^{-1} [3], 445(60) s^{-1} [4], 470(29) s^{-1} [5] and 409(40) s^{-1} [6]. These measurements, while consistent with each other, are not very precise, with errors in the 6–10 % range. However, there is hope to have this situation clarified by the MuSun Collaboration [7], which is performing at present an experiment at the Paul Scherrer Institut, with the goal of measuring Γ^D with a precision of 1 %.

Published under licence by IOP Publishing Ltd

Theoretical work on reactions (1) and (2) is just as extensive, and a brief summary is given in the introduction of Ref. [8]. Here we will discuss the results presented in Refs. [8, 9], and we will compare them with the most recent studies of Ando *et al.* [10] and Ricci *et al.* [11] for reaction (1), and Gazit [12] for reaction (2). We also will comment on the study of Ref. [13].

We start by briefly reviewing the formalism used in the calculation for the muon capture processes, discussed at length in Refs. [8, 13]. The muon capture on deuteron and ^{3}He is induced by the weak interaction Hamiltonian [14], $H_W = \frac{G_V}{\sqrt{2}} \int d\mathbf{x}\, l_\sigma(\mathbf{x}) j^\sigma(\mathbf{x})$, where G_V, the Fermi coupling constant, is 1.14939 $\times 10^{-5}$ GeV^{-2} [15], and l_σ and j^σ are the leptonic and hadronic current densities, respectively. The transition amplitude can be written as

$$\begin{aligned} T_W(f, f_z; s_1, s_2, h_\nu) &\equiv \langle nn, s_1, s_2; \nu, h_\nu \,|\, H_W \,|\, (\mu, d); f, f_z\rangle \\ &\simeq \frac{G_V}{\sqrt{2}} \psi_{1s}^{\rm av} \sum_{s_\mu s_d} \langle \frac{1}{2} s_\mu, 1 s_d | f f_z\rangle\, l_\sigma(h_\nu, s_\mu)\, \langle \Psi_{\mathbf{p}, s_1 s_2}(nn) | j^\sigma(\mathbf{q}) | \Psi_d(s_d)\rangle \ , \end{aligned} \tag{3}$$

for muon capture on deuteron, $\mathbf{p}$ being the nn relative momentum, and [13]

$$\begin{aligned} T_W(f, f_z; s_3', h_\nu) &\equiv \langle ^3{\rm H}, s_3'; \nu, h_\nu \,|\, H_W \,|\, (\mu, ^3{\rm He}); f, f_z\rangle \\ &\simeq \frac{G_V}{\sqrt{2}} \psi_{1s}^{\rm av} \sum_{s_\mu s_3} \langle \frac{1}{2} s_\mu, \frac{1}{2} s_3 | f f_z\rangle\, l_\sigma(h_\nu, s_\mu)\, \langle \Psi_{^3{\rm H}}(s_3') | j^\sigma(\mathbf{q}) | \Psi_{^3{\rm He}}(s_3)\rangle \ , \end{aligned} \tag{4}$$

for muon capture on ^{3}He. In order to account for the hyperfine structure in the initial system, the muon and deuteron or ^{3}He spins are coupled to states with total spin $f = 1/2$ or $3/2$ in the deuteron case, and $f = 0$ or 1 in the ^{3}He case. In Eqs. (3) and (4) we have defined with s_μ (h_ν) the z-component of the muon spin (muon neutrino helicity), and with s_1, s_2, s_d, s_3, s_3' the analogous z-components of the two neutron, deuteron, ^{3}He and ^{3}H spins. The Fourier transform of the nuclear weak current has been introduced as

$$j^\sigma(\mathbf{q}) = \int d\mathbf{x}\, e^{i\mathbf{q}\cdot\mathbf{x}} j^\sigma(\mathbf{x}) \equiv (\rho(\mathbf{q}), \mathbf{j}(\mathbf{q})) \ , \tag{5}$$

with the leptonic momentum transfer $\mathbf{q}$ defined as $\mathbf{q} = \mathbf{k}_\mu - \mathbf{k}_\nu \simeq -\mathbf{k}_\nu$, $\mathbf{k}_\mu$ and $\mathbf{k}_\nu$ being the muon and muon neutrino momenta. The function $\psi_{1s}^{\rm av}$ has been introduced to take into account the initial bound state of the muon in the atom and the charge distribution of the nucleus. It is typically approximated as [14] $|\psi_{1s}^{\rm av}|^2 = \frac{(\alpha\, \mu_{\mu d})^3}{\pi}$ for muon capture on deuteron, and [13] $|\psi_{1s}^{\rm av}|^2 = \mathcal{R}\, \frac{(2\,\alpha\, \mu_{\mu ^3{\rm He}})^3}{\pi}$ for muon capture on ^{3}He, where α is the fine structure constant ($\alpha = 1/137$), $\mu_{\mu d}$ and $\mu_{\mu^3{\rm He}}$ are the reduced masses of the (μ, d) and $(\mu,^3{\rm He})$ systems, and the factor $\mathcal{R}$ approximately accounts for the finite extent of the nuclear charge distribution [14]. This factor has been explicitly calculated in Ref. [8] using the Argonne v_{18} (AV18) [16] or the chiral N^{3}LO (N3LO) [17] two-nucleon interaction, augmented, respectively, by the Urbana IX (UIX) [18] or N^{2}LO (N2LO) [19] three-nucleon interaction, and has been found within a percent of 0.98, the value commonly found in the literature [14].

In the case of muon capture on deuteron, the final state wave function $\Psi_{\mathbf{p}, s_1 s_2}(nn)$ is expanded in partial waves, and the calculation is restricted to total angular momentum $J \leq 2$ and orbital angular momentum $L \leq 3$, i.e., in a spectroscopic notation, to $^1S_0$, $^3P_0$, $^3P_1$, $^3P_2$–$^3F_2$ and $^1D_2$. Standard techniques [20, 14] are then used to carry out the multipole expansion of the weak charge, $\rho(\mathbf{q})$, and current, $\mathbf{j}(\mathbf{q})$, operators. Details of the calculation can be found in Ref. [8]. Here we only note that all the contributing multipole operators selected by parity and angular momentum selection rules are included.

The total capture rate for the two reactions under consideration is then defined as

$$d\Gamma = 2\pi\delta(\Delta E)\overline{|T_W|^2} \times ({\rm phase\, space}) \ , \tag{6}$$

where $\delta(\Delta E)$ is the energy-conserving δ-function, and the phase space is $d\mathbf{p}\, d\mathbf{k}_\nu/(2\pi)^6$ for muon capture on deuteron and just $d\mathbf{k}_\nu/(2\pi)^3$ for muon capture on ^{3}He. The following notation has been introduced: (i) for muon capture on the deuteron

$$\overline{|T_W|^2} = \frac{1}{2f+1} \sum_{s_1 s_2 h_\nu} \sum_{f_z} |T_W(f, f_z; s_1, s_2, h_\nu)|^2 \;, \tag{7}$$

and the initial hyperfine state has been fixed to be $f = 1/2$; (ii) for muon capture on ^{3}He

$$\overline{|T_W|^2} = \frac{1}{4} \sum_{s_3' h_\nu} \sum_{f f_z} |T_W(f, f_z; s_3', h_\nu)|^2 \;, \tag{8}$$

and the factor 1/4 follows from assigning the same probability to all different hyperfine states.

After carrying out the spin sums, the differential rate for muon capture on deuteron ($d\Gamma^D/dp$) and the total rate for muon capture on ^{3}He (Γ_0) are easily obtained [8]. In order to calculate the total rate Γ^D for muon capture on deuteron, $d\Gamma^D/dp$ is plotted versus p and numerically integrated.

The initial and final $A = 2$ and 3 nuclear wave functions entering in Eqs. (3) and (4) have been obtained from the AV18 or the N3LO two-nucleon potential, in combination with, respectively, the UIX or chiral N2LO three-nucleon potentials in the case of $A = 3$. The hyperspherical harmonics (HH) expansion method has been used to solve the A-body bound and scattering problem. For a review, see Refs. [21, 22, 23]. We have used the same method in the context of $A = 2$ systems, for which of course wave functions could have been obtained by direct solution of the Schrödinger equation. A detailed discussion for the $A = 2$ wave functions is given in Ref. [8].

The weak current consists of polar- and axial-vector components, derived within two different frameworks, the "standard nuclear physics approach" (SNPA) and chiral effective field theory (χEFT). The first one goes beyond the impulse approximation, by including meson-exchange current contributions and terms arising from the excitation of Δ-isobar degrees of freedom. The second approach includes two-body contributions derived in heavy-baryon chiral perturbation theory within a systematic expansion, up to N^{3}LO [24, 25]. It should be noticed that, when the transition operator matrix elements are calculated using phenomenological wave functions, it should be viewed as a hybrid χEFT approach. Both SNPA and hybrid χEFT frameworks have been used in studies of weak *pp* and *hep* capture reactions in the energy regime relevant to astrophysics [24, 26, 20]. A detailed discussion of the weak current models is given in Ref. [8]. Here we briefly review their main characteristics.

The polar weak current operator is related to the isovector piece of the electromagnetic current via the conserved-vector-current (CVC) hypothesis. In SNPA, no free parameters are present in the model for the electromagnetic current, which is able to reproduce the trinucleon magnetic moments to better than 1 % [8], as well as a large variety of electromagnetic observables [27, 28, 29]. In the case of hybrid χEFT, the vector current can be decomposed into four terms [25]: the soft one-pion-exchange (1π) term, vertex corrections to the one-pion exchange ($1\pi C$), the two-pion exchange (2π), and a contact-term contribution. Their explicit expressions can be found in Ref. [25]. All the 1π, $1\pi C$ and 2π contributions contain low-energy constants (LECs) estimated using resonance saturation arguments, and Yukawa functions obtained by performing the Fourier transform from momentum- to coordinate-space with a Gaussian regulator characterized by a cutoff Λ. This cutoff determines the momentum scale below which these χEFT currents are expected to be valid, i.e., Λ=500–800 MeV [24]. The contact-term electromagnetic contribution is given as sum of two terms, isoscalar and isovector, each one with a LEC in front, fixed to reproduce the experimental values of $A = 3$ magnetic moments. The resulting LECs are given in table V of Ref. [8].

The two-body axial current operators in SNPA can be divided into two classes: the operators of the first class are derived from π- and ρ-meson exchanges and the $\rho\pi$-transition mechanism. These mesonic operators give rather small contributions [8]. The operators in the second class are those that give the largest two-body contributions, and are due to Δ-isobar excitation [26, 20]. In particular, in the dominant N-to-Δ-transition axial current, the N-to-Δ axial coupling constant is retained as a parameter and is determined by fitting the experimental Gamow-Teller matrix element in tritium β-decay (GT$^{\rm EXP}$). It is important to note that the value of this parameter depends on how the Δ-isobar degrees of freedom are treated. Here, the two-body Δ-excitation axial operator is derived in the static Δ approximation, using first-order perturbation theory (PT). This approach is considerably simpler than that adopted in Ref. [20], where the Δ degrees of freedom were treated non-perturbatively, within the so-called transition-correlation operator (TCO) approach, by retaining them explicitly in the nuclear wave functions [30]. The results for the N-to-Δ coupling constant obtained within the two schemes differ by more than a factor of 2 [20], but the results for the observables calculated consistently within the two different approaches are typically within 1 % of each other. Note that, since in the SNPA model for the vector current there are no free parameter, the TCO approach is used, to avoid the overestimation of Δ-isobar contributions, as in PT, and to reach the agreement with experiment for the magnetic moments, mentioned above.

The two-body axial current operator in χEFT consists of two contributions: a one-pion exchange term and a (non-derivative) two-nucleon contact-term. The explicit expressions for these terms can be found in Ref. [24]. According to Ref. [11], in addition to the contributions of Ref. [24, 8], a term should be added to the leading axial two-body current, in order to satisfy the partially-conserved-axial-current (PCAC) hypothesis. This term has been constructed in Ref. [31], and recently reviewed in Ref. [9], where its explicit expression can be found. It will be called, as in Refs. [11, 31], "potential current". While the coupling constants which appear in the one-pion exchange term are fixed by πN data, the LEC which determines the strength of the contact-term (d_R) has been fixed by reproducing GT$^{\rm EXP}$. The values of d_R for Λ=500–800 MeV are given in table V of Ref. [8], when no potential currents are considered, and $d_R = 1.54(8)$ for $\Lambda = 600$ MeV, when potential currents are included and the AV18/UIX Hamiltonian model is used [9].

2. Results

The results for the doublet capture rate Γ^D of reaction (1) and the total capture rate Γ_0 for reaction (2) are listed in tables 1 and 2, respectively. Both models for the nuclear weak transition operator presented in the previous section have been used, labeled SNPA and χEFT, respectively. The nuclear wave functions have been calculated with the AV18 [16] or the N3LO [17] two-nucleon interactions in table 1, and with the AV18/UIX [18] and N3LO/N2LO [19] two- and three-nucleon interactions in table 2. In the SNPA calculation, two different values for the single-nucleon axial coupling constant g_A are considered, g_A=1.2654(42), taken from Ref. [32] and widely used in studies of weak processes [20, 24, 13, 33], and g_A=1.2695(29), the latest determination quoted by the Particle Data Group (PDG) [34]. This allows for an estimation of the theoretical uncertainty arising from this source. In the χEFT calculation, for consistency with the work of Refs. [24, 10], the older value for g_A has been used. Within each Hamiltonian model, the parameters present in the SNPA and χEFT axial current models have been fitted to reproduce GT$^{\rm EXP}$. Furthermore, the LECs in the χEFT weak vector current have been fitted to reproduce the $A = 3$ magnetic moments. The results obtained including the χEFT potential current contributions, with $\Lambda = 600$ Mev, are given in the last rows of tables 1 and 2, using the AV18 and AV18/UIX, respectively.

Inspection of table 1 shows that the $^1S_0$ contribution is the leading one, but $L \geq 1$ contributions are significant and account for ~ 37 % of the total rate. By comparison between the first and second row of the table, we conclude that there is no difference in the results, within uncertainties, when the older value for g_A, g_A=1.2654(42), or the most recent one, g_A=1.2695(29), is used. This reflects the fact that the N-to-Δ axial coupling constant has been constrained by $\mathrm{GT}^{\mathrm{EXP}}$ in both cases. The cutoff dependence of the χEFT results is very weak: at the three representative values for Λ, they agree with each other within the theoretical uncertainties. Also the model dependence due to the interaction model is very weak, as can be seen comparing the χEFT(AV18) and χEFT(N3LO) results for Λ=600 MeV. Finally, the potential currents arising from PCAC do not give appreciable contributions: the differences in Γ^D, obtained with and without these currents (rows labelled "χEFT(AV18)-with potential currents" and "χEFT(AV18)", respectively) for Λ = 600 MeV, are $\leq$ 0.1%, well below the theoretical uncertainties.

In conclusion, a total capture rate in the range

$$\Gamma^D = (389.7 \div 394.3)\,\mathrm{s}^{-1}\ , \tag{9}$$

can be conservatively ascribed to reaction (1), as obtained by adding (subtracting) to the largest (smallest) value of table 1 the theoretical uncertainties. This result is in agreement with the measurements of Refs. [3, 4, 6], but not with that of Ref. [5]. The two most recent theoretical studies of reaction (1), besides the ones just reviewed, are those of Ando *et al.* [10] and Ricci *et al.* [11]. The study of Ref. [10] goes much along the lines of the study of the *pp* and *hep* reactions of Ref. [24]. The Γ^D for reaction (1) has been calculated within the hybrid χEFT approach, using the AV18 potential. The LEC d_R present in the two-body axial current has been fixed by reproducing $\mathrm{GT}^{\mathrm{EXP}}$. The calculation, however, retaines only the S-wave contribution in the nn final scattering state (the $^1S_0$ state), and higher partial-wave contributions have been estimated based on Ref. [35]. Ando *et al.* have found Γ^D=386 s^{-1}, and $\Gamma^D(^1S_0)$=245(1) s^{-1}, the theoretical error being related to the experimental uncertainty in $\mathrm{GT}^{\mathrm{EXP}}$. The difference with the present results of Eq. (9) and table 1 is very small, and has been traced back [8] to the inclusion in the weak vector current of the $1\pi C$, 2π and contact-term contributions, not present in Ref. [10].

The calculation of Ref. [11] has been performed within the SNPA, using the Nijmegen I and Nijmegen 93 [36] Hamiltonian models, and deriving two-body meson-exchange currents from the hard pion chiral Lagrangians of the $N\Delta\pi\rho\omega a_1$ system. The final results for Γ^D are in the range of 416–430 s^{-1} (see table 1 of Ref. [11]), depending on the potential used, with $\Gamma^D(^1S_0)$=254–268 s^{-1}. The results of Ref. [11] are therefore significantly larger than those listed here, presumably because these authors have not constrained their weak current to reproduce $\mathrm{GT}^{\mathrm{EXP}}$ and the isovector magnetic moment of the trinucleons.

The results for reaction (2) listed in table 2, allow to draw similar conclusions about the dependence on g_A and on the χEFT cutoff. Furthermore, the χEFT results obtained with the AV18/UIX Hamiltonian model are in excellent agreement with those reported in SNPA. Also we observe that the χEFT (AV18/UIX) and χEFT (N3LO/N2LO) results differ by 8 s^{-1}, or less than 1 %. and the PCAC potential currents contributions are well below the theoretical uncertainty. In view of this, we quote conservatively a total capture rate for reaction (2) in the range

$$\Gamma_0 = (1471 \div 1497)\,\mathrm{s}^{-1}\ , \tag{10}$$

by keeping the lowest and upper bounds in the values of table 2, similarly to what has been done for reaction (1) (see Eq. (9)). The 1% spread due to model dependence is a consequence of the procedure adopted to constrain the weak current. These results are in very good agreement with the experimental datum of Ref. [2]. Comparing with the previous studies of Refs. [13] and [12], we notice that the approach used in Ref. [13] is quite similar to the SNPA presented here. In fact, the AV18/UIX has been used to calculated the nuclear wave functions, within the HH method.

However, the model for the weak transition operator is the same as the one presented here, except that PT, in the static approximation, is used to calculate both the two-body vector and axial currents arising from Δ excitation. The study of Ref. [12] used the hybrid χEFT approach, and obtained the nuclear wave functions with the Effective Interaction HH method [37]. The values for Γ_0 of Refs. [13] and [12] are 1484(8) s^{-1} and 1499(16) s^{-1}, respectively, in excellent agreement with the ones presented here.

Table 1. Total rate for muon capture on deuteron, in the doublet initial hyperfine state, in s^{-1}. The different partial wave contributions are indicated. The numbers among parentheses indicate the theoretical uncertainty arising from the adopted fitting procedures. Such uncertainty is not indicated when less than 0.1 s^{-1}. The AV18 and N3LO interactions have been used to calculate the deuteron and nn wave functions. The label "SNPA" and "χEFT" refer to the model used for the weak transition operator. Two different values for the single-nucleon axial coupling constant g_A in SNPA(AV18) and three values for the cutoff Λ in χEFT(AV18) are used, as explained in the text. The last row shows the results obtained including the χEFT two-body potential currents, arising from the PCAC hypothesis.

SNPA (AV18)	$^1S_0$	$^3P_0$	$^3P_1$	$^3P_2$	$^1D_2$	$^3F_2$	Total
g_A=1.2654(42)	246.6(7)	20.1	46.7	71.6	4.5	0.9	390.4(7)
g_A=1.2695(29)	246.8(5)	20.1	46.8	71.8	4.5	0.9	390.9(7)
χEFT (AV18)	$^1S_0$	$^3P_0$	$^3P_1$	$^3P_2$	$^1D_2$	$^3F_2$	Total
$\Lambda = 500$ MeV	250.0(8)	19.9	46.2	71.2	4.5	0.9	392.7(8)
$\Lambda = 600$ MeV	250.0(8)	19.8	46.3	71.1	4.5	0.9	392.6(8)
$\Lambda = 800$ MeV	249.7(7)	19.8	46.4	71.1	4.5	0.9	392.4(7)
χEFT (N3LO)	$^1S_0$	$^3P_0$	$^3P_1$	$^3P_2$	$^1D_2$	$^3F_2$	Total
$\Lambda = 600$ MeV	250.5(7)	19.9	46.4	71.5	4.4	0.9	393.6(7)
χEFT (AV18)-with potential currents	$^1S_0$	$^3P_0$	$^3P_1$	$^3P_2$	$^1D_2$	$^3F_2$	Total
$\Lambda = 600$ MeV	250.1(8)	20.2	46.1	71.3	4.5	0.9	393.1(8)

3. Conclusions

We have reviewed the recent results of Refs. [8, 9], where total rates for muon capture on deuteron and ^{3}He have been calculated within a consistent approach, based on realistic (conventional and chiral) interactions and weak currents consisting of vector and axial-vector components with one- and many-body terms. Two different approaches have been used to derive these operators: the first one goes beyond the impulse approximation, by including meson-exchange current contributions and terms arising from the excitation of Δ-isobar degrees of freedom. This approach, labeled SNPA, has been widely and successfully used in studies of electroweak processes (see for instance Refs. [26, 20, 13, 28, 23]). The second approach includes two-body contributions, beyond the impulse approximation, derived within a systematic χEFT expansion, up to N^{3}LO. It is a hybrid χEFT approach, since matrix elements of weak operators derived in χEFT have been evaluated between wave functions obtained from realistic potentials.

The only parameter in the SNPA nuclear weak current model is present in the axial current and is determined by fitting the experimental value for the triton half-life. In the case of the χEFT approach, three LECs appear, one in the axial-vector and two in the vector component. Of these, one LEC appearing in the isoscalar electromagnetic contact-term does not contribute

Table 2. Total rate for muon capture on ^{3}He, in s^{-1}. The numbers in parentheses indicate the theoretical uncertainties due to the adopted fitting procedure. The triton and ^{3}He wave functions are obtained from the AV18/UIX and N3LO/N2LO Hamiltonians. The label "SNPA" and "χEFT" refer to the model used for the weak transition operator. Two different values for the single-nucleon axial coupling constant g_A in SNPA(AV18/UIX) and three values for the cutoff Λ in χEFT(AV18/UIX) are used, as explained in the text. The last row shows the results obtained including the χEFT two-body potential currents, arising from the PCAC hypothesis.

SNPA (AV18/UIX)	Γ_0
g_A=1.2654(42)	1486(8)
g_A=1.2695(29)	1486(5)
χEFT (AV18/UIX)	Γ_0
Λ = 500 MeV	1487(8)
Λ = 600 MeV	1488(9)
Λ = 800 MeV	1488(8)
χEFT (N3LO/N2LO)	Γ_0
Λ = 600 MeV	1480(9)
χEFT (AV18/UIX)-with potential currents	Γ_0
Λ = 600 MeV	1488(9)

here. The three LECs are fixed to reproduce, respectively, triton half-life and the $A = 3$ magnetic moments. In the χEFT case, the contributions of the two-body axial current, obtained imposing the PCAC hypothesis ar the two-body level, are found to be negligible.

The results for the rate of the considered muon capture reactions are summarized in Eqs. (9) and (10). The very accurate experimental datum of Ref. [2] for the total rate in muon capture on ^{3}He is very well reproduced. For the muon capture on deuteron, a precise measurement, which will become available in the near future [7], will discriminate between the theoretical results of Ref. [11] on one side, and those of Ref. [10] and the present one, on the other side. Finally, we remark that the dependence of the results presented here on the input Hamiltonian model, or on the model for the nuclear transition operator, is weak, of the order of 1 %. This weak model dependence is a consequence of the procedure adopted to constrain the weak current.

Before concluding, we would like to sketch the main steps which should be made in a fully consistent χEFT calculation. The starting idea, as explained in Ref. [38], is that d_R, the LEC determining the strength of the axial current contact-term, and c_D, the LEC entering, together with c_E, the contact-term three-nucleon interaction at N2LO, are related as

$$d_R = \frac{M_N}{\Lambda_\chi g_A} c_D + \frac{1}{3} M_N (c_3 + 2c_4) + \frac{1}{6} \ . \tag{11}$$

Here M_N is the nucleon mass, and c_3 and c_4 are LECs of the dimension-two πN Lagrangian, already part of the chiral two-nucleon potential at NLO. Furthermore, Λ_χ is the the chiral-symmetry-breaking scale, and is set equal to 700 MeV in the N2LO model. In a fully consistent χEFT calculation, the following steps would be required: (i) set all the LECs present in the interaction and in the current consistently; (ii) calculate the set of values $\{c_D, c_E\}$ for which the triton and ^{3}He binding energies are reproduced; (iii) for each set of $\{c_D, c_E\}$, calculate the GT matrix element of tritium β-decay; (iv) determine which value of $\{c_D, c_E\}$ reproduce GT$^{\rm EXP}$; let us call this values $\{\overline{c_D}, \overline{c_E}\}$. (v) For the chosen $\{\overline{c_D}, \overline{c_E}\}$, determine the LEC entering the weak vector current, fitting the tritium and ^{3}He magnetic moments. Now that all the LECs are fixed, reactions (1) and (2) can be studied and the total capture rate results are χEFT predictions. Work along these lines is currently underway.

Journal of Physics: Conference Series **336** (2011) 012026 doi:10.1088/1742-6596/336/1/012026

References

[1] Adelberger E G *et al.* 2011 *Rev. Mod. Phys.* **83** 195
[2] Ackerbauer P *et al.* 1998 *Phys. Lett. B* **417** 224
[3] Wang I-T *et al.* 1965 *Phys. Rev.* **139** B1528
[4] Bertin A *et al.* 1973 *Phys. Rev. D* **8** 3774
[5] Bardin G *et al.* 1986 *Nucl. Phys. A* **453** 591
[6] Cargnelli M *et al.* 1986 *Workshop on fundamental μ physics*, (Los Alamos, LA 10714C); 1989 *Nuclear Weak Process and Nuclear Structure, Yamada Conference XXIII*, ed M Morita, H Ejiri, H Ohtsubo, and T Sato (Word Scientific, Singapore), p 115
[7] Andreev V A *et al.* [MuSun Collaboration], *Preprint* arXiv:1004.1754
[8] Marcucci L E *et al.* 2011 *Phys. Rev. C* **83** 014002
[9] Marcucci L E and Piarulli M (2011) *Few-Body Syst.* **49** 35
[10] Ando S *et al.* 2002 *Phys. Lett. B* **533** 25
[11] Ricci P, Truhlík E, Mosconi B, and Smejkal J 2010 *Nucl. Phys. A* **837** 110
[12] Gazit D 2008 *Phys. Lett. B* **666** 472
[13] Marcucci L E, Schiavilla R, Rosati S, Kievsky A, and Viviani M 2002 *Phys. Rev. C* **66** 054003
[14] Walecka J D 1995 *Theoretical Nuclear and Subnuclear Physics* (New York: Oxford University Press)
[15] Hardy J C *et al.* 1990 *Nucl. Phys. A* **509** 429
[16] Wiringa R B, Stoks V G J, and Schiavilla R 1995 *Phys. Rev. C* **51** 38
[17] Entem D R and Machleidt R 2003 *Phys. Rev. C* **68** 041001
[18] Pudliner B S, Pandharipande V R, Carlson J, and Wiringa R B 1995 *Phys. Rev. Lett.* **74** 4396
[19] Navrátil P 2007 *Few-Body Syst.* **41** 117
[20] Marcucci L E, Schiavilla R, Viviani M, Kievsky A, and Rosati S 2000 *Phys. Rev. Lett.* **84** 5959; Marcucci L E, Schiavilla R, Viviani M, Kievsky A, Rosati S, and Beacom J F 2000 *Phys. Rev. C* **63** 015801
[21] Kievsky A *et al.* 2008 *J. Phys. G: Nucl. Part. Phys.* **35** 063101
[22] Viviani M *et al.* 2006 *Few-Body Syst.* **39** 159
[23] Marcucci L E, Kievsky A, Girlanda L, Rosati S, and Viviani M 2009 *Phys. Rev. C* **80** 034003
[24] Park T-S *et al.* 2003 *Phys. Rev. C* **67** 055206
[25] Song Y-H, Lazauskas R, and Park T-S 2009 *Phys. Rev. C* **79** 064002
[26] Schiavilla R *et al.* 1998 *Phys. Rev. C* **58** 1263
[27] Carlson J and Schiavilla R 1998 *Rev. Mod. Phys.* **70** 743
[28] Marcucci L E, Viviani M, Schiavilla R, Kievsky A, and Rosati S 2005 *Phys. Rev. C* **72** 014001
[29] Marcucci L E, Pervin M, Pieper S C, Schiavilla R, and Wiringa R B 2008 *Phys. Rev. C* **78** 065501
[30] Schiavilla R, Wiringa R B, Pandharipande V R, and Carlson J 1992 *Phys. Rev. C* **45** 2628
[31] Mosconi B, Ricci P, and Truhlík E 2005 *Eur. Phys. J. A* **25** 283
[32] Adelberger E *et al.* 1998 *Rev. Mod. Phys.* **70** 1265
[33] Schiavilla R and Wiringa R B 2002 *Phys. Rev. C* **65** 054302
[34] Amsler C *et al.* [Particle Data Group] 2008 *Phys. Lett. B* **667** 1
[35] Tatara N, Kohyama Y, and Kubodera K 1990 *Phys. Rev. C* **42** 1694
[36] Stoks V G J, Klomp R A M, Terheggen C P F and de Swart J J 1994 *Phys. Rev. C* **49** 2950
[37] Barnea N, Leidemann W, and Orlandini G 2000 *Phys. Rev. C* **61** 054001; 2001 *Nucl. Phys. A* **693** 565; Barnea N and Novoselsky A 1997 *Ann. Phys. (N.Y.)* **256** 192
[38] Gazit D, Quaglioni S, and Navrátil P 2009 *Phys. Rev. Lett.* **103** 102502

Transversity studies with protons and light nuclei

Sergio Scopetta

Dipartimento di Fisica, Università degli Studi di Perugia,
and INFN, sezione di Perugia,
via A. Pascoli, 06100 Perugia, Italy

E-mail: sergio.scopetta@pg.infn.it

Abstract. A general formalism to evaluate time-reversal odd transverse momentum dependent parton distributions (T-odd TMDS) is reviewed and applied to the calculation of the leading-twist quantities, i.e., the Sivers and the Boer-Mulders functions. Two different models of the proton structure, namely a non relativistic constituent quark model and the MIT bag model, have been used. The results obtained in both frameworks fulfill the Burkardt sum rule to a large extent. The calculation of nuclear effects in the extraction of neutron single spin asymmetries in semi-inclusive deep inelastic scattering off transversely polarized ^{3}He is also illustrated. In the kinematics of JLab, it is found that the nuclear effects described by an Impulse Approximation approach are under theoretical control.

1. Introduction

The partonic structure of transversely polarized hadrons is still rather poorly known [1]. As a matter of fact, despite its leading-twist character, the transversity distribution is chiral-odd in nature and not accessible therefore in inclusive esperiments. Due to going-on and forthcoming important measurements in various Laboratories, this issue is nevertheless one of the most widely studied by the hadronic Physics Community, in particular by the Italian one, which is providing crucial contributions to a better understanding of this fascinating subject (see, for example, Refs. [2] and references therein).

Semi-inclusive deep inelastic scattering (SIDIS) is one of the proposed processes to access the parton structure of transversely polarized hadrons. The theoretical description of semi-inclusive processes implies a more complicated formalism, accounting for the transverse motion of the quarks in the target [3, 4, 5]. In particular, the non-perturbative effects of the intrinsic transverse momentum $\vec{k}_T$ of the quarks inside the nucleon may induce significant hadron azimuthal asymmetries [6, 7].

The Sivers and the Boer-Mulders functions were defined in this scenario [8, 9]. Transverse Momentum Dependent pdfs (TMDs) are the set of functions that depend on the intrinsic transverse momentum of the quark, in addition to the dependences, typical of the PDs, on the Bjorken variable and on the momentum transfer Q^2. Their number is fixed counting the scalar quantities allowed by hermiticity, parity and time-reversal invariance. However, the existence of final state interactions (FSI) allows for time-reversal odd functions [10]. In effect, by relaxing this constraint, one defines two additional functions, namely, the Sivers and the Boer-Mulders (BM) functions. In SIDIS, the Sivers function is involved in the description of the single spin asymmetry, measured when an unpolarized beam scatters transeversely polarized targets. The

Published under licence by IOP Publishing Ltd

single spin asymmetries are obtained constructing the difference of semi-inclusive cross-sections with different transverse polarization of the target with respect to the momentum transfer. The BM function appears in the azimuthal asymmetry in unpolarized SIDIS. The latter object refers to the detection of the produced hadron at different angles with respect to the plane containing the momentum transfer and the hadron momentum. All the involved quantities have been conventionally defined in Ref. [11]. According to the latter convention, the Sivers function, $f_{1T}^{\perp q}(x, k_T)$ [8], and the Boer-Mulders function, $h_1^{\perp q}(x, k_T)$ [9], are formally defined through the following expressions[1]:

$$\begin{aligned} f_{1T}^{\perp q}(x, k_T) &= -\frac{M}{2k_x} \int \frac{d\xi^- d^2\vec{\xi}_T}{(2\pi)^3}\, e^{-i(xp^+\xi^- - \vec{k}_T\cdot\vec{\xi}_T)} \\ &\times \frac{1}{2} \sum_{S_y=-1,1} S_y \, \langle PS_y|\overline{\psi}_q(\xi^-, \vec{\xi}_T)\, \mathcal{L}^\dagger_{\vec{\xi}_T}(\infty, \xi^-)\, \gamma^+\, \mathcal{L}_0(\infty, 0)\psi_q(0,0)|PS_y\rangle \\ &+ \text{ h.c.}\,, \end{aligned} \tag{1}$$

taking the proton polarized along the y axis; and

$$\begin{aligned} h_1^{\perp q}(x, k_T) &= -\frac{M}{2k_x} \int \frac{d\xi^- d^2\vec{\xi}_T}{(2\pi)^3}\, e^{-i(xp^+\xi^- - \vec{k}_T\cdot\vec{\xi}_T)} \\ &\times \frac{1}{2} \sum_{S_z=-1,1} \langle PS_z|\overline{\psi}_q(\xi^-, \vec{\xi}_T)\, \mathcal{L}^\dagger_{\vec{\xi}_T}(\infty, \xi^-)\, \gamma^+\gamma^2\gamma_5\, \mathcal{L}_0(\infty, 0)\psi_q(0,0)|PS_z\rangle \\ &+ \text{ h.c.}\,, \end{aligned} \tag{2}$$

where $\vec{S}$ is the spin of the target hadron. The normalization of the covariant spin vector is $S^2 = -1$, M is the target mass, $\psi_q(\xi)$ is the quark field and $\mathcal{L}_{\vec{\xi}_T}$ is the gauge link.[2] The gauge link contains a scaling contribution which makes the T-odd TMDs non vanishing in the Bjorken limit, as it has been shown in Refs. [12, 13, 14].

The difference between the two functions is clearly seen comparing Eq. (1) and Eq. (2). The BM function counts transversely polarized quarks, hence the Dirac operator $\gamma^+\gamma^2\gamma_5$ in Eq. (2), in an unpolarized proton. On the other hand, the Sivers function counts the unpolarized quarks, hence the Dirac operator γ^+ in Eq. (1), in a transversly polarized proton, denoted by the transverse component S_y in the proton state in Eq. (1). If there were no scaling contribution of the gauge link, the two T-odd functions, $f_{1T}^{\perp q}(x, k_T)$ and $h_1^{\perp q}(x, k_T)$, would be identically zero.

2. Quark model calculations

Several quark model calculations of the Sivers and BM functions have been performed in the past years (see, i.e., Refs. [15, 16, 17, 18]) and their phenomenology keeps attracting interest, as it is demonstrated by very recent important contributions [19, 20, 21].

The general formalism for the evaluation of time-reversal odd TMDs in quark models, presented in Ref. [22, 23, 24], is reviewed here below. It is based on an impulse approximation analysis, where the Final State Interactions are introduced through a One Gluon Exchange mechanism, as depicted in figure 1. The used quark models, i.e., the simplest version of the MIT bag model [25], and a non relativistic constituent quark model (NRCQM), do not contain explicit gluonic degrees of freedom. In the Sivers and BM functions the Dirac operators determine the

[1] $a^\pm = (a_0 \pm a_3)/\sqrt{2}$.

[2] The gauge link is defined as $\mathcal{L}_{\vec{\xi}_T}(\infty, \xi^-) = \mathcal{P}\exp\left(-ig \int_{\xi^-}^{\infty} A^+(\eta^-, \vec{\xi}_T)\, d\eta^-\right)$, where g is the strong coupling constant. This definition holds in covariant (non singular) gauges, and in SIDIS processes, as the definition of T-odd TMDs is process dependent.

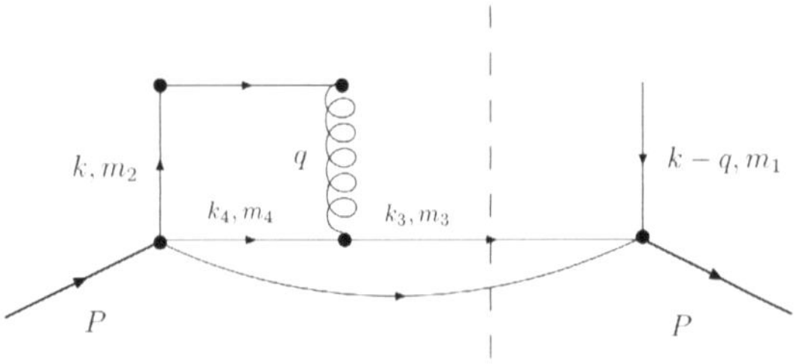

Figure 1. The process taken into account, together with its h.c., in the present evaluation of the T-odd TMDs.

spin structure of the interaction, $V_{m_1 m_2 m_3 m_4}$ between the quarks with initial spins m_2 and m_4 and final spins m_1 and m_3, respectively. For instance, in the MIT bag model [25], one gets, for the Sivers [BM] function

$$
\begin{aligned}
f_{1T}[h_1]^{\perp q}(x, k_T) &= -2\Im[\,]\, i g^2 \frac{M\, E_P}{k_x} \int \frac{d^2\vec{q}_T}{(2\pi)^5} \int \frac{d^3 k_3}{(2\pi)^3} \\
&\times \sum_{m_1, m_2, m_3, m_4} C_{qf[h]}^{m_1 m_2, m_3 m_4}\, V(\vec{k}, \vec{k}_3, \vec{q}_T)_{m_1 m_2, m_3 m_4} \quad . \qquad (3)
\end{aligned}
$$

The imaginary part is taken only for the Sivers function calculation, performed changing the transversely polarized nucleon states from a transversity to a helicity basis. The coupling of the spins at the quark level appears clearly from the expressions

$$
\begin{aligned}
C_{qf[h]}^{m_1 m_2, m_3 m_4} &= \frac{1}{2} \sum_S C_{qS}^{m_1 m_2, m_3 m_4} \\
&= \sum_\beta T_{ij}^a T_{kl}^a \langle PS_z = 1 | b_{qm_1}^{i\dagger} b_{qm_2}^{j} b_{\beta m_3}^{k\dagger} b_{\beta m_4}^{l} | PS_z = -1 \rangle \\
&\left[\frac{1}{2} \sum_{S_z = 1, -1} \sum_\beta T_{ij}^a T_{kl}^a \langle PS_z | b_{qm_1}^{i\dagger} b_{qm_2}^{j} b_{\beta m_3}^{k\dagger} b_{\beta m_4}^{l} | PS_z \rangle \right] \quad , \qquad (4)
\end{aligned}
$$

whose calculation has been performed assuming $SU(6)$ symmetry for the proton state, for $q = u, d$. The interaction $V(\vec{k}, \vec{k}_3, \vec{q}_T)$ has been evaluated using the properly normalized fields for the quark in the bag [25], given in terms of the quark wave function in momentum space, which reads

$$
\varphi_m(\vec{k}) = i\sqrt{4\pi}\, N\, R_0^3 \begin{pmatrix} t_0(|\vec{k}|)\chi_m \\ \vec{\sigma} \cdot \hat{k}\, t_1(|\vec{k}|)\, \chi_m \end{pmatrix} \quad , \qquad (5)
$$

with the normalization factor N given in Ref. [25] The interaction is then

$$
V(\vec{k}, \vec{k}_3, \vec{q}_T)_{m_1 m_2, m_3 m_4} = \frac{1}{q^2} \varphi_{m_1}^\dagger(\vec{k} - \vec{q}_T)\, \gamma^0 \gamma^+ \Gamma^{f[h]}\, \varphi_{m_2}(\vec{k}) \varphi_{m_3}^\dagger(\vec{k}_3)\, \gamma^0 \gamma^+\, \varphi_{m_4}(\vec{k}_3 - \vec{q}_T)\, , (6)
$$

where $\Gamma^{f[h]} = 1$ or $\gamma^2\gamma_5$ for, respectively, the Sivers and the Boer-Mulders function. The expressions in the NRCQM are similar to the one in the bag model. They read, being Ψ the intrinsic proton wave function,

$$
\begin{aligned}
f_{1T}[h_1]^{\perp q}(x,k_T) &= -2ig^2\,\frac{M\,E_P}{k_x}\int\frac{d^2\vec{q}_T}{(2\pi)^5}\sum_{m_1,m_2,m_3,m_4} \\
&\int d\vec{k}_1\,d\vec{k}_3\,(2\pi)^3\delta(k_1^+-xP^+)\delta(\vec{k}_{1\perp}+\vec{q}_\perp-\vec{k}_\perp) \\
&\frac{1}{2}\sum_S\sum_\beta\sum_{ijkl}\delta_{\rm QN}\;\Psi^\dagger_{r\,S}\left(\vec{k}_1,\{m_1,i,q\};\,\vec{k}_3,\{m_3,k,\beta\};\,-\vec{k}_3-\vec{k}_1,m_n\right) \\
&\times\;\; T^a_{ij}T^a_{kl}V(\vec{k}_1,\vec{k}_3,\vec{q})_{m_1m_2,m_3m_4} \\
&\times\;\; \Psi_{r\,S}\left(\vec{k}_1+\vec{q},\{m_2,j,q\};\,\vec{k}_3-\vec{q},\{m_4,l,\beta\};\,-\vec{k}_3-\vec{k}_1,m_n\right)\,,
\end{aligned}
\tag{7}
$$

where $\sum_S=\sum_{S_z=-1,1}$ or $\sum_{S_y=-1,1}S_y$ for the BM and the Sivers function, respectively. The interaction here reads

$$
V(\vec{k}_1,\vec{k}_3,\vec{q})_{m_1m_2m_3m_4} = \frac{1}{q^2}\,\bar{u}_{m_1}(\vec{k}_1)\,\gamma^+\Gamma^{f[h]}\,u_{m_2}(\vec{k}_1+\vec{q})\,\bar{u}_{m_3}(\vec{k}_3)\,\gamma^+\,u_{m_4}(\vec{k}_3-\vec{q}) \quad , \tag{8}
$$

with $u(\vec{k})$ the four-spinor of the free quark states, arising from the Impulse Approximation analysis

In Refs. [22, 23, 24], the T-odd TMDs have been evaluated in both the MIT bag model and the NRCQM assuming initially an $SU(6)$ symmetry for the proton.

In the former case, the qualitative results are understood calculating the coefficients (4), in terms of which it is possible to realize what happens to the quark spins in the FSI process in a perfectly transparent way. In the latter case, one first has to re-express Eq. (7) in terms of the proton state. In spectroscopic notation and with the Jacobi coordinates, one has, using SU(6) symmetry

$$
|^2S_{1/2}\rangle_S = \frac{e^{-(k_\rho^2+k_\lambda^2)/\alpha^2}}{\pi^{3/2}\alpha^3}|\chi\rangle_S\,, \tag{9}
$$

where $|\chi\rangle_S$ is the standard $SU(6)$ vector describing the spin-flavor structure of the proton. In the bag model calculation, a non-zero Sivers function arises through the interference between the upper and lower components of the wave function. Something similar happens in the CQM. This time the interference is between the upper and lower components of the free spinors appearing as a consequence of the use of the Impulse Approximation (see Eqs. (6, 8) for the interaction and Refs. [22, 23, 24] for details). Two different spin combinations contribute to the Sivers function. One of them comes from the spin-flipping of the quark interacting with the photon, i.e. the Y term in the MIT bag calculation

$$
f_{1T}^{\perp q}(x,k_T) = \frac{g^2}{2}\frac{ME_P}{k^x}C^2\int\frac{d^2q_\perp}{(2\pi)^2}\frac{1}{q^2}[C_q^{-+}Y(\vec{q}_\perp,k_T)+C_q^{+-}U(\vec{q}_\perp,k_T)]\,, \tag{10}
$$

with C a normalization factor, C_q a weighting spin-flavor-color factor resulting from the matrix elements (4) and where $Y/U(\vec{q}_\perp,k_T)$ include the momentum dependent part.[3]

On the other hand, there are more spin combinations for the BM function.[4] The first reason is that both non-flipping and double-flipping terms are important. The second reason is the sum

[3] See Eqs. (8-9) of Ref. [23].

[4] See Eq. (13) of Ref. [24].

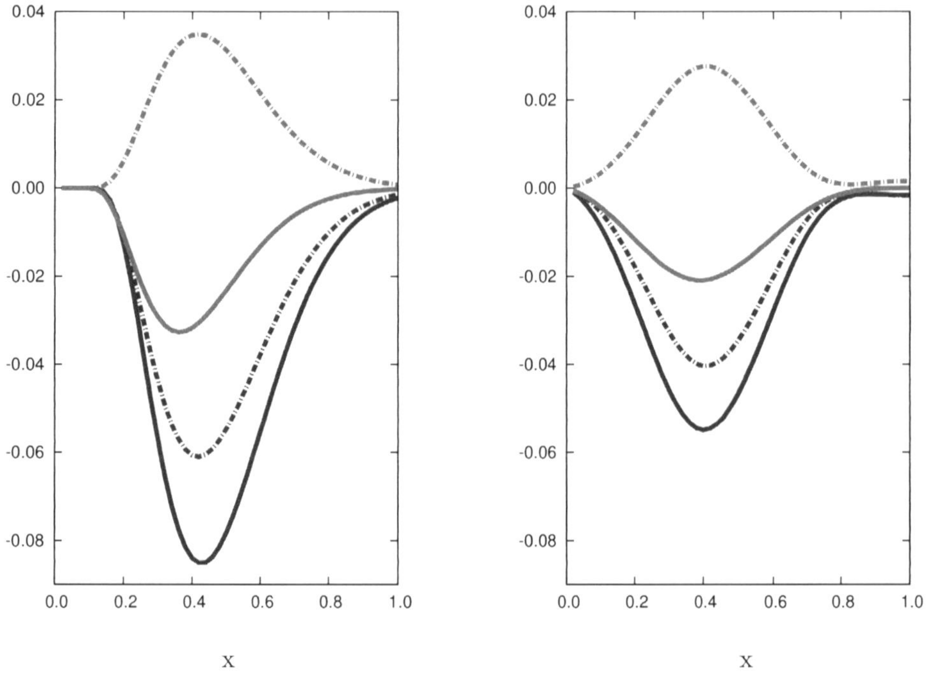

Figure 2. (Color online) Comparison of the moments, $f(h)_{1T}^{\perp(1)u(d)}(x)$, Eq. (11), of the Sivers (red) and Boer-Mulders (blue) functions in the SU(6) NR Constituent Quark Model (left) and the MIT bag model (right). Short-dashed curves represent the d-quark distributions; full curves the u-quark ones.

over the two spin states, i.e. $S_z = -1, 1$. Due to the spin-flavor-color coefficients, i.e., due to the $SU(6)$ symmetry assumption, the non-flipping term is more important than the double-flipping contribution. In effect, the latters are governed by the product of the two lower components of the bag wave function which encodes the most relativistic contribution arising in the MIT bag model. They turn out to be a few orders of magnitude smaller than the dominant ones, arising from the interference between the upper and lower parts of the bag wave function. This also happens if a proper non relativistic reduction of the gauge link, suitable for CQM calculations, is performed, justifying then the non relativistic approximation.

In figure 2 the "first moments" of the Sivers and the Boer-Mulders functions, i.e. the quantities

$$f_{1T}^{\perp(1)\mathcal{Q}}(x) = \int d^2\vec{k}_T \frac{k_T^2}{2M^2} f_{1T}^{\perp\mathcal{Q}}(x, k_T) \ , \tag{11}$$

are shown for u and d quarks in both the CQM and MIT bag model, using $\alpha_s(\mu_0^2)/(4\pi) \simeq 0.13$ [27].

In figure 3 it is shown, in the MIT bag model case, the effect of neglecting the possible contribution of a spin-flip of the spectator quark (the one with initial spin projection m_4 in figure 1), as was done in a previous calculation [15]. The proper inclusion of this term is crucial for the fulfillment of the Burkart Sum Rule, as it is explained here below.

The above-described formalism has been easily extended, in the NR case, to the model of Isgur and Karl [26], allowing for a weak SU(6) breaking. The detailed procedure and the final expressions of the Sivers function in this model can be found in Ref. [22]. To evaluate numerically the Sivers and BM functions, the strong coupling constant g (i.e. $\alpha_s(Q^2)$) has to be fixed. The

Journal of Physics: Conference Series **336** (2011) 012027 doi:10.1088/1742-6596/336/1/012027

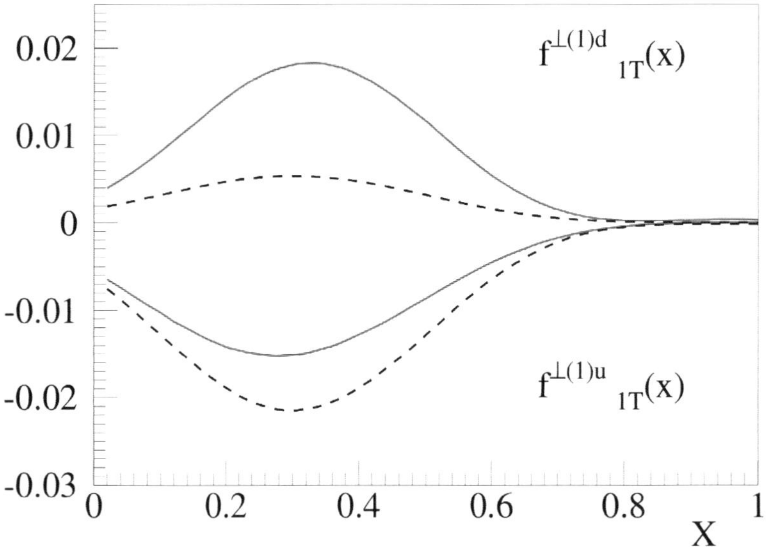

Figure 3. (Color online) The quantity $f_{1T}^{\perp(1)u(d)}(x)$, Eq. (11). Dashed curve: the results in the MIT bag model at μ_0^2, considering that a spin-flip can occurr only for the interacting quark. Full curve: the same, evaluated properly taking into account also the possible spin-flip of the spectator quark.

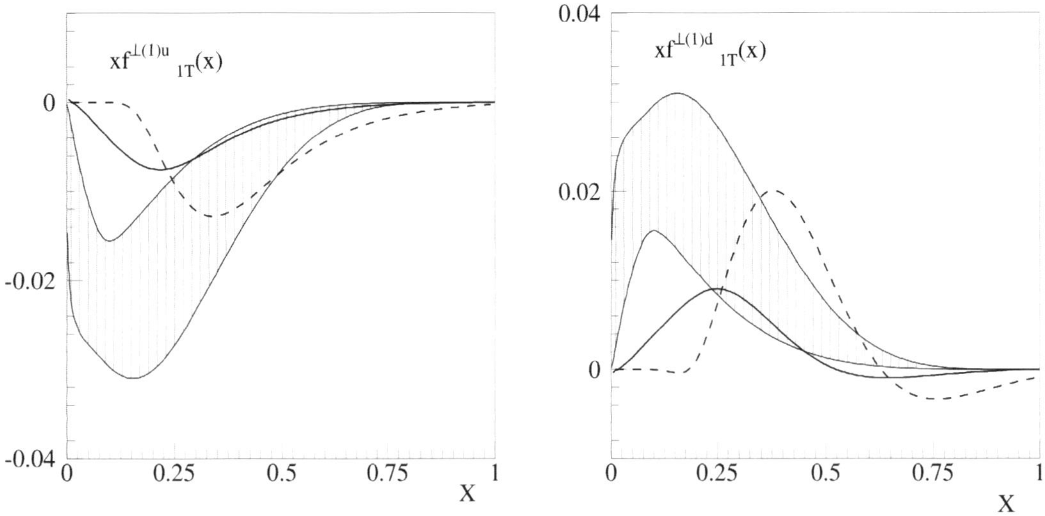

Figure 4. Left (right): the quantity $f_{1T}^{\perp(1)u(d)}(x)$, Eq. (11). Dashed curve: the result in the Isgur and Karl model at μ_0^2. Full curve: the evolved distribution at NLO. Patterned area: the $1-\sigma$ range of the best fit of the HERMES data proposed in Ref. [28].

prescription of Ref. [27]. is used to fix μ_0^2, according to the amount of momentum carried by the valence quarks in the model. Here, assuming that all the gluons and sea pairs in the proton are produced perturbatively according to NLO evolution equations, in order to have $\simeq 55\%$ of the momentum carried by the valence quarks at a scale of 0.34 GeV^2 one finds that $\mu_0^2 \simeq 0.1$ GeV^2 if $\Lambda_{QCD}^{NLO} \simeq 0.24$ GeV. This yields $\alpha_s(\mu_0^2)/(4\pi) \simeq 0.13$ [27]. The results of the present approach for the first moments of the Sivers function are given by the dashed curves in figure 4, where they are compared with a parameterization of the HERMES data, taken at $Q^2 = 2.5$ GeV^2. The patterned area represents the $1-\sigma$ range of the best fit proposed in Ref. [28]. The results are close, in magnitude, to the data, although they have a different shape: the maximum (minimum) is predicted at larger values of x. Actually μ_0^2 is much lower, $Q^2 = 2.5$ GeV^2. For a proper comparison, the QCD evolution from the model scale to the experimental one would be necessary. Unfortunately, the Sivers function is a TMD PDs and the evolution of this class of functions is, to a large extent, unknown. In order to have an indication of the effect of the evolution, a NLO evolution of the model results has been performed, assuming, for $f_{1T}^{\perp(1)Q}(x)$, the same anomalous dimensions of the unpolarized PDFs. From the final result (full curve in figure 2), one can see that the agreement with data improves dramatically and the trend is reasonably reproduced at least for $x \geq 0.2$. Although the performed evolution is not exact, the procedure highlights the necessity of evolving the model results to the experiment scale and it suggests that the present results could be consistent with data, still affected by large errors. A discussion concerning the use of perturbative methods to relate the scale of model calculations to the scale of experimental data, and its impact on the calculation of T-odd TMDs, has been recently presented in Ref. [29] The deep understanding of the QCD evolution of TMDs is progressing fast [30, 31, 32, 33, 34].

Let us see now how the results of the calculation compare with the Burkardt sum rule [35], which follows from general principles. The Burkardt Sum Rule (BSR) states that, for a proton polarized in the positive y direction, $\sum_{Q=u,d}\langle k_x^Q\rangle = 0$ with

$$\langle k_x^Q\rangle = -\int_0^1 dx \int d\vec{k}_T \frac{k_x^2}{M} f_{1T}^{\perp Q}(x, k_T) \,, \tag{12}$$

and must be satisfied at any scale. Within our scheme, at the scale of the model, it is found $\langle k_x^u\rangle = 10.85$ MeV, $\langle k_x^d\rangle = -11.25$ MeV and, in order to have an estimate of the quality of the agreement of our results with the sum rule, we define the ratio $r = |\langle k_x^d\rangle + \langle k_x^u\rangle|/|\langle k_x^d\rangle - \langle k_x^u\rangle|$ obtaining $r \simeq 0.02$, so that one can say that the calculation fulfills the BSR to a precision of a few percent. One should notice that the agreement which is found is better than that found in previous model calculations, where the Sivers function for the flavor u was found to be proportional to that for the flavor d. For the MIT bag model, where the translational invariance is broken and the wave functions are not exact momentum eigenstates, the situatuion is slightly worse and the Sum Rule turns out to be violated of $\simeq 5$ % (one should realize anyway that, in a previous calculation without spin flip of the spectator quark, the violation was found to be $\simeq$ 60 %).

The recent calculation of Ref. [18], using overlaps of Light Cone Wave Functions in the Light Front of Dynamics, where the number of particles and the momentum sum rules are satisfied at the same time, fulfills the Burkart Sume Rule exactly, showing the importance of using exact momentum eigenstates in checking the transverse momentum of the quarks.

3. The Sivers function from neutron (^{3}He) targets

The experimental scenario which arises from the analysis of SIDIS off transversely polarized proton and deuteron targets is puzzling [36, 37]. With the aim at extracting the neutron information to shed some light on the problem, a measurement of SIDIS off transversely polarized

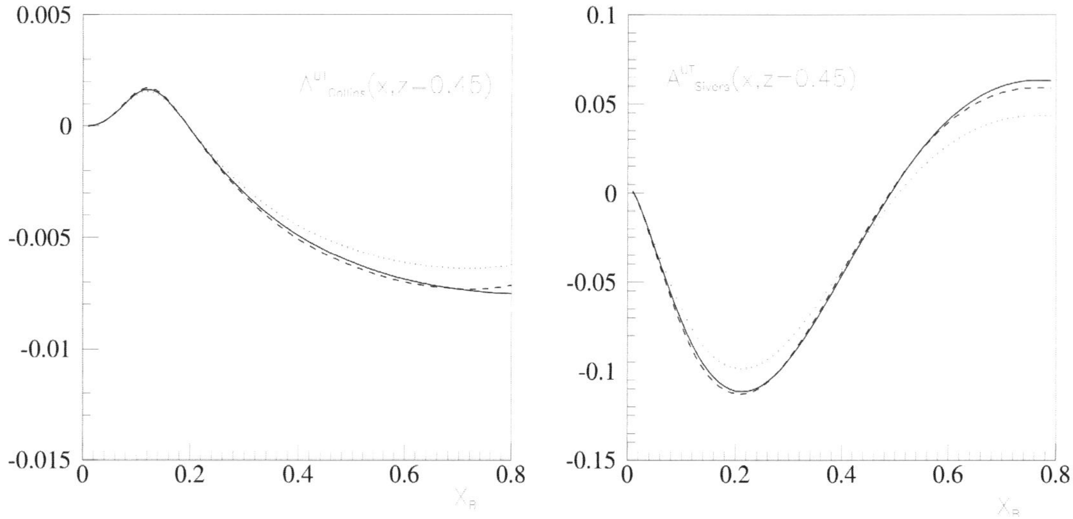

Figure 5. Left (right): the model neutron Collins (Sivers) asymmetry for π^- production (full) in JLab kinematics, and the one extracted from the full calculation taking into account the p_p (dashed), or neglecting it (dotted). The results are shown for z=0.45 and $Q^2 = 2.2$ GeV^2, typical values in the kinematics of the JLab experiments.

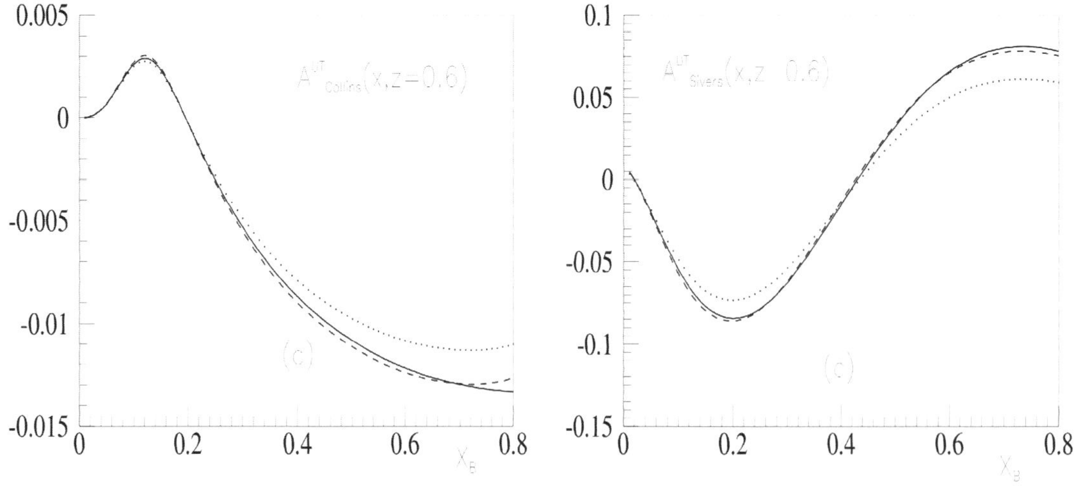

Figure 6. Left (right) The same as in figure 5, but at z=0.6, the maximum value accessed in the JLab experiments.

^{3}He has been addressed [38], and an experiment, planned to measure azimuthal asymmetries in the production of leading $\pi^\pm$ from transversely polarized ^{3}He, has been just completed at Jefferson Lab, with a beam energy of 6 GeV [39]. Another experiment will be performed after the 12 GeV upgrade of JLab [40, 41]. Here, a realistic analysis of SIDIS off transversely polarized ^{3}He, presented in Ref. [42], is sommarized. The formal expressions of the Collins and Sivers contributions to the azimuthal Single Spin Asymmetry (SSA) for the production of leading pions off ^{3}He have been derived, in impulse approximation (IA), including also the initial transverse momentum of the struck quark. The final equations are rather involved and they are not reported here. They can be found in [42]. The same quantities have been then evaluated in the kinematics of the planned JLab experiments. Wave functions [43] obtained within the AV18 interaction [44] have been used for a realistic description of the nuclear dynamics, using overlap integrals evaluated in Ref. [45], and the nucleon structure has been described by proper parameterizations of data [46] or suitable model calculations [47]. The crucial issue of extracting the neutron information from ^{3}He data will be now thoroughly discussed. As a matter of facts, a model independent procedure, based on the realistic evaluation of the proton and neutron polarizations in ^{3}He [48], called respectively p_p and p_n in the following, is widely used in inclusive DIS to take into account effectively the momentum and energy distributions of the polarized bound nucleons in ^{3}He. It is found that the same extraction technique can be applied also in the kinematics of the proposed experiments, although fragmentation functions, not only parton distributions, are involved, as it can be seen in figures 5 and 6. In these figures, the free neutron asymmetry used as a model in the calculation, given by a full line, is compared with two other quantities. One of them is:

$$\bar{A}_n^i \simeq \frac{1}{d_n} A_3^{exp,i} , \qquad (13)$$

where i stands for "Collins" or "Sivers", $A_3^{exp,i}$ is the result of the full calculation, simulating data, and d_n is the neutron dilution factor. The latter quantity is defined as follows, for a neutron n (proton p) in ^{3}He:

$$d_{n(p)}(x,z) = \frac{\sum_q e_q^2 f^{q,n(p)}(x)\, D^{q,h}(z)}{\sum_{N=p,n}\sum_q e_q^2 f^{q,N}(x) D^{q,h}(z)} \qquad (14)$$

and, depending on the standard parton distributions, $f^{q,N}(x)$, and fragmentation functions, $D^{q,h}(z)$, it is experimentally known (see [42] for details). $\bar{A}_n^i$ is given by the dotted curve in the figures. The third curve, the dashed one, is given by

$$A_n^i \simeq \frac{1}{p_n d_n}\left(A_3^{exp,i} - 2p_p d_p A_p^{exp,i}\right) , \qquad (15)$$

i.e. ^{3}He is treated as a nucleus where the effects of its complicated spin structure, leading to a depolarization of the bound neutron, together with the ones of Fermi motion and binding, can be taken care of by parameterizing the nucleon effective polarizations, p_p and p_n. One should realize that Eq. (13) is the relation which should hold between the ^{3}He and the neutron SSAs if there were no nuclear effects, i.e. the ^{3}He nucleus were a system of free nucleons in a pure S wave. In fact, Eq. (13) can be obtained from Eq. (15) by imposing $p_n = 1$ and $p_p = 0$. It is clear from the figures that the difference between the full and dotted curves, showing the amount of nuclear effects, is sizable, being around 10 - 15 % for any experimentally relevant x and z, while the difference between the dashed and full curves reduces drastically to a few percent, showing that the extraction scheme Eq. (15) takes safely into account the spin structure of ^{3}He, together with Fermi motion and binding effects. This important result is due to the peculiar kinematics of the JLab experiments, which helps in two ways. First of all, to favor pions from current fragmentation, z has been chosen in the range $0.45 < z < 0.6$, which means that only high-energy pions are observed. Secondly, the pions are detected in a narrow cone around the direction of the momentum transfer. As it is explained in [42], this makes nuclear effects in the fragmentation functions rather small. The leading nuclear effects are then the ones affecting the parton distributions, already found in inclusive DIS, and can be taken into account in the usual way, i.e., using Eq. (15) for the extraction of the neutron information. In the figures, one should not take the shape and size of the asymmetries too seriously, being the obtained quantities strongly dependent on the models chosen for the unknown distributions [47]. One should instead consider the difference between the curves, a model independent feature which is the most relevant outcome of the present investigation. The main conclusion is that Eq. (15) will be a valuable tool for the data analysis of the experiments [39, 40].

While the analysis of Ref. [42] has been performed assuming the experimental set-up of the experiment described in Ref. [39], but using DIS kinematics, a further analysis is being carried on [49], to investigate possible nuclear effects related to the finite values of the momentum and energy transfers, Q^2 and ν, in the actual experiment. In Ref. [49], the description of Ref. [42] will be also improved, implementing a relativistic Light Front treatment to evaluate the nuclear polarized spectral function. Besides, the problem of possible effects beyond IA, such as final state interactions, will be addressed, and more realistic models of the nucleon structure, able to predict reasonable figures for the experiments, will be included in the general scheme.

Acknowledgments
I thank the organizers of the Conference for the kind invitation, A Courtoy, F Fratini and V Vento for a fruitful and pleasent collaboration, A Del Dotto and G Salmè for relevant discussions. This work is supported in part by the INFN-CICYT agreement.

References

[1] Barone V, Drago A and Ratcliffe P 2002 *Phys. Rep.* **359** 1
[2] Ciullo G, Contalbrigo M, Hasch D, Lenisa P 2009 *Proc. of the Second Workshop on Transverse Polarizatin Phenomena in Hard Processes, "Transversity 2008", Ferrara, Italy, 28-31 May 2008*, (Singapore: World Scientific)
[3] Bacchetta A, Diehl M, Goeke K, Metz A, Mulders P.J. and Schlegel M 2007 *J. High Energy Phys.* JHEP **0702** 093
[4] D'Alesio U and Murgia F 2008 *Prog. Part. Nucl. Phys.* **61** 394
[5] Pasquini B, Cazzaniga S and Boffi S 2008 *Phys. Rev. D* **78** 034025
[6] Mulders P J and Tangerman R D 1996 *Nucl. Phys. B* **461** 197 [Erratum-ibid. 1997 *B* **484** 538]
[7] Cahn R N 1978 *Phys. Lett. B* **78** 269
[8] Sivers D.W. 1990 *Phys. Rev. D* **41** 83; 1991 *Phys. Rev. D* **43** 261
[9] Boer D. and Mulders P.J. 1998 *Phys. Rev. D* **57** 5780
[10] Brodsky S J, Hwang D S and Schmidt I 2002 *Phys. Lett. B* **530** 99
[11] Bacchetta A, D'Alesio U, Diehl M and Miller C A 2004 *Phys. Rev. D* **70** 117504
[12] Brodsky S J, Hoyer P, Marchal N, Peigne S and Sannino F 2002 *Phys. Rev. D* **65** 114025
[13] Collins J C 2002 *Phys. Lett. B* **536** 43
[14] Belitsky A V, Ji X and Yuan F 2003 *Nucl. Phys. B* **656** 165
[15] Yuan F 2003 *Phys. Lett. B* **575** 45
[16] Gamberg L P , Goldstein G R and Schlegel M 2008 *Phys. Rev. D* **77** 094016
[17] Bacchetta A, Conti F and Radici M 2008 *Phys. Rev. D* **78** 074010
[18] Pasquini B and Yuan F 2010 *Phys. Rev. D* **81** 114013
[19] Barone V, Melis S and Prokudin A 2010 *Phys. Rev. D* **81** 114026
[20] Boer D 2011 On a possible node in the Sivers and Qiu-Sterman functions *Preprint* arXiv:1105.2543 [hep-ph]
[21] Bacchetta A and Radici M 2011 Constraining quark angular momentum through semi-inclusive measurements *Preprint* arXiv:1107.5755 [hep-ph]
[22] Courtoy A, Fratini F, Scopetta S and Vento V 2008 *Phys. Rev. D* **78** 034002
[23] Courtoy A, Scopetta S and Vento V 2009 *Phys. Rev. D* **79** 074001
[24] Courtoy A, Scopetta S and Vento V 2009 *Phys. Rev. D* **80** 074032

[25] Jaffe R L 1975 *Phys. Rev. D* **11** 1953
[26] Isgur N and Karl G 1978 *Phys. Rev. D* **18** 4187; 1979 *Phys. Rev. D* **19** 2653 [Erratum-ibid. 1981 *D* **23** 817]
[27] Traini M, Mair A, Zambarda A and Vento V 1997 *Nucl. Phys. A* **614** 472
[28] Efremov A V, Goeke K, Menzel S, Metz A and Schweitzer P 2005 *Phys. Lett. B* **612** 233; Collins J C, Efremov A V, Goeke K, Menzel S, Metz A and Schweitzer P 2006 *Phys. Rev. D* **73**, 014021
[29] Courtoy A Scopetta S and Vento V 2011 *Eur. Phys. J. A* **47** 49
[30] Ceccopieri F A and Trentadue L 2008 *Phys. Lett. B* **660** 43
[31] Kang Z B and Qiu J W 2009 *Phys. Rev. D* **79** 016003
[32] Zhou J Yuan F and Liang Z T 2009 *Phys. Rev. D* **79** 114022
[33] Vogelsang W and Yuan F 2009 *Phys. Rev. D* **79** 094010
[34] Cherednikov I O, Karanikas A I and Stefanis N G 2010 *Nucl. Phys. B* **840** 379
[35] Burkardt M 2004 *Phys. Rev. D* **69** 091501; *Phys. Rev. D* **69** 057501
[36] Airapetian A *et al.* [HERMES Collaboration] 2005 *Phys. Rev. Lett.* **94** 012002
[37] Alexakhin V Y *et al.* [COMPASS Collaboration] 2005 *Phys. Rev. Lett.* **94**, 202002
[38] Brodsky S J and Gardner S 2006 *Phys. Lett. B* **643** 22
[39] Qian X *et al.* 2011 Single Spin Asymmetries in Charged Pion Production from Semi-Inclusive Deep Inelastic on a Transversely Polarized ^{3}He Target *Preprint* arXiv:1106.0363v3 [nucl-ex]
[40] Gao H *et al.* 2011 *Eur. Phys. J. Plus* **126**, 2
[41] Del Dotto A, Master Thesis, Università di Roma "La Sapienza", 2011 (unpublished)
[42] Scopetta S 2007 *Phys. Rev. D* **75** 054005
[43] Kievsky A, Viviani M and Rosati S 1994 *Nucl. Phys. A* **577** 511
[44] Wiringa R B, Stocks V G J and Schiavilla R 1995 *Phys. Rev. C* **51** 38
[45] Kievsky A, Pace E, Salmè G, and Viviani M 1997 *Phys. Rev. C* **56** 64 Pace E, Salmè G, Scopetta S and Kievsky A 2001 *Phys. Rev. C* **64** 055203
[46] Anselmino M, Boglione M, D'Alesio U, Kotzinian A, Murgia F and Prokudin A 2005 , *Phys. Rev. D* **71** 074006; 2005 *Phys. Rev. D* **72** 094007 [2005 Erratum-ibid. *D* **72** 099903]
[47] Amrath D, Bacchetta A and Metz A 2005 *Phys. Rev.D* **71** 114018
[48] Ciofi degli Atti C, Scopetta S, Pace E and Salmè G 1993 *Phys. Rev. C* **48** 968
[49] Del Dotto A, Salmè G and Scopetta S, in preparation

Journal of Physics: Conference Series **336** (2011) 012028 doi:10.1088/1742-6596/336/1/012028

A hybrid-chiral soliton model with broken scale invariance for nuclear matter

V Mantovani Sarti[1] **and A Drago**[1]

[1]Department of Physics, University of Ferrara, ITALY

E-mail: smantovani@fe.infn.it, drago@fe.infn.it

Abstract. We present a model for describing nuclear matter at finite density based on quarks interacting with chiral fields, σ and π. The chiral Lagrangian also includes a logarithmic potential, associated with the breaking of scale invariance. We provide results for the soliton in vacuum and at finite density, using the Wigner-Seitz approximation. We show that the model can reach higher densities respect to the Linear-σ model, up to $\rho \approx 3\rho_0$ for $m_\sigma = 1200$ MeV.

1. Introduction

The phase diagram of nuclear matter can be in principle more complicated than the one based on MIT-bag-like models which predict a direct transition from hadronic matter to quark-gluon plasma (QGP). In the region of high densities and relatively low temperatures a variety of phases can exist in which chiral-symmetry breaking is realized in different ways. On the other hand, to study systems at finite density by using chiral Lagrangians is not a trivial task: for instance models based on the Linear σ-model fail to describe nuclear matter already at $\rho \sim \rho_0$. In Ref. [1] the authors conclude that the failure of the σ model is due to the restrictions on the scalar field dynamics imposed by the Mexican hat potential. In Ref. [2] they use a non-linear realization of chiral symmetry in which a scalar-isoscalar effective field is introduced, as a chiral singlet, to simulate intermediate range attraction. In this way the dynamics of the chiral singlet field is no more regulated by the Mexican hat potential and accurate results for finite nuclei can be found.

Another possible solution to the problem of studying systems at finite density in chiral models is to use a linear realization but with a new potential, which includes terms not present in the Mexican hat potential. A possible guideline in building such a potential is scale invariance.

In QCD scale invariance is spontaneously broken due to the presence of the parameter Λ_{QCD} coming from the renormalization process. Formally, the non conservation of the dilatation current is strictly connected to a not vanishing gluon condensate [3], [4], [5], [6]:

$$\langle \partial_\mu j^\mu_{QCD} \rangle = \frac{\beta(g)}{2g} \langle F^a_{\mu\nu}(x) F^{a\mu\nu}(x) \rangle. \tag{1}$$

In the approach of Schechter, Migdal, and Shifman [7] a scalar field representing the gluon condensate is introduced and its dynamics is regulated by a potential chosen so that it reproduces (at mean-field level) the divergence of the scale current that in QCD is given by Eq. (1). The

Published under licence by IOP Publishing Ltd

potential of the dilaton field is therefore determined by the equation:

$$\theta^{\mu}_{\mu} = 4V(\phi) - \phi\frac{\partial V}{\partial \phi} = 4\epsilon_{vac}\left(\frac{\phi}{\phi_0}\right)^4 \tag{2}$$

where the parameter ϵ_{vac} represents the vacuum energy. To take into account massless quarks a generalization was proposed in Ref. [8], so that also chiral fields contribute to the trace anomaly. In this way the single scalar field of Eq. (2) is replaced by a set of scalar fields $\{\sigma, \boldsymbol{\pi}, \phi\}$.

It has already been shown that an hadronic model based on this dynamics provides a good description of nuclear physics at densities about ρ_0 and describes the gradual restoration of chiral symmetry at higher densities [9].

The new idea we develop in this work is to interpret the fermions as quarks, to build the hadrons as solitonic solutions of the fields equations and finally to explore the properties of the soliton at finite density using the Wigner-Seitz approximation.

The structure of the paper is the following. In Sec. 2 we describe the model we are using, in Sec. 3 we discuss the results for the single soliton, first in vacuum and then at finite density. Finally, in Sec. 4 we present our conclusions and future outlooks.

2. The model

The basic approach of a hybrid chiral model is to couple quarks to mesons in a chirally invariant way. We consider such a Lagrangian:

$$\mathcal{L} = \bar{\psi}\left(i\gamma^{\mu}\partial_{\mu} - g_{\pi}(\sigma + i\boldsymbol{\pi}\cdot\boldsymbol{\tau}\gamma_5)\right)\psi + \frac{1}{2}(\partial_{\mu}\sigma\partial^{\mu}\sigma + \partial_{\mu}\boldsymbol{\pi}\cdot\partial^{\mu}\boldsymbol{\pi}) - V(\phi_0, \sigma, \pi). \tag{3}$$

Here ψ is the quark field, σ and π are the chiral fields and ϕ is the dilaton field which, in the present calculation, is kept frozen at its vacuum value ϕ_0. The potential is given by:

$$\begin{aligned} V(\phi_0, \sigma, \pi) \;=\; & -\frac{1}{4}B\phi_0^4 - \frac{1}{2}B\delta\phi_0^4 Log\frac{\sigma^2 + \boldsymbol{\pi}^2}{\sigma_0^2} \\ & + \frac{1}{2}B\delta\frac{\phi_0^4}{\sigma_0^2}\left(\sigma^2 + \boldsymbol{\pi}^2 - \frac{\sigma_0^2}{2}\right) \\ & - \frac{1}{4}\epsilon_1\left[\frac{4\sigma}{\sigma_0} - 2\left(\frac{\sigma^2 + \boldsymbol{\pi}^2}{\sigma_0^2}\right) + 2\right] \\ & - \frac{1}{4}B\phi_0^4(\delta - 1) + \epsilon_1 \end{aligned} \tag{4}$$

where the logarithmic term generates from (2). The constants in the last line of Eq. (4) ensure that the vacuum energy is zero.
The constants B and ϕ_0 can be fixed by choosing value for the mass of the glueball and the vacuum energy ϵ_{vac}, while $\delta = 4/33$ is provided by the QCD beta function and it corresponds to the relative weight of the fermionic and of the gluonic degrees of freedom.

The first two terms of the potential are responsible for the breaking of scale invariance, while the term in the third line explicitly breaks the chiral invariance of the Lagrangian. The Euler-Lagrangian equations that follow from Lagrangian (3) are:

$$\begin{aligned} &[i\partial\!\!\!/ - g_{\pi}(\sigma + i\boldsymbol{\pi}\cdot\boldsymbol{\tau}\gamma_5)]\psi = 0, \\ &\partial_{\mu}\partial^{\mu}\sigma = -g\bar{\psi}\psi - \frac{\partial V}{\partial\sigma}, \\ &\partial_{\mu}\partial^{\mu}\boldsymbol{\pi} = -ig\bar{\psi}\boldsymbol{\tau}\gamma_5\psi - \frac{\partial V}{\partial\boldsymbol{\pi}}. \end{aligned} \tag{5}$$

The values of the parameters used for calculations in vacuum are listed in table 1.

Table 1. Values of the parameters.

Quantity	Value
$\lvert\epsilon_{vac}\rvert^{1/4}$ (MeV)	236
g	5
m_σ (MeV)	550
B	24.37
$\epsilon_1^{1/4}$ (MeV)	114
δ	4/33
σ_0 (MeV)	93
ϕ_0 (MeV)	175.23

2.1. The hedgehog ansatz

We are working in the mean-field approximation, where mesons are described by time-independent, classical fields and where powers and products of these fields are replaced by powers and products of their expectation values. The quark spinor in the spin-isospin space is:

$$\psi = \frac{1}{\sqrt{4\pi}} \begin{pmatrix} u(r) \\ iv(r)\boldsymbol{\sigma}\cdot\hat{\boldsymbol{r}} \end{pmatrix} \frac{1}{\sqrt{2}}(|u\downarrow\rangle - |d\uparrow\rangle) \tag{6}$$

Once fixed the form of the quark wave function the self-consistents ansatze for the meson fields are:

$$\langle\hat{\sigma}\rangle = \sigma(r), \quad \langle\hat{\boldsymbol{\pi}}_a\rangle = r_a\pi(r)$$

where $\sigma(r)$, $\pi(r)$ are radial functions of r.

3. Results

3.1. Single soliton in vacuum

Solving the vacuum case requires the following boundary conditions for the fields:

$$\begin{aligned} u'(0) = v(0) = 0, \\ \sigma'(0) = \pi(0) = 0 \end{aligned} \tag{7}$$

while at infinity (in practice at a value $r = R = 4$ fm), the boundary conditions read:

$$\begin{aligned} &\sigma(R) = \sigma_0, \ \pi(R) = 0, \\ &\frac{v(R)}{u(R)} = \sqrt{\frac{-g\sigma(R)+\varepsilon}{-g\sigma(R)-\varepsilon}}. \end{aligned} \tag{8}$$

A test for convergence of the solution comes from another way of expressing the energy obtained by Rafelski [10] by integrating out the fermionic fields:

$$E_{Raf.} = \int_a^R d^3r \left[4\left(V - \sigma\frac{\partial V}{\partial\sigma} - \pi\frac{\partial V}{\partial\pi} \right) \right]. \tag{9}$$

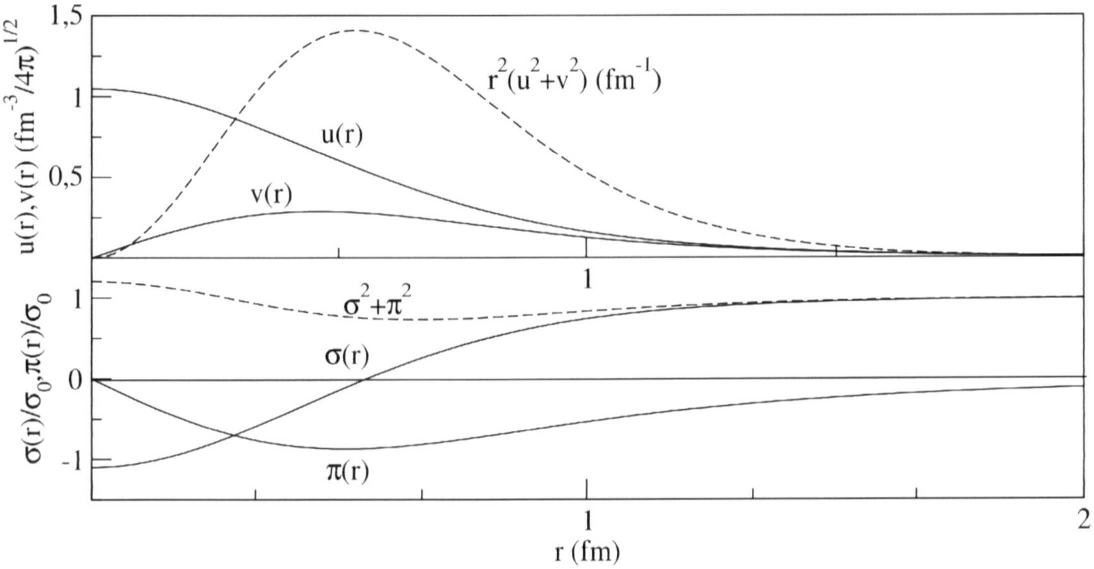

Figure 1. Fields in vacuum for the solution $g = 5, m_\sigma = 550$ MeV.

Our solutions satisfy this consistency test up to a precision of the order of 10^{-3}.

In table 2 we present the static properties of the nucleon at Mean Field level and we compare them with experimental values and in table 3 we show the decomposition of the soliton total energy in various contributions and the comparison with the Linear-σ model [11].

Table 2. Various nucleon properties at mean field level in the present work and comparison with experimental values.

Quantity	MFA	Exp.
$M\,(MeV)$	1175.6	1085
$\langle r_e^2\rangle_{I=0}$	$(0.73\,fm)^2$	$(0.72\,fm)^2$
$\langle r_m^2\rangle_{I=1}$	$(0.91\,fm)^2$	$(0.80\,fm)^2$
g_A	1.27	1.26

Table 3. Contributions to the soliton total energy in the Logarithmic model and comparison with Linear-σ model. All quantities in MeV.

Quantity	Log. Model	Linear-σ Model
Quark eigenvalue	83.1	107.4
Quark kinetic energy	1138	1056.9
E_σ (mass+kin.)	334.5	320.3
E_π (mass+kin.)	486	373.1
Potential energy $\sigma - \pi$	105.7	120.7
$E_{q\sigma}$	−101.4	−62.3
$E_{q\pi}$	−787	−673.2
Total energy	1175.6	1136.2

3.2. Single soliton at finite density

In order to describe a soliton system at finite density we use the Wigner-Seitz approximation. This approach is very common for this kind of calculations and it has already been widely applied to both non-linear [12, 13, 14] and Linear-σ models [15].
Specifically the Wigner-Seitz approximation consists of replacing the cubic lattice by a spherical symmetric one where each soliton sits on a spherical cell of radius R with specific boundary conditions imposed on fields at the surface of the sphere. The specific configuration of the meson fields, each one centered at a lattice point, generates a periodic potential in which the quarks move.
In particular now the spinor of quark fields must satisfy the Bloch theorem:

$$\psi_{\boldsymbol{k}}(r) = e^{i\boldsymbol{k}\cdot\boldsymbol{r}}\Phi_{\boldsymbol{k}}(r), \tag{10}$$

where $\boldsymbol{k}$ is the crystal momentum (for the ground state is equal to zero) and $\Phi_{\boldsymbol{k}}(r)$ is a spinor that has the same periodicity of the lattice.
We should explain in detail our choice of boundary conditions, because there are various sets of possible boundary conditions [15, 14]. In particular we relate the choice of our boundary conditions to the parity $\boldsymbol{r} \to -\boldsymbol{r}$ operation; respect to this symmetry the lower component $v(r)$ of quark spinor, the pion $\pi(r)$ and the $\rho(r)$ fields are odd, so all of these fields should vanish at R:

$$v(R) = \pi(R) = \rho(R) = 0. \tag{11}$$

For the remaining fields and the upper Dirac component we apply the usual "flatness" condition on the boundary:

$$u'(R) = \sigma'(R) = \omega'(R) = A'_S(R) = A'_T(R) = 0. \tag{12}$$

Basically the calculation is based on solving the set of coupled fields equations in a self-consistent way for each step in R; we start from the vacuum value, $R = 4$ fm, and we slowly decrease the cell radius down to the smallest radius for which self-consistent solutions can be obtained.

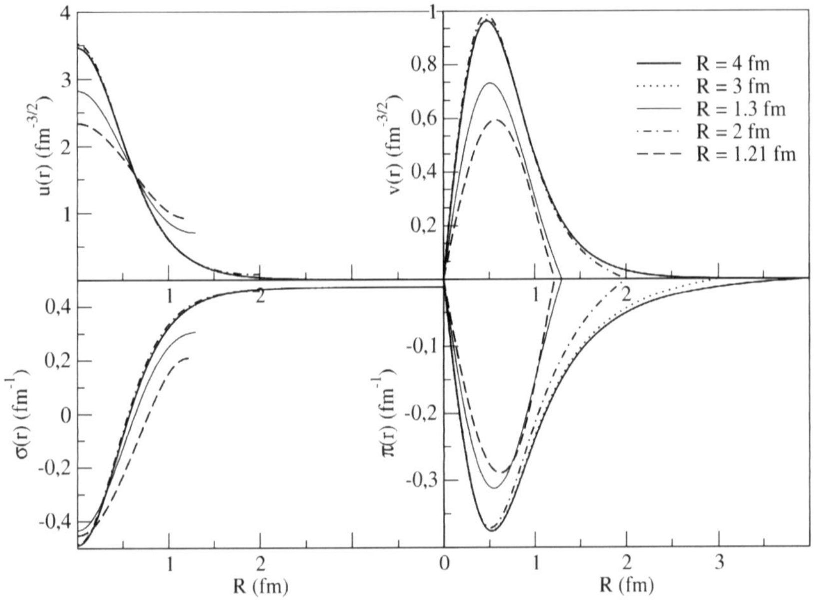

Figure 2. Upper and lower components of the Dirac spinor, σ and pion fields for different values of the cell radius R.

In figure 2 we plot the Dirac and the chiral fields for different values of R; up to $R = 2$ fm, the solutions do not change significantly, but as the cell radius shrinks to lower values, we see that all the fields are deeply modified by the finite density. In figure 3 we show the results, for the total energy of the soliton, in the present model and in the Linear-σ model. For a fixed value of the σ mass, the logarithmic model do reach higher densities; as the σ mass raises, the system can get to lower R because chiral fields are stuck on the chiral circle.

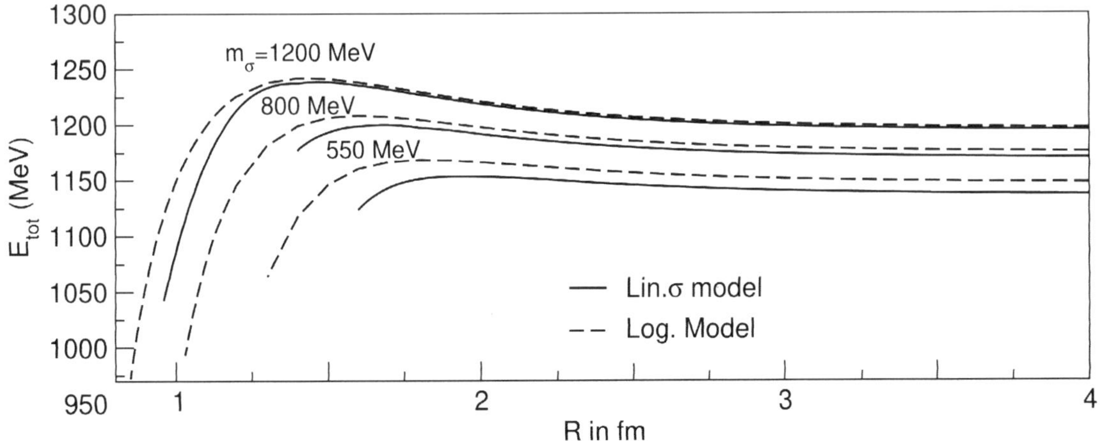

Figure 3. Total energy of the soliton as a function of cell radius R for the Linear-σ model [15] for different values of m_σ and present model.

4. Conclusions

We used a Lagrangian with quarks degrees of freedom based on chiral and scale invariance to study how the soliton behaves in vacuum and at finite density. At zero density the interplay between quarks and chiral fields leads to a soliton of mass $M = 1175.6 MeV$, not too far from the experimental value given by the medium value 1085 MeV of $M_n + M_\Delta$. We employed the Wigner-Seitz approximation to describe the dense system and we compared our results to the Linear-σ model [15]. In this approach we showed that the new potential, which includes the scale invariance, allows the system to reach higher densities at fixed m_σ. This improvement is totally due to the different dynamics between chiral fields, given by the logarithmic potential.

The present work will be extended in several directions. First of all the finite density approach will be applied to a model including vector mesons, in order provide the necessary repulsion at large density. Next we will provide a precise and accurate calculation of the eigenvalues band in the soliton crystal in order to see how the system is affected by these effects [15]. Finally, the model can also be studied also at finite temperature, including the dynamics of the dilaton field. It is interesting to note that when this model has been investigated at finite temperature assuming the fermions to be hadrons [9], a phase diagram similar to the one proposed by McLerran and Pisarski [16] was obtained. In principle the approach based on the Wigner-Seitz scheme should be able to recover a scenario similar to the one discussed in [9], but starting from more fundamental ingredients.

Acknowledgments

It is a pleasure to thank B Y Park and V Vento for many stimulating discussions, M Birse and J McGovern for useful comments and tips on calculations.

References

[1] Furnstahl R J and Serot B D 1993 *Phys. Rev. C* **47** 2338
[2] Furnstahl R J, Serot B D and Tang H B 1996 *Nucl. Phys. A* **598** 539
[3] Heide E K, Rudaz S and Ellis P J 1994 *Nucl. Phys. A* **571** 713
[4] Carter G W and Ellis P J 1998 *Nucl. Phys. A* **628** 325
[5] Carter G W, Ellis P J and Rudaz S 1997 *Nucl. Phys. A* **618** 317
[6] Carter G W, Ellis P J and Rudaz S 1996 *Nucl. Phys. A* **603** 367
[7] Schechter J 1980 *Phys. Rev. D* **21** 3393
[8] Heide E K, Rudaz S and Ellis P J 1992 *Phys. Lett. B* **293** 259
[9] Bonanno L and Drago A 2009 *Phys. Rev. C* **79** 045801
[10] Rafelski J 1977 *Phys. Rev. D* **16** 1890
[11] Birse M C and Banerjee M K 1985 *Phys. Rev. D* **31** 118
[12] Hahn D and Glendenning N K 1987 *Phys. Rev. C* **36** 1181
[13] Glendenning N K 1986 *Phys. Rev. C* **34** 1072
[14] Amore P and De Pace A 2000 *Phys. Rev. C* **61** 055201
[15] Weber U and McGovern J A 1998 *Phys. Rev. C* **57** 3376
[16] McLerran L and Pisarski R D 2007 *Nucl. Phys. A* **796** 83

Journal of Physics: Conference Series **336** (2011) 012029 doi:10.1088/1742-6596/336/1/012029

The role of Rutherford, the father of nuclear physics

Carlo Bernardini

Dipartimento di Fisica, Università La Sapienza, Roma, Italy

E-mail: `Carlo.Bernardini@roma1.infn.it`

Abstract. In this article, I will give a short account of Rutherford's contributions to the birth and early developments of nuclear physics.

1. Introduction: annus mirabilis

Up until the most recent centenary celebrations in physics, 1905 was called the *annus mirabilis* mostly because of two turning points: relativity and quantization. No doubt, these are remarkable changes in the development of the foundations of physics. But it is true that even considering the exceptional advent of the phenomenological models of microscopic objects it introduced a new unprecedented picture of the world. Only in 1906 did Jean Perrin give a persuasive proof of the existence of atoms with no mention of their internal structure; the first speculative model of atomic structure was Thomson's pancake of a spherical positive charge distributed in space with electrons oscillating inside at special frequencies corresponding to the spectral lines. Probably, this period was one of the most exciting for physicists throughout the world.

2. Lord Rutherford of Nelson

Ernest Rutherford (1871 – 1937) came to the Cavendish Laboratory (UK) from his native New Zealand thanks to a grant; he spent a short time in and around that period at McGill University in Canada. In 1900 he made an extremely important discovery [1, 2]: the data of radioactive decays can be ordered by assuming that each emitting species has a characteristic "probability per unit time" of generating "Becquerel rays". Intrinsically casual phenomena enter the classical scene dominated by deterministic mechanical theories. Accurate measurements showed at this point that the time distribution of decays is Poissonian; this was clearly reported in the famous text *Radiation from radioactive substances* by Rutherford, Chadwick and Ellis ([3], p. 172). I think that this excellent and deep sample of the "change of paradigm" is not sufficiently underlined in the current teaching of physics. Rutherford did not interrupt his analysis at this result: he was a true Galilean and immediately moved to "sensate experience", to sensible tests for investigating the positive charges in atoms. The most plausible route to this end is scattering. Despite his reluctance to use theoretical calculations, he himself made the exercise of computing the coulomb scattering between two point charges; since then, we call his result the Rutherford cross section [4]. He deliberately neglected the electron cloud (he only knew that electrons were a small part of the atomic mass). But he immediately got confirmation of his ideas: an amount of back-scattered alpha particles compatible with his calculations showed that the atoms have a heavy nearly point-like central nucleus. This marked the birth of the so called planetary model

Published under licence by IOP Publishing Ltd

Journal of Physics: Conference Series **336** (2011) 012029 doi:10.1088/1742-6596/336/1/012029

that was quantized by Niels Bohr and Arnold Sommerfeld after a short period of time (1913). It may be useful to list the problems on which Rutherford worked since then (with more or less success):

- Isotopes: the meaning of A (the mass number) and Z (the atomic number) and nuclear stability.
- Beta decay and the nuclear emission of energetic electrons.
- The mechanism of alpha decay.
- The experiments performed by German physicists.
- The sources of subatomic particles.
- Inelastic reactions.
- Nuclear electromagnetic radiation.

Following on the classification of the years prior to 1930, in epistemological terms, Sir Ernest was a "classical realist"; very conservative indeed. He is extremely diffident of theoretical physicists. There is a lucky accident in his favour: the Rutherford cross section has no quantization problems in the sense that the classical "collision parameter method" produced exactly the same results as the dominating first order Born approximation amplitude of the quantum theory of scattering.

3. A short review of the problems

The existence of isotopes is clear. The number A (the "mass") is very close to an integer multiple of the hydrogen mass (proton = hydrogen nucleus). The measured electron mass in cathode rays was known to be 2000 times smaller than that of a proton. An unstable nucleus emitted electrons in beta decay. It was very natural to think that a nucleus consisted of A protons and $A - Z$ electrons; altogether $2A - Z$ particles (fermions). The nuclear electrons should be tightly bound to protons in states unpredicted by current atomic models. Theory was in serious difficulty, for many reasons.

i) It could not account for the electron-proton bound state;
ii) Why should these states be unstable so as to produce beta decay?
iii) Why was the spectrum of beta decay electrons continuous up to a maximum energy?

These problems occupied nearly thirty years of nuclear physics. In 1928, Franco Rasetti noticed that according to the theory of Raman molecular spectra a system of $2A - Z$ fermions should have a half-integer spin in the case of nitrogen ($A = 14, Z = 7$), contrary to the experimental evidence obtained by Rasetti himself. The imagination of physicists exploded; but the idea of a neutron is not so easily conjectured up. Some nuclear reaction events in which neutrons are emitted were confused with gamma ray emission. As far as the electron spectra are concerned, Bohr, Kramers and Slater proposed resurrecing an idea by Exner according to which a microscopic energy conservation might be a statistical law; even Schrödinger proposed this, years later. Many physicists had been infected by the seemingly open mindedness of theoretical physics, notwithstanding the difficulties of the Heisenberg representation of the mathematics of the time; that mathematics was mostly based on a deep understanding of partial differential equations, determining the immediate success of the Schrödinger approach especially after the interpretation of the "probability amplitudes" by Max Born. The tunnel effect, as proposed by George Gamow in 1928, gave a convincing explanation of alpha decay as a completely non-classical phenomenon in the penetration of forbidden potential barriers. Gamow imagined that an attractive short range nuclear potential added to the Coulomb repulsion formed a barrier of finite thickness; then he computed by the Schrödinger equation the probability of the transition in which an alpha particle is emitted by a nucleus (A, Z) leaving a nucleus $(A - 4, Z - 2)$ and a fast helium nucleus. This was immediately a great success; Condon and Gurney reached the same conclusion just few weeks later. Also the Geiger and Nuttall law, known since 1912, and giving a simple relation between the lifetime and the alpha energy, was easily reproduced by the model.

Let me add a short comment on what the Italians (I already mentioned the important contribution by Rasetti) were doing at the time. Rutherford hesitated in believing such a possibility as the tunnel effect. He had devised a model in which the alpha particles were like satellites around an unstable nucleus to which they were bound by polarization forces; he had presented such a model at the conference held in Como in 1927 to celebrate Alessandro Volta. Giovanni Gentile jr., who had just completed the physics course at university and worked with a student, Ettore Majorana, showed that the Rutherford satellite model was unsustainable. Gentile also wrote a popular book [5], *Fisica Nucleare*, issued in 1937 which is still very up to date. Rutherford had excellent collaborators: among them, James Chadwick. At that time in Germany, Bothe and Becker, using electric counters, had explored the possible existence of energetic gamma rays. They actually found a neutral component; but the assumption that it was a gamma ray required that it have too large an energy. Chadwick repeated the German experiment and showed clearly that what was emitted was a "neutral proton", to be rebaptized "neutron" shortly afterwards. One of the first to understand thet there was a neutron in these events was Ettore Majorana.

4. Amaldi about the Cavendish group

Edoardo Amaldi wrote and published in 1984 an important review [6]: *From the discovery of the neutron to the discovery of nuclear fission.* In that report, the paragraphs from 1.4 to 1.6 are dedicated to the Cavendish group, particularly to Rutherford and Chadwick, but also to Feather and Dee. On June 3, 1920, Rutherford gave a "Bakerian Lecture" to the Royal Society; on that occasion, he mentioned the neutron for the first time as a short for the electron-proton tightly bound state. Amaldi recounts that throughout the twenties the word neutron was improperly used many times, always referring to an exotic bound state. So did Nernst, in 1898 speaking of electrolytic solutions, as referred by Johannes Stark, in a conjecture about radioactivity. Many people were impressed by the regularity of isotopic masses as evidenced by Aston using mass spectrometers. I also list some amusing speculations in the literature:

- The famous Dutch lawyer Antonius Johannes van den Broek was very interested in the periodic system of Mendelejev and speculated on electrons and protons. In 1921, Van den Broek proposed a model with neutral particles constituted by alpha particles bound with two electrons; this possibility was reconsidered by Rutherford in 1927.
- Working on the Aston data, the Chicago chemist William Draper Harkins proposed in 1915 a model in which beryllium was a bound state of a neutron and two alpha particles.
- Dimitri Ivanenko in the USSR was perhaps the first to conceive of a model with protons and neutrons as genuine different nucleons.

When Chadwick produced the experimental evidence of the neutron, in 1932, the theoretical difficulties of the electron-proton bound state appeared unsurmountable. The notion of the De Broglie wave length of a particle indicated that the nuclear size implied relativistic energies for the bound electrons. For instance, even for a large nucleus with $A = 210$ the bound electron should possess 210 MeV. Nevertheless, Werner Heisenberg and his assistant Guido Beck insist on e-p bound states. The most impressive changes will occur in the thirties.

- The Heisenberg-Majorana exchange forces, based on concepts developed for molecular dynamics, were accepted by the nuclear community as fundamental forces.

- Pauli proposed the neutrino and shortly after Fermi invented the theory of weak interactions. A new dedicated mathematics was invented by Pascual Jordan, in which the introduction of creation and destruction operators, made the physics of quantum fields much easier.
- Frederic Joliot and Irene Curie discovered the transmutation reactions of light nuclei.
- Fermi understandood the efficiency of slow neutrons in destabilizing nuclear structures; he failed to understand the most dramatic result: Uranium fission.

Sir Ernest Rutherford, Lord of Nelson, died in October 1937. Forgive my too short account of his enormous history: he was the reference head of a community which is one of the most impressive in the history of science. Thus, the father of nuclear physics.

References

[1] Rutherford E 1900 *Philosophical Magazine* ser. 5 **49** 1; *ibid.* 161
[2] Rutherford E and Soddy F 1902 *Philosophical Magazine* ser. 6 **4** 370; *ibid.* 569
[3] Rutherford E, Chadwick and Ellis C D 1919 *Radiation from radioactive substances* (Cambridge University Press)
[4] Rutherford E 1911 *Philosophical Magazine* ser. 6 **21** 669
[5] Gentile G jr 1937 *Fisica Nucleare* (Edizioni Roma)
[6] Amaldi E 1984 *Physics Reports* **111** 1

Journal of Physics: Conference Series **336** (2011) 012030 doi:10.1088/1742-6596/336/1/012030

Investigating biological matter with theoretical nuclear physics methods

Pietro Faccioli

Dipartimento di Fisica, Università degli Studi di Trento, Via Sommarive 14, Povo (Trento), I-38123 Italy.
INFN, Gruppo Collegato di Trento, Via Sommarive 14, Povo (Trento), I-38123 Italy.

E-mail: faccioli@science.unitn.it

Abstract. The internal dynamics of strongly interacting systems and that of bio-polymers such as proteins display several important analogies, despite the huge difference in their characteristic energy and length scales. For example, in all such systems, collective excitations, cooperative transitions and phase transitions emerge as the result of the interplay of strong correlations with quantum or thermal fluctuations. In view of such an observation, some theoretical methods initially developed in the context of theoretical nuclear physics have been adapted to investigate the dynamics of biomolecules. In this talk, we review some of our recent studies performed along this direction. In particular, we discuss how the path integral formulation of the molecular dynamics allows to overcome some of the long-standing problems and limitations which emerge when simulating the protein folding dynamics at the atomistic level of detail.

1. Introduction

Proteins are macromolecules involved in virtually all the biochemical processes which take place inside the cell, such as e.g. catalysis, charge transport, signal transduction. Clearly, understanding the physical principles which drive the internal dynamics of individual proteins and shape the protein-protein interaction would have countless implications in molecular biology, drug design and nano-biotechnology. In particular, a central open challenge at the interface of physics, biochemistry and molecular biology concerns the prediction of the protein folding pathways i.e. the sequence of conformational transitions which these molecules perform in order to reach their stable and biologically active configuration (see e.g. figure 1) .

From a theoretical perspective, the problem of determining the time evolution of proteins has been mostly investigated by means of Molecular Dynamics (MD) simulations, i.e. through the direct integration of the classical equations of motion of all the atoms in the system. From the conceptual point of view, such an approach should provide a reasonable approximation, since quantum corrections are expected to be quite small. On the other hand, from the practical point of view, the applicability of classical MD simulations is strongly limited by their intrinsic computational complexity. Indeed, even using computational resources comparable with those employed in the largest contemporary lattice QCD simulations, it is possible to follow the dynamics over time intervals which do not exceed a few hundreds ns [1]. Unfortunately, the dynamics which drives the protein phenomenology occurs at time scales which are several orders of magnitude larger, typically ranging from few ms to several seconds.

Published under licence by IOP Publishing Ltd

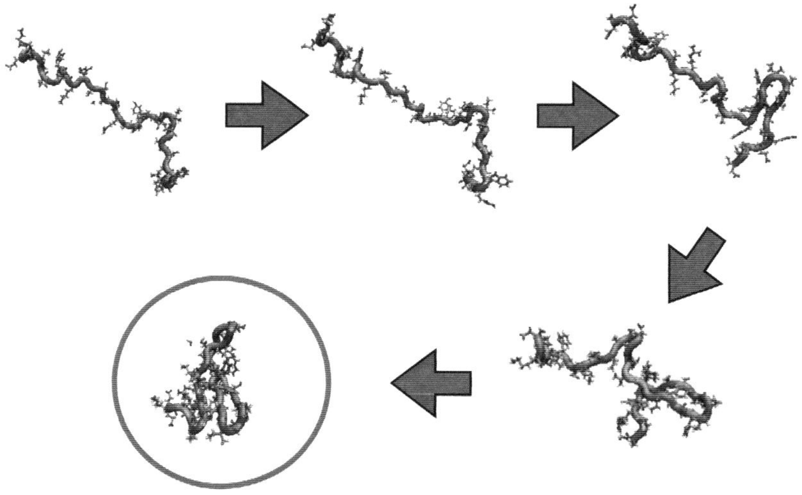

Figure 1. (Color online) Schematic representation of a protein folding pathway. This trajectory corresponds to a biological instanton and was obtained using the Dominant Reaction Pathways approach (see discussion in section 4.

The decoupling of the microscopic time scales from the time scales at which conformational reactions occur is a common feature in many bio-molecular systems. It is due to the fact that most conformational transitions are cooperative and involve the overcoming of free-energy barriers. As a result, in a MD simulation, most computational time is wasted to describe thermal oscillations inside the initial meta-stable state.

Given the huge gap between the time scales at which the relevant physics takes place and the time intervals which can be covered by MD simulations, it is natural to expect that a real breakthrough in our comprehension of biological matter at the fundamental level is more likely to come from new ideas and new theoretical approaches, rather than from the next several generations of supercomputers.

In the next sections, we discuss how functional integral approaches originally developed in the context of nuclear theory can provide an extremely powerful framework to investigate the dynamics of proteins as well as other complex molecular systems. Before introducing the mathematical details of this theory, it is instructive to discuss some basic aspects of protein structure and phenomenology and to highlight several analogies with the dynamics of strongly interacting systems.

2. QCD and protein dynamics: so far yet so close

From the biochemical point of view, proteins are poly-peptide chains, i.e. weakly branched polymers formed by 20 different types of monomers called amino-acids. Typical globular proteins consist of about 100 amino-acids, but some chains can contain as many as $\sim 30,000$ of such monomers.

From the physical point of view [2], proteins are very different from essentially all other known polymers: indeed, while standard hetero-polymers are prototypes of disordered mesoscale systems, proteins are characterized by their unique "ability" to spontaneously and reversibly fold into a well-defined stable configuration (denominated the *native state*), in which they perform their biological functions (see e.g. figure 2). Such a remarkable feature is the result of the evolutionary pressure, which has selected only a very specific subset in the space of all possible

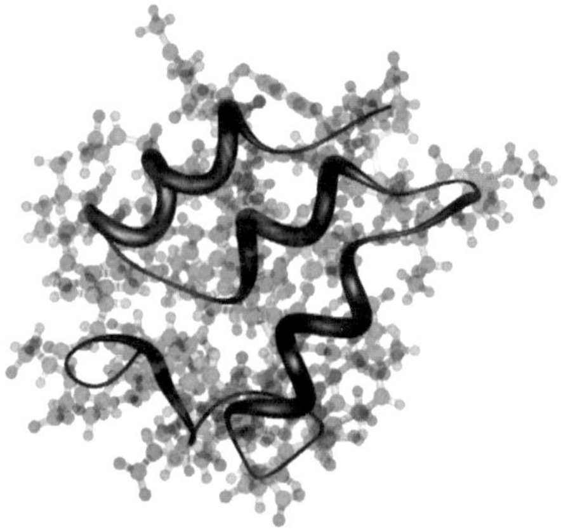

Figure 2. (Color online) Ball-and-stick representation of the native state of a globular protein. The tube represents the back-bone of the chain.

amino-acid sequences.

The native state of a protein is characterized by an extremely low entropy, since it is defined by the small thermal oscillations around a single conformation. In this sense, proteins may be regarded as non-symmetric mesoscale crystals. If the temperature of the water surrounding the molecule is increased, or if a denaturant agent (typically urea) is added to the solvent, then proteins can reach a much more entropic phase called the *molten globule*. In this state, the chain is still collapsed, but is free to explore the different compact configurations, in analogy with a liquid drop. Finally, at even higher temperatures or denaturant concentrations, proteins swell and start to behave as flexible random coils (see right panel of figure 3).

Interestingly, the phase diagram of proteins in solution displays several analogies with that of strongly interacting matter (see left panel of figure 3). Indeed, the low-temperature and low-density phase of QCD is characterized by a very low entropy, since only colorless hadrons are present in the spectrum. On the other hand, above the deconfinement phase transition, the entropy is much larger, since also quantum states in color multiplets can contribute to the partition function.

In addition, flavor symmetry implies that the hadronic spectrum can be approximatively organized into SU(3) multiplets. Hence, since quantum states belonging to the same multiplet can be transformed into one another by a flavor rotation, the number of truly distinct hadronic wave-functions is actually much smaller than the number of hadrons recorded in the particle data book. Similarly, it has been experimentally observed that nearly identical native structures are adopted by poly-peptide chains characterized by very different amino-acid sequences. This is to say that the number of native structures which can be realized is much smaller than the number of sequences in the data bank of known proteins.

These analogies are not accidental. They reflect the fact that the dynamics of both hadrons and proteins is characterized by the interplay of large correlations and quantum or thermal fluctuations. It is natural to address the question whether some of the powerful theoretical techniques developed to describe the non-perturbative internal dynamics of hadrons may be exported to describe the non-perturbative internal dynamics of proteins. In the next section, we shall see that this intuition can be made rigorous, i.e. that it is possible to formulate the dynamics of biomolecular systems in terms of an effective quantum many-body problem.

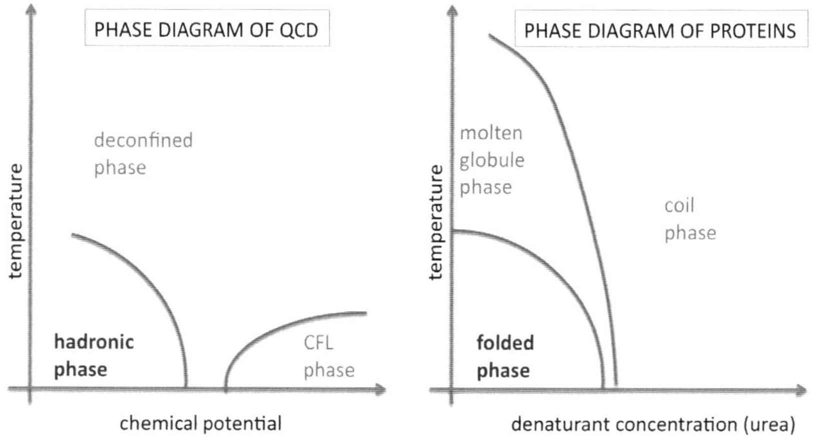

Figure 3. (Color online) Comparison between the phase diagram of QCD and of proteins in solution.

3. Path integral formulation of molecular dynamics

The dynamics of biomolecular systems in solution is often described by means of the Langevin equation, which reads

$$m\,\ddot{x} = -\gamma\dot{x} - \nabla U(x) + \xi(t), \tag{1}$$

In this equation, $x = (x_1, x_2, ..., x_N)$ is a macro-vector specifying the coordinates of all the atomic nuclei in the molecule, $U(x)$ is the potential energy and $\xi(t)$ is a stochastic force, which simulates the effects of the random Brownian collisions with the molecules in the solvent. The contemporary models for the potential energy $U(x)$ — which is often improperly called the *force field* — are defined phenomenologically in order to reproduce the results of quantum chemistry calculations of the equilibrium configurations of small molecules [3]. For sake of simplicity and without loss of generality, in the following we present the formalism in the case of one particle in one spacial dimension. The generalization to the multi-dimensional case is straightforward.

The acceleration term on the left hand side of Eq. (1) introduces inertial effects which are damped on a time-scale $t \sim m/\gamma \equiv \tau_D$. For most biomolecular systems, τ_D is of order of the fraction of a *ps*, hence much smaller than the relevant microscopic time-scale associated to the dynamics of the torsional angles, which takes place at the *ns* time-scale.

For $t \gg \tau_D$ the Langevin Eq. (1) reduces to

$$\frac{\partial x}{\partial t} = -\frac{D}{k_B T}\,\nabla U(x) + \eta(t), \tag{2}$$

where $D = \frac{k_B T}{m\gamma}$ is the so-called diffusion coefficient and $\eta(t) = \frac{1}{\gamma}\,\xi(t)$ is a rescaled white noise that satisfy the so-called fluctuation-dissipation relationship:

$$\langle \eta(t)\eta(t')\rangle = 2D\delta(t-t'). \tag{3}$$

It can be shown —see e.g. [4]— that the probability distribution sampled by the stochastic differential Eq. (2) satisfies then the Fokker-Planck (FP) equation:

$$\frac{\partial}{\partial t}\,P(x,t) = D\nabla\left(\frac{1}{k_B T}\,\nabla\,U(x)\,P(x,t)\right) + D\,\nabla^2\,P(x,t). \tag{4}$$

A universal property of all the solutions of the Eq. (4) is that, in the long time limit, they converge to the Boltzmann weight, regardless of their initial conditions.

By performing the substitution

$$P(x,t) \equiv e^{-\frac{1}{2k_BT}U(x)}\ \psi(x,t), \tag{5}$$

the FP Eq. (4) can be formally re-written as a Schrödinger Equation in imaginary time:

$$-\frac{\partial}{\partial t}\psi(x,t) = \hat{H}_{eff}\ \psi(x,t), \tag{6}$$

where the effective "Quantum Hamiltonian" operator reads

$$\hat{H}_{eff} \ = \ -D\nabla^2 + V_{eff}(x), \tag{7}$$

and $V_{eff}(x)$ is an effective potential and is defined as

$$V_{eff}(x) = \frac{D}{4(k_BT)^2}\ \left[(\nabla U(x))^2 - 2k_BT\,\nabla^2 U(x)\right]. \tag{8}$$

Hence, we have shown that the problem of studying the *real time* stochastic dynamics of a protein in solution can be mapped into the problem of determining the *imaginary time* evolution of an effective quantum many-body system.

Based on the analogy with quantum mechanics, it is immediate to obtain a path-integral representation of the solution of (4), subject to the boundary conditions $x(0) = x_i$ and $x(t) = x_f$:

$$\begin{aligned} P(x_f,t|x_i) &= e^{-\frac{1}{2k_BT}(U(x_f)-U(x_i))}\langle x_i|e^{-\hat{H}_{eff}t}|x_f\rangle \\ &= e^{-\frac{1}{2k_BT}(U(x_f)-U(x_i))}\int_{x_i}^{x_f}\mathcal{D}x(\tau)\ e^{-\int_0^t d\tau\ \left(\frac{\dot{x}^2(\tau)}{4D}+V_{eff}[x(\tau)]\right)}. \end{aligned} \tag{9}$$

This equation expresses the fact that the Green's function of the Fokker-Planck equation is formally equivalent to a quantum-mechanical propagator in imaginary time. On the other hand, it is important to stress that the variable t which enters in these equations corresponds to the physical time.

4. The theory of biological instantons

The path integral formulation of the stochastic dynamics of protein is particularly convenient if one is interested in investigating the structure of the protein folding pathways. Indeed, such a formalism allows to focus directly on the non-equilibrium dynamical trajectories which actually reach the native state x_f in a given time interval t. This way, one avoids wasting time to simulate trajectories which remain confined in the initial denatured state. A second important advantage of Eq. (9) is that it assigns a probability density to each folding pathway $x(\tau)$:

$$\text{Prob.}[x(\tau)] \propto \exp[-S_{eff}[x]]. \tag{10}$$

Clearly, the folding pathways which are most likely to be realized by a given protein are the solutions of the "classical" equations of motion generated by the effective action

$$S_{eff}[x] = \int_0^t d\tau\ \left(\frac{\dot{x}^2}{4D} + V_{eff}[x]\right). \tag{11}$$

These trajectories typically overcome barriers of the effective potential $V_{eff}(x)$. Hence, they correspond to instantons in the effective quantum theory. The so-called Dominant Reaction

Pathways (DRP) approach developed by our group [5, 6, 7, 8, 9] is based on the saddle-point approximation of the stochastic path integral (9) and focuses on such "biological instantons". Unlike in QCD, such an approach is not affected by infrared divergences, hence it is completely rigorous. The expansion parameter controlling the accuracy of this approximation is the thermal energy k_BT, which enters in the definition of the diffusion constant D and of the effective potential $V_{eff}(x)$.

In principle, the instantons can be obtained by relaxing numerically a discretized representation of the effective action functional $S_{eff}[x]$, in complete analogy with the cooling procedure developed in lattice field theory. In practice, however, the simultaneous presence of fast and slow internal time scales (which range from fs to ns) makes such a task very challenging. Indeed, the discretized elementary time interval Δt has to be chosen much smaller than the smallest time scale. Hence, computing a single dominant pathway would require to find a minimum of a function with a gigantic number of degrees of freedom: $3N \times N_t$, where N is the number of atoms in the protein —typically from several hundreds to many thousands— and N_t is the number of time discretization steps —typically 10^6—.

Fortunately, a major numerical simplification of this problem can be achieved by exploiting the fact that the dynamical system defined by the effective action (11) is simplectic, i.e. it conserves the "effective energy" [1]

$$E_{eff} = \frac{1}{4D}\dot{x}^2(t) - V_{eff}[x(t)]. \tag{12}$$

Hence, rather than minimizing directly the effective action $S_{eff}[x]$, it is possible to obtain the biological instantons using the Hamilton-Jacobi (HJ) formulation of classical mechanics. In other words, the trajectories obeying the classical equations of motion and subject to the boundary conditions $x(t) = x_f$, and $x(0) = x_i$ are those which minimize the effective HJ functional

$$S_{HJ}[x(l)] = \frac{1}{\sqrt{D}} \int_{x_i}^{x_f} dl \sqrt{E_{eff}(t) + V_{eff}[x(l)]}, \tag{13}$$

where $dl = \sqrt{dx^2}$ is the measure of the distance covered by the system in configuration space, during the transition. Hence, the dominant reaction pathways can be viewed as the geodesic in a space with a curved metric.

The major advantage of the HJ formulation is that it allows to remove the time as the independent variable and replacing it with the curvilinear abscissa l, which has the dimension of a length scale. Since there is no gap in the length scales of molecular systems, the discretization of the HJ is expected to converge extremely much faster than the time discretization of the effective action $S_{eff}[x]$. Indeed, typically 100 slices are sufficient to achieve an accurate representation of the path.

The effective energy E_{eff} is an external parameter which determines the time at which each configuration of a dominant reaction pathway $\bar{x}(\tau)$ is visited, according to the usual HJ relationship

$$t(x) = \int_{x_i}^{x} dl \frac{1}{\sqrt{4D(E_{eff} + V_{eff}[\bar{x}(l)])}}. \tag{14}$$

Typically, one is interested in studying transitions which terminate close the local minima of the potential energy $U(x)$. The residence time in such end-point configurations must be much longer than that in the configurations visited during the transition. From Eq. (14) it follows

[1] Note that E_{eff} has the dimension of a rate, therefore this quantity does not have the physical interpretation of a mechanical energy

that these conditions are verified if $E_{eff} \sim -V_{eff}(x_o)$, where x_o is a configuration in the vicinity of a local minimum of $U(x)$.

In practice, computing the dominant reaction pathway connecting two given configurations x_i to x_f amounts to minimizing a discretized version of the effective HJ functional:

$$S^d_{HJ}[x(l)] = \sum_{n=1}^{N_s-1} \sqrt{\frac{1}{D}\left[E_{eff} + V_{eff}\left(x(n)\right)\right]}\ \Delta l_{n,n+1}, \tag{15}$$

where N_s is the number of path discretization slices. Once such a path has been determined, one can reconstruct the time at which each of the configurations is visited during the transition, using Eq. (14). In particular, the time interval between the n-th and the $(n+1)$-th slice is

$$\Delta t_{n+1,n} = \frac{\Delta l_{n+1,n}}{\sqrt{4D(E_{eff} + V_{eff}[\bar{x}(n)])}}, \tag{16}$$

where $\Delta l_{n+1,n} = \sqrt{(\bar{x}(n+1) - \bar{x}(n))^2}$.

4.1. Thermal fluctuations around the dominant folding pathways

The biological instantons encode information about the reactive dynamics, hence can be used as starting point to compute the folding rate, whose inverse gives the typical time a protein takes to reach the native state, starting from a denatured state. This is a fundamental observable in protein kinetics, which is accessible from fluorescence experiments.

It turns out that, in order to obtain an equation for the rate, it is necessary to go beyond the lowest-order saddle-point approximation, and estimate the contribution of small thermal fluctuations around the dominant paths [8].

This corresponds to evaluating the instanton weight at one-loop level, by functionally expanding the effective action to quadratic order, around the instanton solution $\bar{x}(\tau)$:

$$\begin{aligned} S_{eff}[x] &= S_{eff}[\bar{x}] + \frac{1}{2}\int_0^t d\tau' \int_0^t d\tau'' \frac{\delta^2 S_{eff}[\bar{x}]}{\delta x_i(\tau')\delta x_k(\tau'')}\left(x_i(\tau') - \bar{x}_i(\tau')\right)\left(x_k(\tau) - \bar{x}_k(\tau)\right) + \ldots \\ &\simeq S_{eff}[\bar{x}] + \frac{1}{2}\int_0^t d\tau \int_0^t d\tau' (x_i(\tau') - \bar{x}_i(\tau'))\ \hat{F}^{\tau,\tau'}_{ik}[\bar{x}]\ (x_k(\tau) - \bar{x}_k(\tau)), \end{aligned} \tag{17}$$

where we have introduced the fluctuation operator $\hat{F}[\bar{x}]$, defined as

$$\hat{F}^{\tau,\tau'}_{ik}[\bar{x}] \equiv \frac{\delta^2 S_{eff}[\bar{x}]}{\delta x_i(\tau')\delta x_k(\tau'')} = \left[-\frac{1}{2D}\ \delta_{ik}\ \frac{d^2}{d\tau^2} + \partial_i\partial_k\ V_{eff}[\bar{x}(t)]\right]\delta(\tau - \tau'). \tag{18}$$

The formal one-loop expression for the conditional probability is therefore

$$P(x_f, t|x_i)\ \simeq\ \mathcal{N}\ \frac{e^{-\frac{\beta}{2}(U(x_f)-U(x_i))}}{\sqrt{\det \hat{F}[\bar{x}]}}\ e^{-S_{eff}[\bar{x}]}. \tag{19}$$

Such an expression is clearly divergent and needs to be regularized. To this end, one can multiply and divide by the conditional probability density of a reference system $P_{reg.}(x_f, t|x_i)$. For example, one can use as regulator the conditional probability associated to the diffusion in an external harmonic potential, for which the propagator $P_{reg.}(x_0, t|x_0)$ is analytically known. The result is [8]:

$$P_{DRP}(x_f, t|x_i) = P_{reg.}(x_0, t|x_0)\ e^{-\frac{\beta}{2}(U(x_f)-U(x_i))}\ e^{S^{reg.}_{eff}[\bar{x}_{reg.}] - S_{eff}[\bar{x}]} \sqrt{\frac{1}{\det\left(\hat{F}^{-1}_{reg.}[\bar{x}_{reg.}]\ \hat{F}[\bar{x}]\right)}}, \tag{20}$$

where $\hat{F}_{reg}$ is the fluctuation operator for the reference system used in the regularization.

In practice, the determinant $\det\left(\hat{F}_{reg.}[\overline{x}_{reg.}]\ \hat{F}^{-1}[\overline{x}]\right)$ has to be evaluated numerically from the discretized representation of the fluctuation operators. To obtain such a representation, one needs to express the time derivative using discretized time intervals. It is most convenient to use the intervals $\Delta t_{i,i+1}$ evaluated from the dominant trajectory, according to Eq. (16). The fluctuation operator reads

$$\begin{aligned}\hat{F}[\bar{x}]^{i,j}_{k,m} &= \frac{-1/D}{\Delta t_{m+1,m}+\Delta t_{m,m-1}}\delta_{i,j}\left[\frac{\delta_{k,m+1}}{\Delta t_{m+1,m}}-\delta_{k,m}\left(\frac{1}{\Delta t_{m+1,m}}+\frac{1}{\Delta t_{m,m-1}}\right)+\frac{\delta_{k,m-1}}{\Delta t_{m,m-1}}\right]\\ &+ \frac{\partial^2 V_{eff}(\bar{x}(k))}{\partial x_i \partial x_j}\delta_{k,m},\end{aligned} \tag{21}$$

where the indexes $k,m = 1,\ldots,N_s$ run over the path frames, while the indexes $i,j = 1,\ldots,d$ label the degrees of freedom of the system.

4.2. Improved effective actions

The main numerical advantage of the DRP formalism arises from the possibility of removing the time as a dynamical variable, and replace it with the curvilinear abscissa l. Since there is no gap in the characteristic length scales of molecular systems, the convergence of the discretization of l is usually very fast compare to that of the discretization of t. As a result, using such a formulation, it is possible to gain information about the reaction mechanism at a very low computational cost.

On the other hand, in order to obtain information about the dynamics, one needs to compute the times at which each configuration is visited along the dominant path, using Eq.s (14) and (16). The calculation of the time intervals $\Delta t_{i,i+1}$ is also needed in the discretized representation of the fluctuation operator (21), which enters in the calculation of the order $k_B T$ corrections arising from non-equilibrium stochastic fluctuations around the dominant path.

Clearly, the trick to remove time by switching to the HJ formulation does not help in the evaluation of explicitly time-dependent observables. As a result, in order to achieve an accurate description of the dynamics, or in order properly take into account of the effects of fluctuations, one would need to use a large number of path frames, with a consequent significant increase of the computational cost of a DRP simulation.

Fortunately, the convergence of the calculation of the time intervals in the DRP approach can be greatly improved by adopting the effective stochastic theory (EST) developed in [10]. The main idea of the EST is to exploit the gap in the internal time scales in order to analytically perform the integral over the fast Fourier components of the paths $x(\tau)$ which contribute to the path integral (9), using Wilson's approach. Through such a procedure, the effects of the fast dynamics is rigorously and systematically integrated out by renormalizing the effective potential:

$$V_{eff}(x) \quad \to \quad V_{eff}^{EST}(x) = V_{eff}(x) + V_{eff}^{R}(x). \tag{22}$$

where, to leading order in the effective theory one has

$$V_{eff}^{R}(x) \quad = \quad \frac{D\Delta t_c\ (1-b)}{2\pi^2 b}\nabla^2 V_{eff}(x) + \ldots \tag{23}$$

In such an Eq., Δt_c is a cut-off time scale which must be chosen much smaller than the fastest internal dynamical time scale and b is a parameter which defines the interval of Fourier modes which are being analytically integrated out. Typically, for molecular systems $\Delta t_c \sim 10^{-3}$ps and $b \sim 10^{-2}$ [11]. The dots in Eq. (23) denote higher order correction in an expansion in the ratio of slow and fast time scales (so-called slow-mode perturbation theory) and can be found in the original publication [10].

The EST generates by construction the same long-time dynamics of the original "bare" theory, but has a lower time resolution. In the context of MD simulations, this implies that the EST can be integrated using much larger discretization time steps [11]. In the context of the DRP simulations, the utility of the EST resides in the fact that much fewer path discretization time steps are required in order to achieve a convergent calculation of the time interval from the dominant path, through Eq. (14).

4.3. Generalization to ab-initio quantum chemistry calculations

So far we have discussed the DRP formalism in the specific context of protein folding. On the other hand, this theory applies in general to any physical system described by the over-damped Langevin Eq. (2). In particular, it can be applied to investigate rare chemical reactions taking place in solution [12]. Typically, such reactions involve large modifications of the electronic structure, e.g. associated to the breaking or forming of covalent bonds. Hence, in investigating such processes, a classical description based on some phenomenological expression for the molecular potential energy $U(x)$ entering in the Langevin equation is no longer appropriate, and one must adopt a sub-atomic approach, in which the quantum electronic degrees of freedom are explicitly taken into account.

In the Born-Oppenheimer approximation [3], this is done in two steps. First, the Schrdinger equation for the electronic degrees of freedom is (approximatively) solved holding fixed all the nuclear coordinates. The corresponding ground-state energy is then interpreted as the potential molecular energy $U(x)$ and is used to derive the effective potential $V_{eff}(x)$, which enters the DRP formulas. This approach allows to rigorously take into account of the quantum effects on the electron dynamics, to all order in $\hbar$ and to leading order in the ratio m/M between the mass of the electron m and the typical mass of the atomic nuclei M.

4.4. Quantum corrections to the diffusive dynamics of the atomic nuclei

In the discussion performed so far, the dynamics of the atomic nuclei has been assumed to be entirely classical, and to evolve according to the Langevin Eq. (2). Such an approximation is certainly reliable for most atomic species which are found in biomolecular systems, such as carbon or oxygen. On the other hand, it becomes questionable for the lightest species, notably hydrogen. Hence, in the study of reactions involving a large number of such atoms, the quantum corrections to the stochastic dynamics of nuclei should be consistently taken into account.

To tackle this problem, we have derived the multidimensional generalization [15] of the semiclassical extension of the Langevin Eq. (2) and of the corresponding FP equation (4), in which quantum corrections are systematically included to order $\hbar^2$ [13, 14]. From these equations it is immediate to compute the quantum corrections to the DRP equations [15]. We found that the effective action receives contribution from an additional term in the form (adopting again for simplicity a one-dimensional notation):

$$S^Q_{eff}[x] = \int_0^t d\tau \; V^Q_{eff}[x(\tau)], \tag{24}$$

where

$$V^Q_{eff}(x) \simeq \frac{D\,\lambda}{4(k_BT)^3}|\nabla U(x)|^2\nabla^2 U(x) \tag{25}$$

is the correction to the effective potential and $\lambda = \frac{\hbar^2}{12\, k_B T m_i}$ is a characteristic parameter which determines the size of quantum fluctuations. In conformational transitions driven by the formation of hydrogen bonds, the quantum corrections to the dominant reaction pathways have been found to give raise to sizable effects, see Ref. [15].

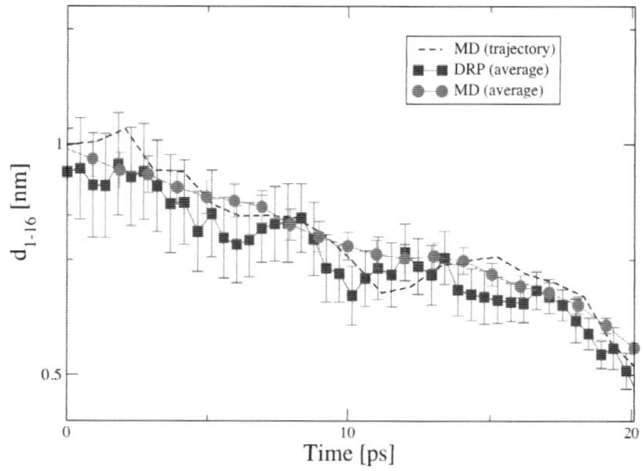

Figure 4. (Color online) Evolution of the end-point distance of the protein during a folding reaction, averaged over folding trajectories obtained in DRP and MD simulations, respectively. The dashed line represents a typical trajectory.

5. Some applications

The DRP approach has been extensively tested and validated on toy systems [7, 8] and simplified protein models [16, 17], and then applied to investigate realistic transitions, including conformational transitions of peptides [6, 18], chemical reactions [12] and realistic protein folding reactions [21]. In this section, we briefly review some of these studies.

5.1. Validation of the DRP theory on a coarse-grained model

The most straightforward way to assess the accuracy of the DRP approach consists in comparing its predictions with those obtained directly from MD simulations. Unfortunately, for realistic models MD simulations of protein folding reactions are not presently feasible.

Hence, we have validated the DRP approach using much simpler coarse-grained models, with protein-like properties. For example, the model used in Ref.[17] displays a folding transition with a well-defined and unique native state. The predictions for the time evolution of the distance between the end-points of the chain (a typical observable in protein folding) obtained in the DRP and MD approaches is compared in figure 4. We see that the biological instanton standpoint provides a realistic description of the reactive dynamics.

5.2. Atomistically detailed protein folding simulations

The computational efficiency of the DRP approach allows to investigate a large class of reactions which cannot be studied on the existing supercomputer using standard MD simulations. Only very few examples of successful MD simulations of protein folding reactions have been reported to date. These studies were either performed on a special purpose supercomputing machine in which all the key MD algorithms are implemented at the hardware level [19] or through very long large-scale simulations performed on several thousands of CPU's, distributed world wide [20].

On the other hand, using the DRP method [21] it was possible to simulate the same reaction at a comparable level of statistics in less than a single day on only 32 CPU's. The DRP results were found to be consistent with those reported in [19], and provided an explanation of the discrepancy between the results of Ref. [19] and Ref. [20]. An illustration of one of the folding

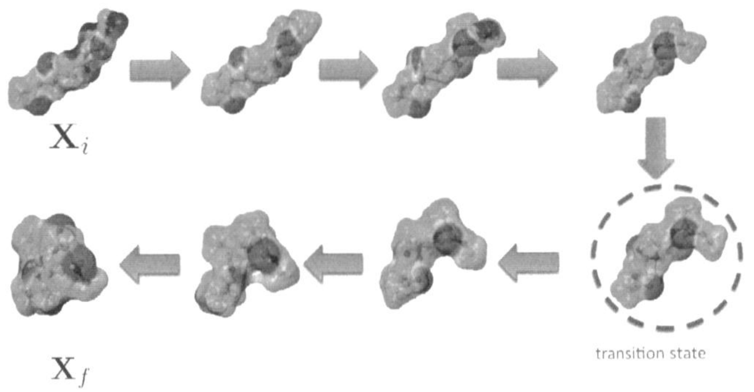

Figure 5. (Color online) Folding of a peptide obtained from quantum DRP calculations. The shaded area represents the distribution of electric charge.

trajectories obtained in the DRP approach is displayed in figure 1. Simulations on much larger systems (unaccessible to MD simulations) are now being performed.

5.3. Folding of a small peptide from ab-initio quantum mechanical simulations

Using the DRP approach we have performed the first study of the folding of a peptide based on *ab-initio* Quantum Mechanical calculations, using the Born-Oppenheimer approach described in section 4.3. This type of analysis could never performed using MD or Car-Parinello approaches [3]. The results of our simulations are schematically represented in Fig. 5. This study allowed for the first time to access the reliability of the existing classical force fields in predicting non-equilibrium reactive trajectories. This is an important and non-trivial issue, since the phenomenological force fields are fitted in order to reproduce quantum mechanical calculations on static configurations, or in equilibrium condition. Hence, there is in principle no guarantee that they should be accurate in also in the regions of phase-space which are mostly visited by the non-equilibrium reactive trajectories. On the other hand, our studies have revealed that classical calculations give results that are indeed in good agreement with those of quantum *ab-initio* simulations. This study also allowed to quantify to what extent the electronic structure of the chain is modified during the reaction.

6. Conclusions

Nuclear physics provides an excellent training ground where developing advanced theoretical approaches to interacting many-particle systems. Indeed, a number of schemes and approximations originally proposed in this context [22] have later become standard tools in the field of strongly correlated electrons, quantum liquids, cold atoms, ...

In particular, in this talk we have reviewed our recent attempt to export functional integral techniques originally derived in QCD to investigate the non-equilibrium diffusive dynamics of biomolecules such as proteins. We have argued that such an approach opens the door to the study of a large number of processes which simply could not be investigated using standard MD simulations.

Interdisciplinary studies of this type may also emphasize the importance of pushing forward the fundamental research in nuclear theory. Indeed, they show that this activity is valuable also in a perspective which goes beyond the traditional perimeter of nuclear physics.

Acknowledgments

I would like to thank all the collaborators which have been involved in the development of the DRP approach, and in particular H Orland, F Pederiva, G Garberoglio, S a Beccara and M Sega.

PF is a member of the Interdisciplinary Laboratory for Computational Science (LISC), a joint venture of Trento University and Bruno Kessler foundation. The numerical simulations were performed on the Aurora supercomputer at LISC. This research was partially funded by the Provincia Autonoma di Trento and by INFN, through the AuroraScience project.

References

[1] Freddolino P L *et al.* 2008 *Biophys. J.* **94** L75
[2] Finkelstein A V and Ptitsyn O B 2002 *Protein Physics: a Course of Lectures* (Academic Press, London)
[3] Leach A R 2001 *Molecular modeling: principle and applications" (2nd ed.)* (Pearson Education, Harlow, England)
[4] Schwabl F 2006 *Statistical Mechanics (2nd ed.)* (Springer-Verlag, Berlin)
[5] Faccioli P, Sega M, Pederiva F and Orland H 2006 *Phys. Rev. Lett.* **97** 108101
[6] Sega M, Faccioli P, Pederiva F, Garberoglio G and Orland H 2007 *Phys. Rev. Lett.* **99** 118102
[7] Autieri E, Faccioli P, Sega M, Pederiva F and Orland H 2009 *J. Chem Phys.* **130** 064106
[8] Mazzola G, a Beccara S, Faccioli P and Orland H 2011 *J. Chem. Phys.* **134** 164109
[9] Elber R and Shalloway D 2000 *J. Chem. Phys.* **112** 5539
[10] Corradini O, Faccioli P and Orland H 2009 *Phys. Rev. E* **80** 061112
[11] Faccioli P 2010 *J. Chem. Phys.* **133** 164106
[12] a Beccara S, Garberoglio G, Faccioli P and Pederiva F 2010 *J. Chem. Phys.* **132** 111102
[13] Ankerhold J, Pechukas P and Grabert H 2001 *Phys. Rev. Lett.* **87** 086802; Ankerhold J and GrabertH 2008 *Phys. Rev. Lett.* **101** 119903 (Erratum); Ankerhold J 2001 *Phys. Rev. E* **64** 060102
[14] Coffey W T, Kalmykov Y P, Titov S V and Mulligan B P 2007 *J. Phys. A* **40** F91; Coffey W T, Kalmykov Y P, Titov S V and Cleary L 2008 *Phys. Rev. E* **78** 031114
[15] a Beccara S, Garberoglio G and Faccioli P 2011 *J. Chem. Phys.* **135** 034103
[16] Faccioli P 2008 *J. Phys. Chem.* **B112** 137560
[17] Faccioli P, Lonardi A and Orland H 2010 *J. Chem. Phys.* **133** 045104
[18] a Beccara S, Faccioli P, Sega M, Garberoglio G, Pederiva F and Orland H 2011 *J. Chem. Phys.* **134** 024501
[19] Shaw D E *et al.* 2010 *Science* **330** 341
[20] Ensign D E and Pande V J 2009 *Biophys. Journ.* **96** L53
[21] a Beccara S, Skrbic T, Covino R and Faccioli P, presently under review.
[22] Negele J W and Orland H 1998 *Quantum many-particle systems* (Westview Press)

Journal of Physics: Conference Series **336** (2011) 012031 doi:10.1088/1742-6596/336/1/012031

Status and perspectives of the INFN Laboratori Nazionali del Sud

M Lattuada

INFN - Laboratori Nazionali del Sud, Via S. Sofia 62, 95123 Catania, Italy
Dipartimento di Fisica e Astronomia, Via S. Sofia 64, 95123 Catania, Italy

E-mail: `lattuada@lns.infn.it`

Abstract. The nuclear research carried out at the Laboratori Nazionali del Sud is mainly based on two ion accelerators, a 15 MV Tandem and a K800 superconductive cyclotron. The large variety of stable and radioactive beams produced and their wide energy range, together with the availability of high performance detection systems, allow for experimental research in the fields of the nuclear structure, reaction mechanisms and nuclear astrophysics. A continuous upgrading of the beam production and detection facilities is performed to mantain the role of the LNS in the panorama of the European nuclear laboratories. In addition the research activity in astroparticle physics led to the construction of two underwater laboratories in front of the sicilian coast. They allow to check and validate submarine technologies in view of the realization of the European project KM3NeT, consisting in the installation of an observatory for high energy cosmic neutrinos in deep sea. All these experimental resources also represent formidable tools for applied researches that cover multidisciplinary fields concerning proton therapy, cultural heritage, seismology, environmental sciences and many others.

1. Introduction

The Laboratori Nazionali del Sud (LNS), one of the four national laboratories of the Istituto Nazionale di Fisica Nucleare (INFN), started its operation at the beginning of eighties, thanks to the installation of a Tandem accelerator with nominal maximum voltage of 13 MV. It was later upgraded to 15 MV and since that time is running without any relevant interruption till today. The second important step in the development of LNS was represented by the acquisition of the K800 superconductive cyclotron (CS) designed and realized by tha LASA group in Milan and transferred to LNS at the beginning of nineties. After the commissioning the CS started to operate as post-accelerator of the Tandem beams, but few years later it was possible to decouple the two accelerators and let them run independently thanks to the construction of ECR ion source producing high charge state ions to be injected into the cyclotron. From that time LNS can offer to the nuclear community a large variety of stable beams with energies ranging from few A·MeV to 80 A·MeV depending on the ions to be accelerated and on the accelerator used for that.

More recently an ISOL low energy radioactive beam facility (EXCYT) was installed, based on the combined use of the two accelerators and in addition just few years ago the CS beam started to be used also for in-flight production of intermediate energy unstable beams. These two facilities allowed to extend the LNS nuclear research to unstable beams induced reactions at both low and intermediate energies.

Published under licence by IOP Publishing Ltd

Figure 1. (Color online) The LNS Tandem.

Figure 2. (Color online) The LNS superconductive cyclotron and its operative diagram.

The field of the LNS activities has been expanded in the last decade with the NEMO project, an initiative aimed at the design and construction of a submarine observatory for high energy cosmic neutrinos to be deployed in the Mediterranean seafloor, just in front of the sicilian coast. Today the permanent LNS staff consists of about 110 INFN people, including researchers, technologists, administratives and technicians. In addition about 40 researchers from other public institutions, mainly from the University of Catania, are associated to the LNS activities, together with a huge number of undergraduate and PhD students, post doc fellows and young temporary researchers that lead the total number of people daily working at LNS above 200. An idea of the impact of LNS on the scientific community can be also given by the number of users that is around 300 per year, one third of them belonging to foreign research institutions.

2. Beam production facilities

The two LNS accelerators are shown in figures 1 and 2. Figure 2 also reports the operative diagram of the cyclotron, showing the ion beams accelerated up to now and their respective

energies. It can be seen that the maximum attainable energy is 80 A·MeV for the lightest beams and it has been obtained for protons and ^{12}C. For the heaviest elements the maximum energy is 25 A·MeV and ^{197}Au beams with this energy have been produced, indeed.
Today the cyclotron is endowed with two ECR sources. One is working at room temperature (Cesar) and the second one (Serse) is based on superconductive technologies and works at liquid helium temperature.
Figure 3 shows a layout of the LNS accelerator hall and experimental rooms. The main elements of the facility EXCYT are also schematically shown.
EXCYT is a first generation ISOL facility using the cyclotron to produce the primary beam and the Tandem as post accelerator.The cyclotron beam is produced at intermediate energy and delivered to a production target thick enough to stop the beam. The large number of different species produced in the beam-target interaction are then extracted by heating the target and transported to a charge exchange channel that provides them with a negative ionization. A multiple stage separator along the line allows to get a clean exotic beam of the requested species that can be injected into the Tandem, accelerated and sent to the experimental hall where the nuclear processes induced by this beam can be studied.

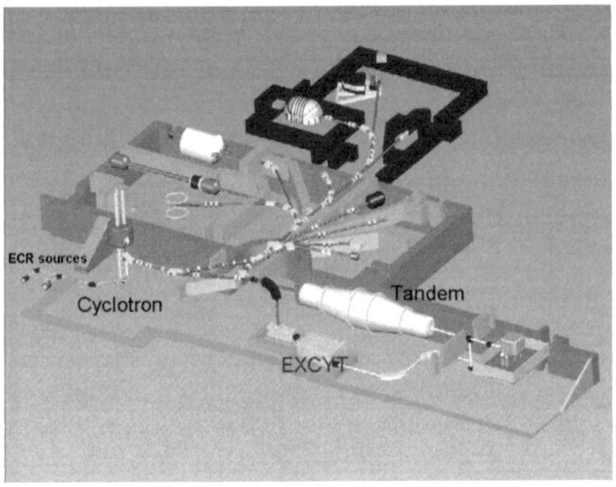

Figure 3. (Color online) Layout of the accelerator area and experimental rooms.

The ^{8}Li was chosen as the first EXCYT beam and it was produced by using a 45 A·MeV ^{13}C primary beam and a graphite target. Owing to the low efficiency of the charge exchange process (around 3%) the final beam intensities attainable by EXCYT are low, if compared to the usual stable beam intensities. In fact the maximum intensity obtained for the first ^{8}Li beam was about $7{\cdot}10^4$ particles/s. It implies that special care has to be taken when planning an experiment with the EXCYT beams, in order to compensate the low intensity in some way, for instance using 4π detector arrays.
Following the requests of the LNS users, the development of a ^{15}O beam is now under way, for which one expects a charge exchange efficiency more favourable than for ^{8}Li by a factor close to 10.
Radioactive beams at intermediate energy are also produced at LNS using the well known technique of in-flight projectile fragmentation, where the primary beam is again provided by the cyclotron. A thin production target is placed just after the extraction line of the CS. The interaction of the primary beam with this target produces many different ions moving in the forward direction together with the residual beam and approximately with the same velocity. By a proper setting of the two 45 degree dipoles used for the beam transport, particles with a given magnetic rigidity are selected and separated from the other reaction products and from

the residual beam. Due to the limited selectivity of these magnets, actually a 'cocktail' beam is transported along the line to the experimental room where the set up of the experiment is mounted. In this last part of the line each beam particle has to cross a tag detector consisting in a thin double side strip silicon detector, that gives a measure of its energy loss and position. This information, together with the measurement of the time of flight on a given path, allows the complete identification of each single beam particle and measurement of its energy. In the off-line analysis of the experiment, mass, charge and energy of the projectile can be assigned to each event, so that in principle multiple experiments are feasible with these beams.
Intensities of the order of 10^4 particles/s have been achieved during the first runs of the facility. After a careful study of the transport beam line, new focusing elements have been installed and a new articulated diagnostics system has been implemented on the line leading to the CHIMERA room, so that a gain by a factor around 25 - 30 is now expected in the intensitiy of the radioactive beams when delivered to CHIMERA.

3. Detection facilities

The development of the beam production facilities has been accompanied by the parallel development of detection facilities.
The first big 4π detector designed and realized at LNS was MEDEA (see figure 4), an array of 180 BaF_2 crystals that can detect and identify light particles and gamma rays. Later the detector has been coupled to a superconductive solenoid (SOLE), able to separate the very forward emitted reaction products from the residual beam. MACISTE, a zero degree array of telescopes, each one made of a ionization chamber and a plastic detector, completes the apparatus, allowing the detection and clear identification of the heaviest reaction products emitted in the forward direction. Looking at a possible development in the employment of the detector, an experimental study is taking place to verify the response of the MEDEA crystals to neutrons.

Figure 4. (Color online) The Medea-Sole-Maciste detector.

Big efforts of several researchers belonging to LNS, to the Catania INFN unit and to other italian and foreign groups led to the realization of CHIMERA, a 4π charged particle detector made of 1200 telescopes each one consisting of a CsI(Tl) E-stage and a Silicon ΔE stage (figure 5).

The detector has been, and still presently is, widely used by large international collaborations to study the physics of collisions at the Fermi energy. A very recent upgrade of the detector concerns the implementation of a hardware pulse height analysis of the signals coming from both the E and ΔE stages. This analysis, applied to the signals coming from the Silicon stage,

Figure 5. (Color online) The CHIMERA 4π detector.

allows to identify the low energy ions that stop in the such stage. On the other hand, the most penetrating particles that leave in the ΔE detector a signal indistinguishable from the background can be efficiently identified by a pulse shape analysis of the signal produced in the E stage. This important upgrading of the detector extended the field of operation of CHIMERA on both the low and the high energy sides.

Figure 6. (Color online) The Magnex spectrometer.

MAGNEX (figure 6) is a high resolution, large acceptance (50 msr) magnetic spectrometer, that was specially designed for both inclusive and coincidence experiments at Tandem energy with stable and unstable beams. Its main elements are a quadrupole magnet, vertically focusing the particles, and a horizontally focusing dipole. The focal detectors is made of a gas chamber that measures the energy loss of the ejectiles and a wall of Silicon detectors for the measure of their residual energy. Adding the measurement of the flight time one gets a complete information on emission angle, energy, charge and mass of the ejectiles.
Thanks to an agreement between LNS and the Nuclear Physics Institute of Orsay, the neutron detector EDEN has been temporarily transferred to LNS and coupled to MAGNEX, forming a unique experimental complex able to perform exclusive measurements specially useful for studies on the formation and decay of neutron-rich systems.

Of course besides these big devices, LNS owns a number of smaller modular detectors (CLAD, Hodo-Big, Hodo-small, Trasma, etc.), whose geometry can be changed according to the requirements of each specific experiment.

4. Research with accelerators

In the past years most of the beam time of the two accelerators (around 75% on average) has been allocated to nuclear physics experiments, whilst about 25% of the beam time has been used for application studies and interdisciplinary activities.
It is impossible here to recall all the excellent results deduced from the several experiments performed with the LNS beams in these last 30 years. It must be underlined that, even if only some examples are mentioned in the sequel, there is a huge amount of studies in nuclear structure and in reaction mechanisms that should deserve to be mentioned as well. Moreover it is to be acknowledged here the work of the small but active group of nuclear theorists operating at LNS. Most of the experimental studies in the field of intermediate energy collisions have been developed in harness with them, who have interpreted in the most suitable way the role of theorists in a nuclear laboratory.
Nuclear research with stable and unstable Tandem beams concerns the nuclear astrophysics, the spectroscopy of neutron-rich nuclei, the study of dissipative processes in the nuclear interaction, the structure of the light nuclei and its effects on the reaction mechanisms and so on.

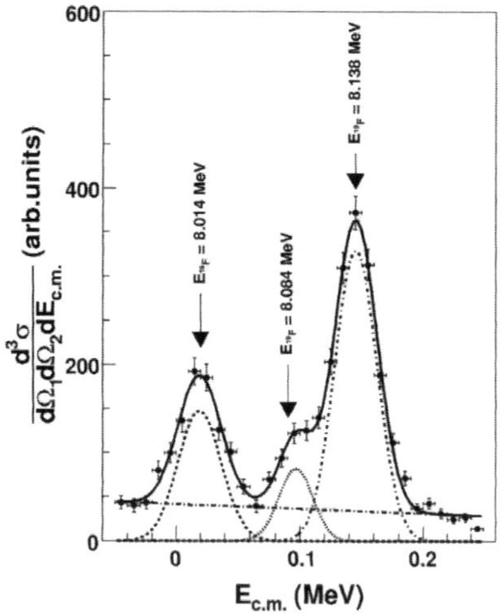

Figure 7. The $^{18}O(p,\alpha)$ excitation function as deduced by the Trojan Horse method (data from Ref. [1]).

Nuclear astrophysics represents one of the leading research topic at LNS. The development of the Trojan Horse method and its application to several nuclear systems allowed the indirect extraction of astrophysically relevant quantities concerning the interaction between charged particles, otherwise unaccessible by direct measurements because ot the huge Coulomb barrier to incident energy ratio. Such measurements can be of crucial impact on our knowledge of the processess occurring in stars. For instance the recent discovery of a resonance at 20 keV [1][2] in the $^{18}O(p,\alpha)^{15}N$ reaction (figure 7) implies a 40 % increase of the value of the astrophysical factor with respect to the value included in the NACRE compilation for that process.
In the field of the nuclear structure interesting results have been recently achieved from the comparison of the scattering cross sections of Be isotopes on ^{64}Zn, measured at LNS with a ^{9}Be

beam and at ISOLDE with ^{10}Be and ^{11}Be beams. It has been observed that the halo structure of ^{11}Be heavily affects the scattering cross section [3] which is strongly suppressed with respect to the scattering cross section of the other two isotopes (figure 8). This effect has been related to the increased contribution of absorption processes that in halo nuclei can occur even at large distances.

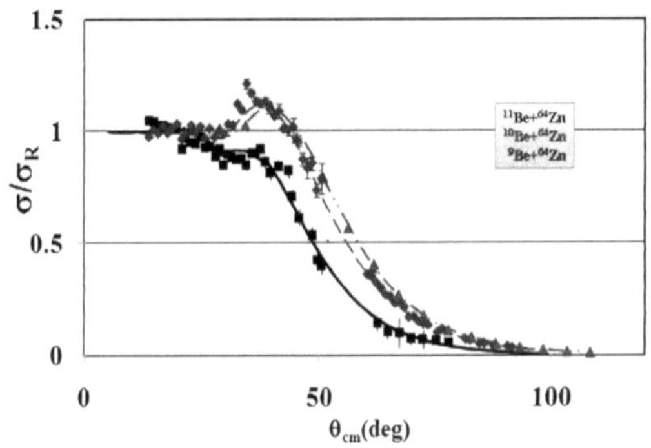

Figure 8. (Color online) Comparison of the ^{9,10,11}Be+^{64}Zn scattering angular distributions (data from Ref. [3]).

As already mentioned, the EXCYT experiments have to take into account the low beam intensities that presently can be produced. So the first experiments using the ^{8}Li beam have been planned keeping in mind this limitation. As a clear example, one can mention the measure of the scattering excitation function of the ^{8}Li+α system that was performed using the resonant scattering method on thick gas target.

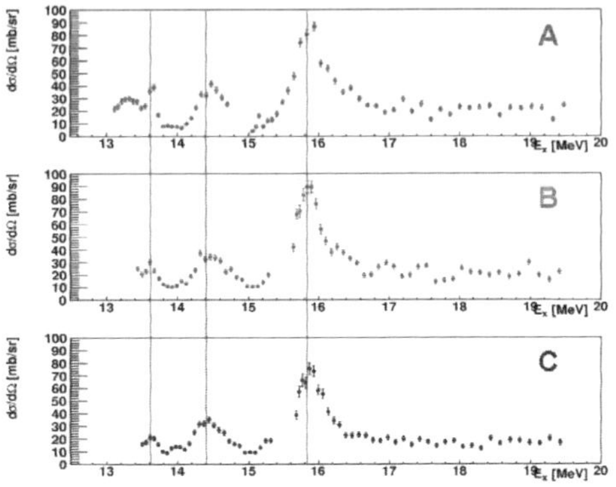

Figure 9. (Color online) ^{8}Li+α scattering excitation functions at three angles (unpublished data).

The method consists in delivering the beam into a reaction chamber filled with He gas at

a pressure high enough to stop the beam and detecting the recoil α-particles. The gas acts as target and as degrader at the same time and one can measure the excitation function in a wide energy range, with a single beam energy, saving a lot of beam time. The goal of the experiment was the search for resonances in the scattering cross section that could be related to population of cluster states in ^{12}B with α-^{8}Li configuration.
The resonances showed in the preliminary excitation functions at different angles and reported in figrue 9 need to be better understood by a comparison with cluster model calculations, before drawing definitive conclusions.
The first campaigns of experiments with the beams produced by the CS concerned the study of the collisions and the behavior of the nuclear matter at intermediate energy. The de-excitation modes of the GDR in medium mass nuclei, the study of dissipative processes, the nuclear caloric curve and the phase transition in nuclei were among the subjects of the experimental campaigns with the cyclotron. Afterwards, the availability of the powerful detector CHIMERA gave the opportunity to perform exclusive experiments with multiple coincidences, thanks to the high granularity of the detector. The unambiguous reconstruction of the final state for each event allows to investigate in greater details the physics at the Fermi energy, including the different aspects of the projectile multifragmentation process that is typical of these energies [4],[5] and the role of the entrance channel isospin on the dynamics of collisions [6].
As an example of recent results, it has been found that in semi-peripheral three-body reactions, besides the fragments produced by the sequential decay, the most neutron-rich intermediate mass fragments (IMF) are emitted in the very first stage of the collision (<40 fm/c), from a region of overlap of the two colliding nuclei. Furthermore, it was possible to establish a kind of emission cronology for the IMFs, in the sense that with increasing IMF charge the corresponding emission time scale also increases to values as large as 300 fm/c or more [5].
This is shown in figure 10, where the correlation between the relative velocities of the IMFs and the projectile-like or target-like fragments is reported for the system ^{124}Sn+^{58}Ni at 35 A·MeV.

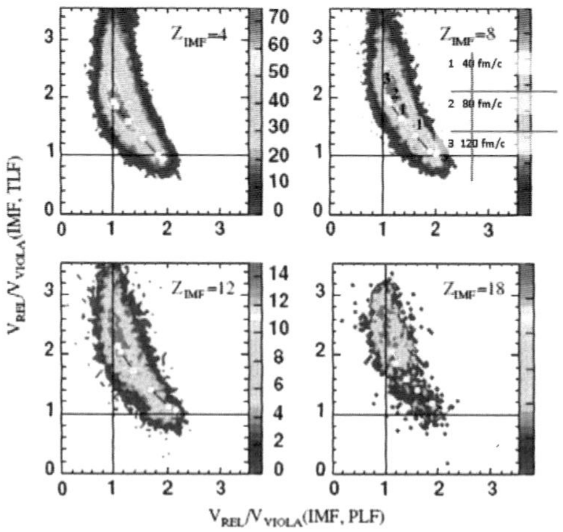

Figure 10. (Color online) IMF production yield of the ^{124}N+^{64}Ni at 35 A·MeV reported as a function of the fragment relative velocities (data from Ref. [5]).

Moreover few years ago the study of the emission of two correlated protons from the excited ^{18}Ne represented the first interesting result achieved thanks to the development of in-flight fragmentation beams. The relative energy spectrum of the two coincident protons (figure 11) emitted after the interaction of a secondary ^{18}Ne beam with a lead target showed a peak at the expected p-p relative energy that could reflect the preformation of a di-proton in ^{18}Ne, as effect of the pairing interaction [7].

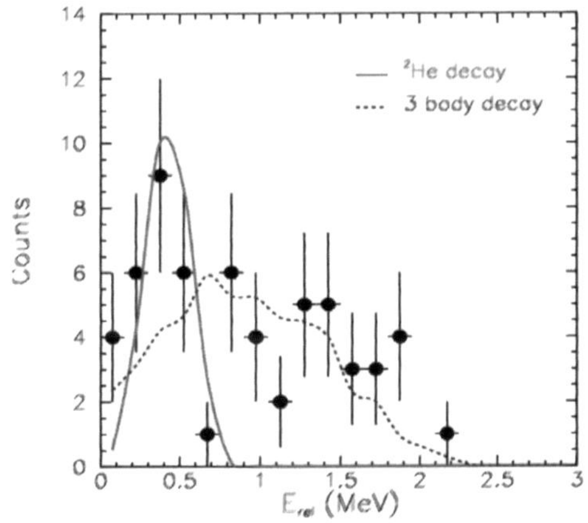

Figure 11. (Color online) Relative energy spectrum of two coincident protons emitted by excited ^{18}Ne (data from ref. [7]).

The two accelerators represent also a useful tool for a number of non-nuclear activities that are performed at LNS. The beams are used, indeed, also for radiobiological studies, researches on the modifications induced by ion beams on the behaviour of superconductive materials, investigation on the radiation hardness of electronic components to be used in space missions. For these measurements LNS has recently received the ESA certification, an acknowledgment of the reliability of the work jointly performed by the LNS staff and a private partner company to set up the system.
Futhermore the CS is also used for therapy purpose in the frame of an agreement involving LNS, the University of Catania and the University Hospital. Since 10 years patients affected by uveal melanoma are treated with a 62 MeV proton beam produced by the CS. Every year five weeks of beam time are devoted to these protontherapy sessions.
Finally, a procedure of elemental analysis of archaeological finds with the Tandem and CS beams has been set up, allowing to deduce, in a non destructive way, information on their internal composition. Ancient coins, paintings, documents and other cultural heritage samples have been processed. Exploiting the wide energy range of the LNS beams such kind of analysis can be performed at different depth, from the surface to the inner layers.

5. Technological research without accelerators

The panorama of the LNS activities also includes a large number of technological researches that do not require the use of accelerators. Among these initiatives one must recall the study of scintillating fibers as cheap, easy-to-use tools for a continuous, efficient monitoring of nuclear waste repositories, an issue that is part of the INFN-Energy strategic project.
Moreover an explicit mention is deserved by the development of portable PIXE-α, XPIXE-α and BSC-XRF systems that allow 'in situ' non destructive analysis of cultural heritage finds. The use of these instruments is, of course, complementary to the analysis technique based on the use of ion beams, already described. In connection with this activity a laboratory has been

installed at LNS for the home production of the ^{210}Po sources that are needed for the operation of the portable instruments.
Finally intense efforts have been devoted to the design and realization of innovative ion sources. The competence acquired during the past years in this field by the LNS source group is worldwide acknowledged and is witnessed by the involvement in a large number of national and international projects. Valuable contributions have been given, and are still given, to the realization of ion sources in italian (LNS, Legnaro National Laboratory, the italian center for hadrotherapy CNAO) and european (GSI, ESS) laboratories.

6. Astroparticle physics

More than ten years ago another scientific initiative was undertaken by the LNS researchers in collaborations with colleagues of other INFN units, in the field of astroparticle physics. It is the NEMO project, that concerns the design and realization of an undersea telescope for high energy cosmic neutrinos to be installed in the Mediterranean Sea, with the goal of extending the present limits of the observable universe.
Today the project NEMO is completely integrated in the European project called KM3NeT, born from the collaboration of researchers from several European countries and included in the first road map of research infrastructures of ESFRI.
The structure of the telescope will consist of a number of towers, about 800 m tall, each one supporting the optical sensors to detect the Cherenkov light emitted by the muons produced by the decay of neutrinos after interaction with the sea water. The towers are expected to cover a water volume of the order of 1 km^3. Calculations and simulations are carried out to determine the parameters of the most efficient configuration of the detector, namely the number of towers, the number of detectors in each tower, the distance between adiacent towers and so on.
In the meanwhile a number of measurement campaigns have been performed to check and compare the characteristics of the proposed sites that could host the telescope. For the high transparency of the water, the low biological activity and a suitable morphology of the seafloor, the Italian collaboration has proposed a site at 3500 m depth, about 100 km offshore the sicilian coast. A parallel R&D actifity has been carried out to test and validate the mechanical and electrical solutions for the construction and deployment of the elements of the telescope on one side, and for the transfer of the electrical power and retrieve of the signals from the detectors on the other side. As a necessary support to this activity, two underwater laboratories have been realized and fully equipped. One is the so called 'test site' located at 2000 m depth, 20 km offshore the Catania port, and the second one is the italian candidate site for the final installation of the observatory. Both of them are connected with the respective on-shore stations by electro-optical cables, 25 and 100 km long respectively.
These infrastructures have attracted the interest of several researchers operating in contexts other than astroparticle physics, since they represent a unique opportunity to perform observations and measurements in deep sea. Thus several multidisciplinary collaborations grew up in fields like seismolgy, geophysics, marine biology, environmental sciences. Moreover the two laboratories partecipate to the European Network of Excellence ESONET and to EMSO, a European large-scale infrastructure composed of seafloor observatories for long-term monitoring of environmental processes.

7. Perspectives and conclusions

LNS was born as a nuclear physics laboratory and is still playing a role in the panorama of the European laboratories, thanks to the reliable operation of its accelerators and to the large variety of stable and unstable beams produced. To mantain this role, improvements and upgrades have already been made in the last years, for instance on the transport line of the fragmentation beams, and others are in progress, for instance on the two ECR ion sources of the cyclotron.

Moreover it is to be checked the interest of users to exploit the beams produced by EXCYT at low energy, that could avoid the bottleneck of the charge exchange channel. In this case a new beam line should be installed downstream the mass separator. Anyway the traditional nuclear research at LNS has to be reinforced, primarily for its contribution to a deeper knowledge of the nature, but also for its wide implications in the development and application of nuclear technologies and methodologies to interdisciplinary researches.
The astroparticle project of the deep sea neutrino observatory is the foreseeable medium term future of the Laboratory and will require increasing attention and dedicated resources. Enormous R&D efforts have been made in the last ten years on the different aspects of the project concerning optical and acoustic sensors, electronics and mechanics, in the frame of the KM3-NeT collaboration. Submarine technologies have been developed and validated by taking advantage of the underwater laboratories installed in front of the sicilian coast, that in turn have stimulated the birth of innovative scientific collaborations in interdisciplinary research fields.
Applications and technological developments represent the third proficuous field of activity at LNS, a very articulated field spanning from radiobiology to medical diagnostics and therapy, from science of materials to seismology. These activities bring additional visibility to the Laboratory and contribute to a harmonic development of fundamental and applied research at LNS.

References

[1] La Cognata M *et al.* 2008 *Phys. Rev. Lett.* **101** 152501
[2] La Cognata M *et al.* 2010 *Astrophys. Journ.* **708** 796
[3] Di Pietro A *et al.* 2010 *Phys. Rev. Lett.* **105** 022701
[4] Pagano A *et al.* 2004 *Nucl. Phys.* **A 734** 504
Papa M *et al.* 2007 *Phys. Rev.* **C 75** 054616
[5] De Filippo E *et al.* 2005 *Phys. Rev.* **C 71** 044602
[6] Planeta R *et al.* 2008 *Phys. Rev.* **C 77** 14610
Amorini F *et al.* 2009 *Phys. Rev. Lett.* **102** 112701
Lombardo I *et al.* 2010 *Nucl. Phys.* **A 834** 458c
[7] Raciti G, Cardella G, De Napoli M, Rapisarda E, Amorini F and Sfienti C 2008 *Phys. Rev. Lett.* **100** 192503

Institute of Physics
Dirac House, Temple Back
Bristol BS1 6BE UK

proceedings
.com